Energy Sustenance and Environmental Safety Evaluation

Energy Sustenance and Environmental Safety Evaluation

Edited by **Lucas Collins**

⬚SYRAWOOD
PUBLISHING HOUSE

New York

Published by Syrawood Publishing House,
750 Third Avenue, 9th Floor,
New York, NY 10017, USA
www.syrawoodpublishinghouse.com

Energy Sustenance and Environmental Safety Evaluation
Edited by Lucas Collins

Printed in the United States of America.

Contents

Preface

This book has been an outcome of determined endeavour from a group of educationists in the field. The primary objective was to involve a broad spectrum of professionals from diverse cultural background involved in the field for developing new researches. The book not only targets students but also scholars pursuing higher research for further enhancement of the theoretical and practical applications of the subject.

Sustenance of energy is a prime concern of the modern world, especially due to the growing population and depleting resources. This book is the collective contribution of internationally renowned scientists and academicians and would help graduate and post graduate students of environmental studies and associated disciplines to gain comprehensive insights into topics like environmental impact of energy, mitigation of environmental impacts, sustainable development, etc. This book will be beneficial for students, researchers and professionals engaged in the fields of energy and environment.

It was an honour to edit such a profound book and also a challenging task to compile and examine all the relevant data for accuracy and originality. I wish to acknowledge the efforts of the contributors for submitting such brilliant and diverse chapters in the field and for endlessly working for the completion of the book. Last, but not the least; I thank my family for being a constant source of support in all my research endeavours.

Editor

Life cycle analysis and environmental effect of electric vehicles market evolution in Portugal

João P. Ribau, Ana F. Ferreira

LAETA, IDMEC, Instituto Superior Técnico, Universidade de Lisboa, 6 Av. Rovisco Pais, 1, 71049-001 Lisboa, Portugal.

Abstract

Fossil fuel dependency in Portugal is represented in around 76% of the total primary energy use, from which almost half is associated to the road transport sector. The reduction of imported fossil energy, pollutants and CO_2 emissions is seen as a solution to a more sustainable energy system. This paper analyzes the market penetration of battery electric vehicles in the road transport sector as an alternative and more efficient technology, considering its maximum share in the transport sector in 2050. The main goal is to evaluate the energy consumption, air pollutants (including CO_2 emissions), and the economic impacts of conventional and electric vehicles in Portugal. The environmental Kuznets effect in the studied factors is also evaluated. Life cycle methodology was applied to the "fuel" production and use stage, and to the materials of the vehicle. Although reducing energy consumption and emissions is essential, the relation of such impact within the region economy is also extremely important. Based on a Kuznets curve hypothesis, some of those impacts were possible to co-relate with the gross domestic product evolution in Portugal. The evolution of the energy source share, energy production efficiency, vehicle type share in the Portuguese light duty vehicle fleet, and technology efficiency, was also considered. Although the electrification of the road sector can potentially lower the fossil fuel importation, the electricity demand should increase. Nevertheless, it is estimated that around 43% of the energy consumption, 47% of CO_2 emissions, and 17%-40% of air pollutants could be reduced with the expected electric vehicle evolution.

Keywords: Life cycle; Kuznets; Electric vehicle; Emissions; Energy consumption.

1. Introduction

Fossil fuels are at the center of global climate changes causing negative environmental impacts worldwide. In 2010, these energy sources accounted for around 76% of the Portuguese total primary energy consumption, being oil (49.1%), coal (7.2%) and natural gas (19.7%) the major fuel sources; and whereas renewable energy sources accounted for around 23%. The road transportation sector which was the sector that consumed more energy represented approximately 37% of the total final energy consumption in Portugal in 2010, and was responsible for about 30% of CO_2 emissions (Eurostat [1] and DGEG [2]). Oil, electricity and natural gas consumption have shown decreases of 2.7%, 4.1% and 5.3%, respectively, due to the increasing implementation of renewable energies and efficiency improvements. The 2003/30/EC European Directive aims to promote the use of biofuels and other renewable fuels instead of diesel or oil for transport purposes in each member state. In long term, this is expected to contribute to the fulfillment of European climate change agreements (Directive 2003/30/EC [3]). The development of alternative vehicle technologies and new energy sources has been performed in the last decades. These are key factors to minimize the environmental and energy issues that the world faces. The strategies to reduce fuel consumption and emissions in conventional vehicles are one step to be taken into account [4]. The gradual electrification of the vehicle is one of the strategies adopted by the automotive industry and the policy makers. Vehicle electrification enables the improvement of urban air quality (no local emissions), the diversification of primary energy sources (electricity can be generated from a wider range of sources, not necessarily from fossil origin), and allows the use of more efficient propulsion technologies (such as regenerative braking and low consumption electric driven components). Several studies already address and compare alternative vehicle technologies such as battery electric vehicles (BEV) and plug-in electric vehicles (PHEV) with conventional vehicles, and also alternative fuels, such as the hydrogen. Ribau [5] uses life cycle methodology to compare different technologies, but it mainly focuses on the vehicle propulsion system, namely different engines for plug-in hybrid vehicles. In that study the energy consumption and CO_2 emissions from the fuel production and vehicle use were considered. Different kinds of engines and battery sizes showed to be more appropriate for different drive styles. Life cycle assessment (LCA) was applied in several studies to evaluate the energy consumption and CO_2 emissions of alternative fuels, like biohydrogen and biodiesel ([6-13]), and alternative vehicles ([8, 14-17]). Baptista [14] developed a model which consists in the analysis of scenarios of alternative fuels and vehicle penetration in road transportation sector in Portugal for the year 2050. However it doesn't focus on air pollutant emissions neither on possible economic impacts of such scenarios, namely on Gross Domestic Product (GDP) and Green Net National Income (GNNI). The analysis of GNNI and genuine savings considering the Kuznets curve in Portugal was performed by Mota [18]. The environmental Kuznets curve is a hypothesized relationship between various indicators of environmental degradation and income per capita. In rapidly growing countries, where little or no change in infrastructures or technology improvements are developed, a proportional growth of energy consumption, pollution and other environmental impacts relatively to the economy growth, is expected. This is also known as the scale effect, in which an economic growth can lead to an "environmental degradation". However, in wealthier countries, where growth rate is slower, and pollution reduction and energy efficiency policies are in effect, a leveling or decreasing of the "environmental degradation" along the economic growth can be developed, leading to the environmental Kuznets effect. In this kind of countries the development of the economy led also to the development of the technology, infrastructures, and services sectors, which usually results in efficiency and pollutant emissions treatment techniques improvement, therefore forcing the environmental degradation to cease or decrease.

In [19], a software was developed to analyze the performance of BEVs from the perspective of economic and environmental impact in the Tokyo area, considering three electricity generation mix options in Japan by 2030. However, the study didn´t considered a Kuznets effect analysis or relate the different indicators studied. Although in [20] the hypothetic Kuznets curve applied to carbon dioxide emissions and economic growth is studied, it didn´t focused other air pollutant emissions, neither in the transport sector. Regarding pollutants only, the Clean Air for Europe report (CAFE [21]) shows the cost-benefit of air quality considering the analysis of air pollutant emissions like $PM_{2.5}$, NH_3, SO_2, NO_x and VOCs and respective costs, from each European (EU25) Member State.

None of the previous studies covers both energy consumption and emissions, and its relation to a country´s economy impact, especially for the road transportation sector. Energy production and emissions have a tremendous impact in a country's importations share and political commitments. Therefore it is with major interest to relate both energy and emissions in Portugal with economic growth factors aiming to analyze from a sustainability point of view.

In this study the main objective is to estimate the influence of electric vehicle penetration in Portugal regarding evolution scenarios to 2050, in terms of energy, CO_2 and air pollutant emissions and its possible economic impacts. The existence of a possible environmental Kuznets curve effect regarding the energy consumption, CO_2 and air pollutant emissions (in light duty vehicle sector in Portugal) accounting the Portuguese GDP evolution, is analyzed. Although one of the objectives is to identify the Kuznets effect, some difficulties are expected in relating some factors that can have concurrent tendencies. One approach taken regarding the pollutant emissions was to assign a GDP and cause-effect dependent price to the emissions based on the GNNI. The energy consumption and emissions evaluation accounted the life cycle of the energy used in the vehicles and the materials used in vehicle fabrication. The evolution of the vehicle technology efficiency, the electricity generation mix, and the Portuguese road vehicle fleet evolution to the year 2050 is accounted.

From a point representing the current location of Portugal in a Kuznets curve (Figure 1) the possibilities of the future direction to take in order to achieve the objective/target, and therefore to decrease the "environmental degradation", were highlighted. The attribute objective* in Figure 1 refers to energy consumption and emissions (which includes CO_2, NO_x, SO_x, VOC, CO, PM and NH_3 emissions) reduction target, due to political commitments, Kyoto protocol, "20-20-20" targets and energy imports reduction targets in Portugal.

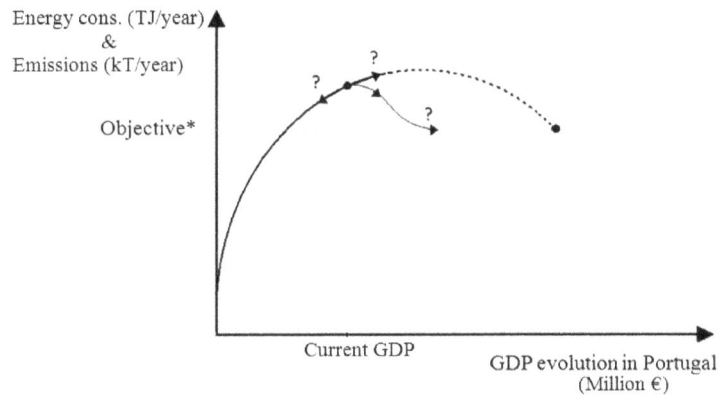

Figure 1. Representation of a possible Kuznets curve considering different scenarios to reach the objective (Energy consumption, Emissions (Particulate matter, CO_2, Greenhouse Gases,..))

2. Economic, energy, and emissions evolution in Portugal

2.1 GDP and population characterization

The GDP growth rate in Portugal in the last 20 years has been rather irregular (Figure 2). At the present time it is difficult to find a consensus in what would be the average GDP growth for the next years. The global crisis has caused a hitherto unseen fiscal expansion and economic uncertainty. The Energy Roadmap 2050, communication from the European Commission assumes an annual average GDP growth rate of 1.7% for EU-27 (European Commission [22]). Although it is easy to find supporters of the opinion that Portugal should have an average GDP growth above European levels in the next decades, it is more difficult to find supporters of the idea that it will be a reality. This ambiguity can be found on the socio-economic development scenarios set for Portugal in Figure 2. These scenarios, conservative and fénix, were based on a report on New Energy Technologies Competitiveness Analysis [23]. The conservative scenario is based on the: i) continuity of the development model of the last 15 years, an investment in non-transactional assets and low economic growth rate; ii) a reduction of the industry sector weight on the GDP and on the other hand an increase of the services sector; iii) a decrease in the population; iv) no changes in the transports. The fénix scenario is based on the: i) rebirth of the economy based on investment and policies for production of added value assets; ii) an increase of the industry sector share on the GDP and a decrease of the services sector share on the other hand, leading to a higher increase of the Gross value added in the industry sector; iii) an increase in the population; iv) new transport policies and habits towards a decrease of short distance traffic, less dependence on individual transport and a reinforcement of the rail transport for goods transport.

For the purpose of this work, the conservative scenario will be followed and therefore an annual growth of 1% for the Portuguese GDP will be assumed. At this point it is believed that the conservative scenario is the most realistic although it can change in some years. Therefore, a sensitivity analysis on the GDP growth rate will also be addressed. Within the same framework, it is of great interest to study the relation of both energy and emissions evolution in Portugal with economic growth factors, aiming to analyze a sustainability point of view and its possible impact in the country importation share and committed policies.

2.2 Energy and emissions characterization

The Portuguese energy consumption profile covers the energy use in the following sectors: industrial, transportation, domestic, electricity and heat, and services. The total energy consumption and CO_2

emissions are shown in Figure 3, regarding GDP evolution during year 1990-2008 (World Bank [24]). Note that the evidenced directions refer to the objective to be achieved, the energy use and CO_2 emissions decreasing, disregarding the GDP tendency (see Figure 1).

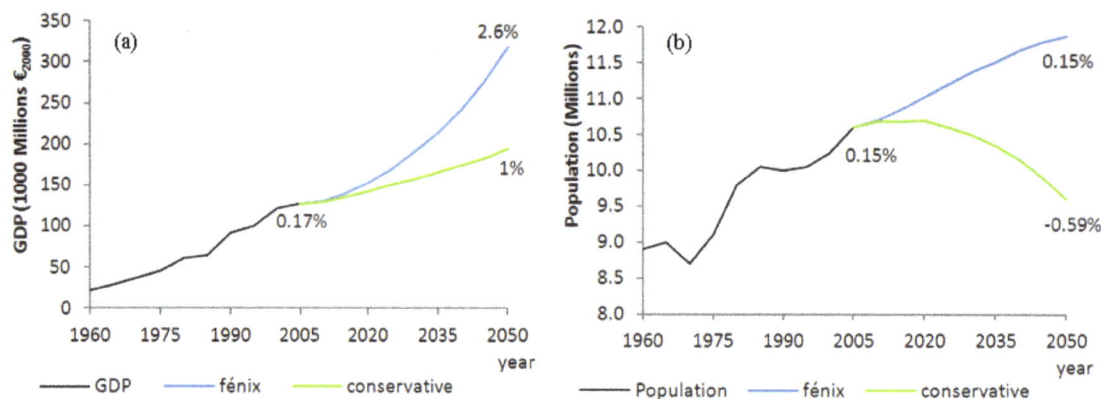

Figure 2. Two scenarios of GDP evolution (a) and population evolution (b) to Portuguese case, 1960-2050 (Adapted from [23])

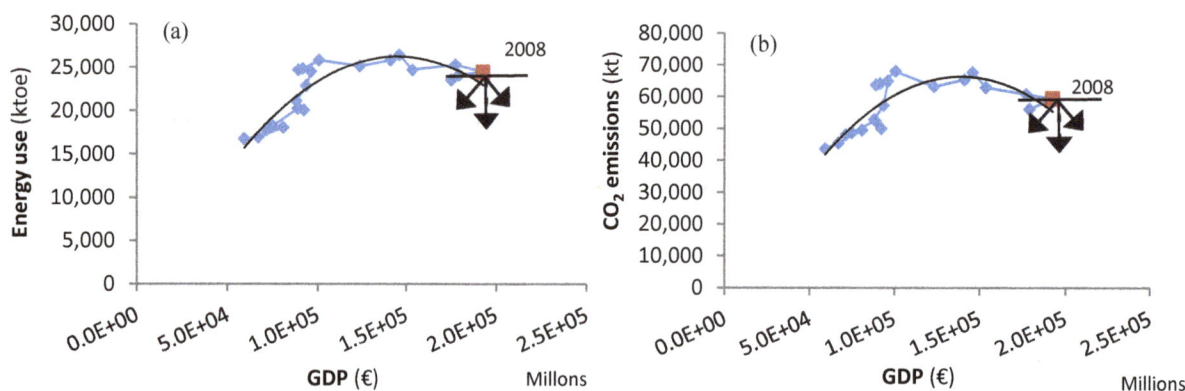

Figure 3. Relation between energy use (thousands of tons of oil equivalent) (a) and CO_2 emissions (thousands of tons) (b), as a function of GDP (current Euro €) evolution in Portugal (Adapted from Word Bank [24] data)

In this work the CO_2 emissions are considered separately from the pollutant emissions. Emissions are originated from the energy use described above and from industrial processes that emit directly and indirectly gaseous compounds (e.g. gasification processes).

It is interesting to see the resemblance between the energy use and the CO_2 emissions evolution (Figure 3), however note that the CO_2 has a more pronounced decrease in the last years (higher GDP values). Although the energy demand has increased along the years, the energy use efficiency also increased due to the technology improvements, therefore lowering the energy use in Figure 3. Besides the technology improvements, more strict political commitments like the Kyoto protocol and 20-20-20 directive had a very important role to lower the CO_2 emissions. The same can be applied to the pollutant emissions (Figure 4).

The pollutant emissions composed by sulphur dioxide (SO_2), nitrogen oxides (NO_x), particulate matter ($PM_{2.5}$), ammonia (NH_3) and volatile organic compounds (VOCs), and their associated impacts are the main responsible for the total damages from air emissions (Figure 4) [25, 26]. Therefore those pollutants were considered in this study. Alike to the energy and CO_2 the same objective also applies to air emissions: to reduce the air pollutants. Despite GDP evolution, this objective is generally being achieved. The use of more improved processes and technologies are the major responsible in the emissions variation. The decrease of SO_x is directly associated to the decrease of coal based industries (e.g. coal power plants). Besides the increasing energy demand, the political commitments and regulation for pollutant emissions, as also technology progress (catalyzers, filters…), inverted or suspended that increase.

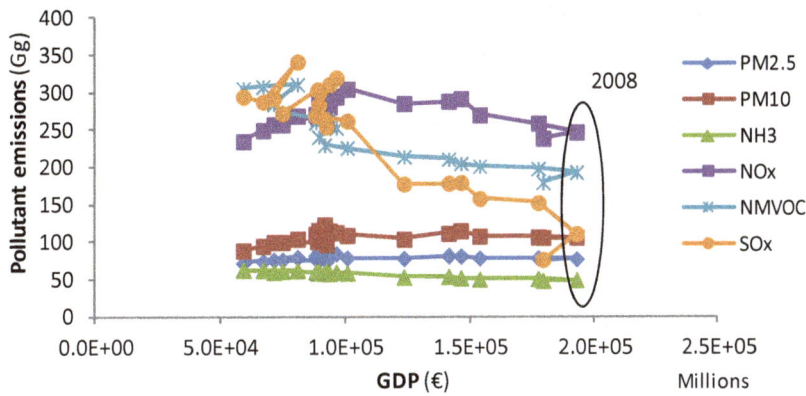

Figure 4. Total damages from air emissions (Gigagrams) as a function of GDP (current Euro €) in Portugal

In Figure 3, the directions highlighted by the black arrows are placed in the current GDP and represent the direction of the objectives since that year beyond, namely the energy use, CO_2 and air pollutant emissions decrease (the same can be assumed for Figure 4). This objective directions regard to the desired scenarios as mentioned in introduction. Assuming that the objective will be achieved one of three scenarios can occur (represented by one possible direction): the energy and emissions decreasing followed by a decline in GDP, a rising GDP, or GDP maintenance. These possibilities are pretended to represent a Kuznet curve shape in the relation of energy/environment and economic factors.

2.3 Green net national income
Unlike conventional accounting, "green accounting" goes beyond welfare depending on just marketed produced goods. Welfare is allowed to depend on health, environmental amenities, pollution levels, or availability of natural resources. These arguments can be seen as alternative forms of consumption, not consumption of conventionally produced goods but of natural resource services, health services, etc. [26, 27]. According to the theory of green accounting, finding a decreasing GNNI implies that in the future there will be a decrease in utility. Thus, according to the definition of sustainability as non-decreasing utility, this would indicate unsustainable development. The GNNI should account at least for the depletion of natural resources (minerals and forests), the health damages from air emissions (SO_2, NH_3, NO_x, VOC, $PM_{2.5}$) and the value of technological progress. The GNNI is defined by the following mathematical equation:

$$GNNI = GNI - CFC - e.E + \left(Q^R - f_R\right)\dot{S} + Q_t \tag{1}$$

Where, GNI is the Gross National Income, CFC is the Consumption of Fixed Capital, (Q^R-f_R), $\dot{S}$ is the value of rents from resource stock depletion, e.E is the welfare cost of emissions (where e is the marginal damage cost of emissions in 2010 per metric ton in Portugal, and E is the amount emissions in metric tons), and Q_t is the time effect [16].
In this study it will only be considered the e.E factor, and other factors will be considered static. In other words only the emissions contribution will be accounted. To calculate the factor e.E in the GNNI, the marginal damage costs (€2010) by air pollutant in Portugal was considered, Table 1 (data from [20]).
In Figure 5 the considered pollutant emissions cost evolution in Portugal is shown. The cost share of each pollutant is directly related to its consequent damage associated cost (Table 1) and its emitted quantity in that year.
Particulate matter is the largest contributor to the total damages from air emissions, followed by SO_2. Both account on average for more than half of the total costs. But whereas, the emissions of SO_2 decrease, the emissions from $PM_{2.5}$ increase in average. In the last years the damages in human health derived from particulate matter, namely the $PM_{2.5}$ has been gaining more attention. As a percentage of GNI, the damages from air emissions have been decreasing. From 1990 to 2005 the best estimate is that the cost of air emissions in Portugal averages 8% of GNI with a decreasing trend [21].
The pollutant emissions accounting are generally done in a local basis. Using the GNNI methodology, pollutant emissions can be related to a country's sustainable development. The energy can be easily

related to the cost of the energy sources and energy importation, and the CO_2 can also be compared to a cost. With GNNI methodology it is possible to attribute a cost to each pollutant emission.

Table 1. Estimates of marginal damage cost by air pollutant in Portugal (€2010/ton)

damage costs (€2010/ton)	Best	Low	High
SO_2	6900	3500	10000
NH_3	7400	3700	11000
NO_x	2200	1300	3200
VOC	1200	500	1600
$PM_{2.5}$	44000	22000	64000

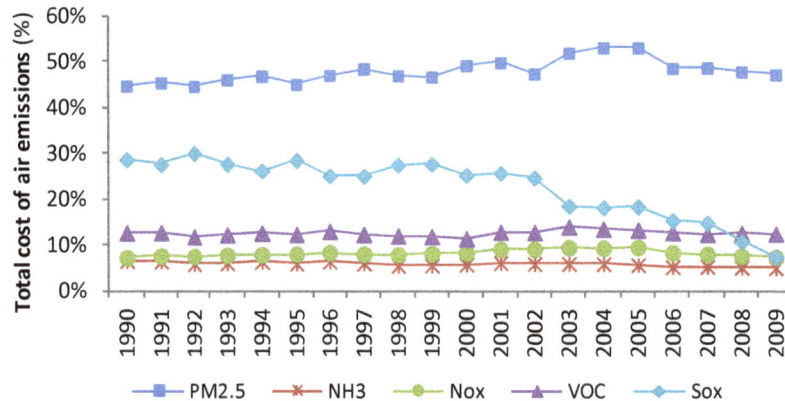

Figure 5. Annual evolution of the cost share of air pollutant emissions in Portugal

3. Road transportation sector and electric vehicle market penetration

The case study considers the Portuguese road transport sector, namely the LDV fleet energy consumption and emissions analysis. Figure 6 shows the evolution of Portuguese LDV fleet and the new diesel vehicle registrations share along GDP from 1990-2008.

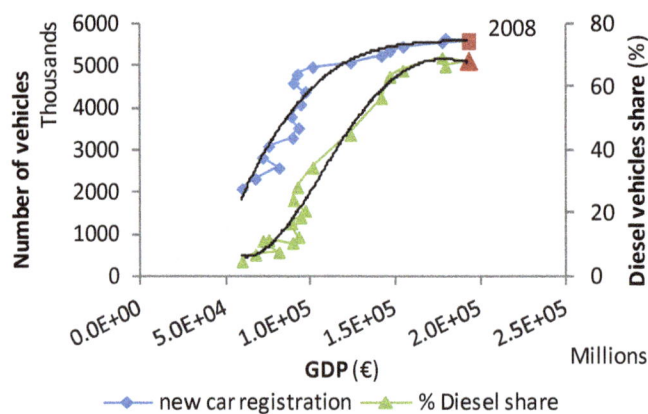

Figure 6. Light duty vehicle fleet in Portugal, the number of vehicles and the market share of the diesel LDV (%)

In the last years, the increasing fuels prices, namely gasoline, led to the diesel market share growth. Although gasoline vehicles are the majority, the share of diesel in the road vehicle fleet has increased. Generally, the road vehicle fleet has grown in recent years; however that tendency is slowing [24]. Figure 7 shows the energy consumption and CO_2 emissions evolution, regarding the road vehicle sector, relatively to the Portuguese GDP [24]. Once again the objective is well highlighted in the presented figures concerning to the decreasing of energy and emissions.

The evolution of GDP and road sector energy consumption per capita of the past two decades is presented in Figure 8 [23, 24], indicating an average growth of both indicators. The energy consumption of the road sector and the emissions after a strong increase slowed down. This follows the new car registration tendency, and the increased efficiency in the vehicles due to technology and regulation actions (Figure 9).

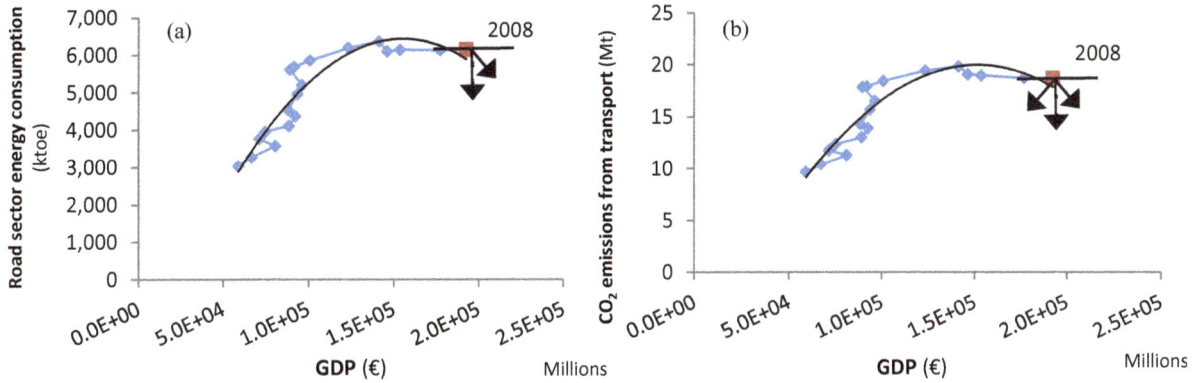

Figure 7. Energy consumption from road transport sector (ktons of oil equivalent) (a) and CO_2 emissions from transport sector (millions of metric tons) (b) as a function of Portuguese GDP

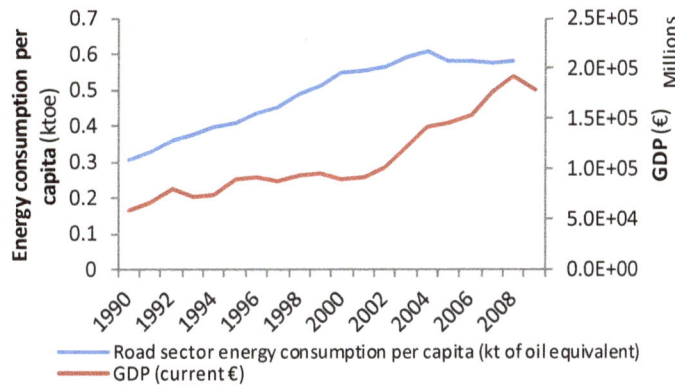

Figure 8. Evolution of road sector energy consumption (ktoe) per capita and GDP 1990-2009 in Portugal (Adapted from [24])

There are several studies concerning alternative vehicle market penetration (MOBI.E [28], and McKinsey&Company [29]). In this study, estimations resulting from a developed model (Baptista [14]) were used. This model, besides estimating the BEVs future market (Figure 9) also considers the efficiency improvement of the vehicle technology. This improvement has a linear progress to 2050, and besides used in energy consumption calculations, it was also used in the same proportion to calculate the emissions [14]. In this scenario the energy consumption (Tank-to-Wheel stage) in gasoline LDVs and diesel LDVs is expected to decrease to around 35.7% by 2050, while BEVs to decrease 25.4%.

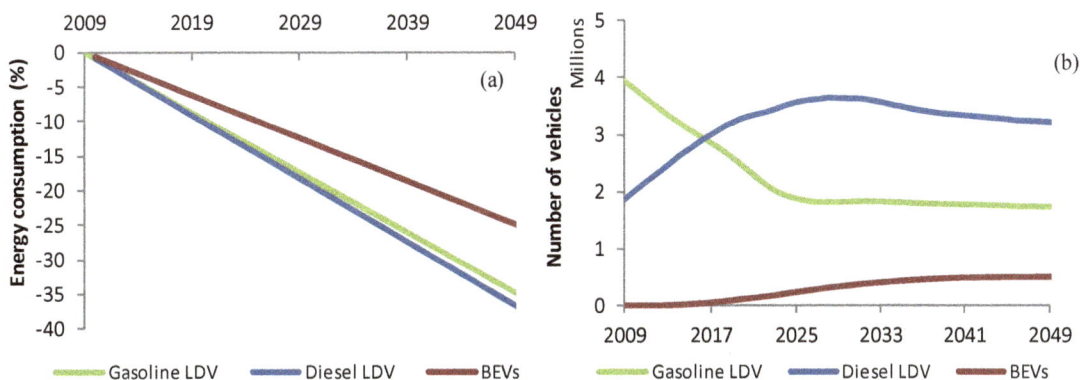

Figure 9. Vehicle energy consumption evolution in TTW (%) (a) and number of vehicles of the Portuguese fleet (b) (Scenario 2010-2050)

4. Life cycle analysis

The evolution trend of conventional and electric vehicle technologies towards 2050 would be evaluated accounting energy consumption, CO_2 and air pollutant emissions. This type of evaluation is performed using the life cycle analysis (LCA) methodology. LCA is an important tool to estimate the energy balance and environmental impact of a system. It can be used also to compare different energy systems including vehicle technologies and production systems ([30]). In this study the Principles of ISO 14040-14044 [31] are followed. The energy and emissions impacts of advanced vehicle technologies and new transportation fuels evaluation were assessed using a specific LCA for fuels and vehicles, the Well-to-Wheel (WTW) and Cradle-to-Grave (CTG) analysis. The WTW analysis is often divided into Well-to-Tank (WTT) and Tank-to-Wheel (TTW) assessment. WTT starts with the fuel feedstock production, followed by fuel production, and ends with the fuel distribution to the pump or vehicle tank, while TTW focus on the fuel utilization at the vehicle operation. The main difference between WTT and TTW lies in the delimitation of the system boundary. In some vehicle analysis studies, such as [8, 30, 32], TTW and WTT are combined, considering the fuel and its application in light duty vehicles. CTG consists in the analysis of the materials used in vehicle and it can be added to WTW analysis. Energy consumption and CO_2 and air pollutant emissions are accounted in WTT, TTW, and CTG. The present work will be focused in LCA, including the energy production (WTT), energy use (TTW) and material used in the vehicle (CTG). An energy cost analysis would be also included in this study.

4.1 Tank-to-wheel

The TTW stage considers the energy consumption and associated CO_2 and other air pollutants emitted by the vehicle/fuel combination. For simulating conventional or alternative vehicle technologies, ADVISOR vehicle simulation software [33] was used. ADVISOR is a micro-simulating tool to estimate the performance, fuel economy, and tailpipe emissions of conventional and new vehicle technologies (hybrid and electric powertrains). This software was used in several studies for vehicle simulation (some already mentioned in Section 1) such as in [30]. Vehicle specifications (detailed in Table A.1 in Appendix A) and a real driving cycle Cascais-Lisboa (specifications in [8]) were the main inputs used in this study. The vehicles chosen for the simulations are based on existing vehicles and available data. They all meet a close value of the power/weight ratio. The BEV, since it doesn't have any combustion engine will not present local emissions on the TTW stage. The main goal of TTW is to compare the BEV operation with the conventional internal combustion engine vehicles. An average of EURO 4 and 5 standards [34] is used to validate the emissions from the conventional gasoline and diesel vehicles (Table 2).

Table 2. Reference values to EURO emissions in light duty vehicles

(g/km)	Tier	Date	CO	THC	NMHC	NO_x	HC+NO_x	PM
Diesel	Euro 4	Jan-05	0.5	-	-	0.25	0.30	0.025
	Euro 5	Sep-09	0.5	-	-	0.18	0.23	0.005
Gasoline	Euro 4	Jan-05	1	0.1	-	0.08	-	-
	Euro 5	Sep-09	1	0.1	0.068	0.06	-	0.005

The energy consumption and the emissions that resulted from the vehicle simulations are presented in Section 6. Note that the evolution of the vehicle technology is accounted as shown in Figure 9.

4.2 Well-to-tank

WTT accounts for the energy consumption and emissions from the primary energy resource extraction through the delivery and process of the fuel to the vehicle's fuel tank (the same applies for electricity). For the WTT analysis the EcoInvent 2.0 database for SimaPro 7.1 software, was adapted for the average Portuguese electricity generation mix, [34] was used to estimate the electricity generation air pollutant emissions.

In this study the electricity is used as "fuel" to the BEV, and its WTT stage data was based in previous works such as [6, 9]. The Portuguese electricity production mix is composed by 49% of non-renewable and 51% of renewable energies (2010 data), with 8 % of energy losses in distribution [30, 35-37]. The resulting energy consumption in order to obtain 1 MJ of electricity generated was 1.02 MJ. A more detailed description of the Portuguese electricity mix is shown in Table A2 (Appendix A).

Following the EcoIvent database the Portuguese electricity generation emits around 174940 kg CO_2/TJ (2004 data). The same database also provides data on the pollutant emissions (Table A3 of Appendix A). However, new and updated values of CO_2 emissions were calculated as 87.190 g CO2/MJ due to improved power plant efficiencies and electricity mix. Then, the emissions from electricity production (Table A3) were proportionally updated regarding updated CO_2 data from [6, 34], as shown in Table 3.

Table 3. Updated CO_2 and pollutant emissions factor for energy production in Portugal 2010

(g/MJ)	CO_2	CH_4	CO	VOC	NO_x	PM	SO_x	NH_3
Gasoline	18.076	0.109	0.016	0.200	0.062	0.007	0.116	7.356E-06
Diesel	9.432	0.099	0.014	0.186	0.046	0.004	0.045	4.439E-06
Electricity	*87.190*	0.133	0.025	0.094	0.227	0.054	0.760	9.968E-05

*Considering 42.8 MJ/kg to diesel and 43.5 MJ/kg to gasoline of LHV.

The evolution scenarios developed in this study were based in the model first developed by [14], which includes also the electricity production efficiency improvement and electricity generation share evolution through the years to 2050 (see Figure 10). The tendency used from the previous model [14] was adjusted to the electricity production and updated by the author in [6] and Table A2 (Appendix A). Table 4 presents the electricity production efficiency and its resulting emissions following the electricity generation mix and plant evolution from Table 3 and Figure 10.

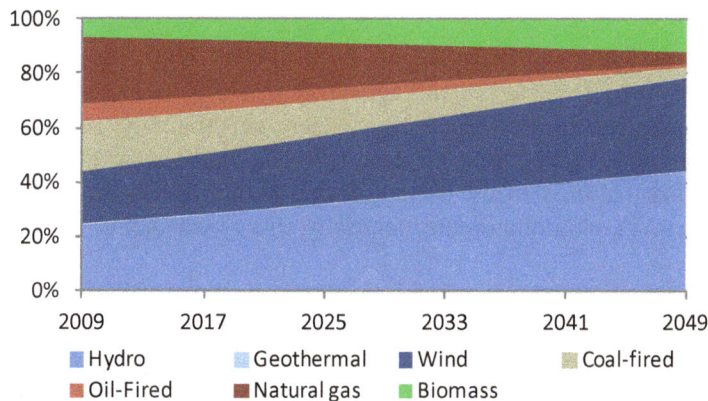

Figure 10. Electricity generation mix evolution scenario 2009-2050

Table 4. Energy and emissions factors of power generation regarding the adjusted electricity mix scenario 2009-2050

	2010	2020	2030	2040	2050
Energy consumption (MJ/MJ)	1.021	0.836	0.651	0.467	0.281
Produced emissions (g/CO_2)	87.19	69.92	52.64	35.37	18.10

The WTT methodology for the gasoline and diesel includes similar processes but different refineries. The following processes were considered: crude extraction, crude transportation, crude refinery and storage and distribution. The WTT data considered in this study for diesel and gasoline is based in [30]. The evolution of the efficiency of diesel and gasoline production from present year to 2050 was not considered to change. In order to obtain 1 MJ of gasoline and diesel fuel, 0.14 MJ and 0.16 MJ of energy is consumed respectively, and 12.5 grams and 14.2 grams of CO_2 respectively (Table A.4 of Appendix A). Pollutant and CO_2 emissions data was also calculated from Eco Invent database [36] (Table A.3) and thereafter updated accounting Portugal present energy efficiency values for fuel production (Table 3)

4.3 Cradle-to-grave

For the CTG stage, the GREET (The Greenhouse Gases, Regulated Emissions, and Energy use in Transportation Model) software from the US Argonne National Laboratory was used, namely GREET 2.7 model [38]. CTG accounted only the materials used in the vehicle. Besides the vehicle power train, body and frame materials, the replacement of consumable elements of the vehicle, such as fluids, tires, batteries, lubricants are also considered (Table A5 of Appendix A).

The total energy and CO_2 emissions of the CTG pathways were distributed along the vehicle lifetime kilometers traveled. In this study it was considered to be 200000 km (Directive 2009/33/CE [39]). Once this study reflects a Portuguese scenario, the Portuguese electricity generation mix evolution (Figure 10) was introduced in this stage also and accounted in the fabrication processes of the materials.

5. Energy cost estimations

The price of oil in international markets highly influences the price of diesel and gasoline. The price of oil is the price with the highest unpredictability in the primary energy market. The estimated oil derivate fuels prices and new road vehicle technologies evolution scenarios were approximated by a linear tendency of growth. Table 5 indicates the estimated prices to the user (based on [40]).

Table 5. Prices estimation of oil, gasoline, diesel, and electricity to the user.

Year	Oil ($/bbl)	Gasoline (€/L)	Diesel (€/L)	Electricity (€cent/KWh)
2010	120	1.69	1.539	15.98
2020	150	1.96	1.827	21.5
2030	200	2.40	2.307	27.5
2040	230	2.66	2.595	33.5
2050	280	3.10	3.075	40.8

The scenarios developed by the European Commission, 2011 indicate an increase in the generation costs for electricity that will have to be reflected in the consumer price. This increase is mostly due to the introduction of new technologies in the electricity generation (namely renewable energies), the introduction of carbon tariffs, fossil fuels price increase (oil, coal and gas) and the construction of new generation facilities. Assuming that most of the users will charge their BEVs at home, the domestic tariff evolution is considered. The taxes evolution on the electricity energy was not considered in this study.

6. Results and discussion

6.1 LCA applied to light duty vehicle estimations

The proposed vehicles, a BEV, a diesel and gasoline internal combustion engine vehicles (detailed in Appendix A, Table A1) were simulated in ADVISOR software. The diesel and gasoline vehicles achieved 2.10 MJ/km and 2.46 MJ/km of energy consumption, and 156 g/km and 179 g/km of CO_2 emissions respectively. The BEV in the same conditions achieved 0.43 MJ/km and zero emissions.

The evolution of TTW energy consumption, CO_2 and air pollutants emissions per vehicle for years 2009-2050 was regarded, and was estimated based on the technology efficiency tendency (Figure 9) for each vehicle. In Table B.1 and Table B2 (Appendix B) the TTW values achieved for the 2009-2050 scenarios are presented. As expected, the vehicle technology improvements lead to the energy consumption and emissions decreasing. The energy consumption and emissions are estimated to decrease around 37% and 25% for conventional and electric vehicle, respectively, by 2050. In order to calculate the total energy consumed in TTW stage, the number of vehicles (Figure 9), the vehicle type share in Portuguese fleet (diesel, gasoline and BEV), and daily travelled distance were accounted. It was considered that LDVs in Portugal travel in average 22 km.day-1 [41].

In the WTT stage the evolution of the electricity generation mix in Portugal to 2050 resulted from an electricity mix scenario for the years 2009-2050 in Section 4.2. The results of WTT energy production efficiency, CO_2 and pollutants emissions for a scenario 2009-2050 are summarized in Tables B3 and B4 (Appendix B). In this scenario the WTT factors in terms of energy and emissions are maintained at 0.140 and 0.160 for gasoline and diesel, respectively. However, for BEV, regarding the electricity production, it's possible to see a reduction of energy and CO_2 emissions from 1.021 MJ/MJ to 0.281 MJ/MJ and 87.192 gCO_2/MJ to 18.096 gCO_2/MJ. The pollutants emissions values relatively to the energy required by the vehicles were also reduced. This reduction in WTT stage is mainly due to the expected power plants efficiency improvements and renewable resources increasing in Portuguese electricity generation sector. Nevertheless, it can be seen that (per MJ) the electricity production is still responsible for larger losses than the diesel or gasoline production. Besides the electric vehicle do not emit local air pollutants (in the usage phase), the energy consumption and emissions associated with the energy production (WTT) are responsible for a major share of the life cycle of this vehicle.

The energy that is consumed in the plants to produce the diesel, gasoline or electricity used in the respective vehicles can be seen as the energy losses during the fuel production (see Appendix B, Table B3 and B4). Multiplying those values (MJ/MJ and g/MJ) by the diesel, gasoline and electricity consumed in the vehicle fleet (MJ/km) (Table B1 and B2) allows us to determine the actual energy consumed (and emissions) in the WTT stage due to the usage of such fuel.

The energy consumption and emissions associated to the materials used in the vehicles are presented in detail in Appendix B, which accounts also with the electricity mix evolution used in the 2009-2050 scenario, since the electricity is the main energy used in material fabrication. Around 0.416 MJ/km and 0.420 MJ/km are regarded to the CTG energy consumption for the gasoline and diesel vehicle respectively, and around 24.3 g/km and 25.1 g/km of CO_2 emissions. Accounting with the evolution scenario, by 2050 is expected that CTG energy consumption should decrease around 16.7% for both conventional vehicles, and around 21% for CO_2 emissions. Although the BEV accounts higher CTG energy consumption and CO_2 emissions, respectively 0.531 MJ/km and 32.0 g/km, its reduction potential by 2050 is also expected to be higher, around 27.7% and 36.7% for energy and emissions respectively.

Figures 11 to 13 show the evolution of the LCA energy consumption and emissions, composed by TTW, WTT, and CTG, of the Portuguese fleet.

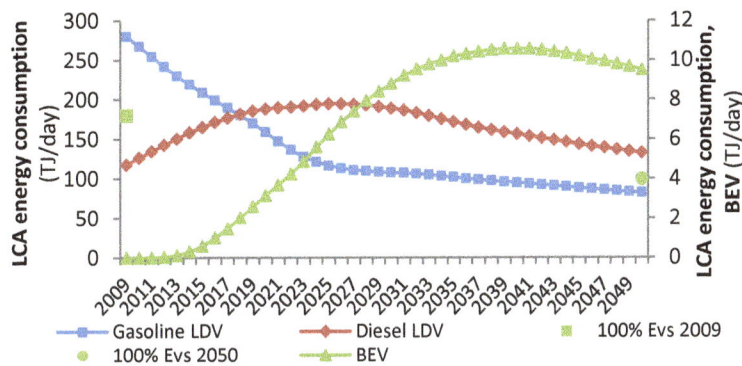

Figure 11. Evolution scenario of LCA energy consumption (TJ/day) regarding energy production efficiency and electricity mix, regarding energy production efficiency, LDV fleet evolution and technology improvements to 2050

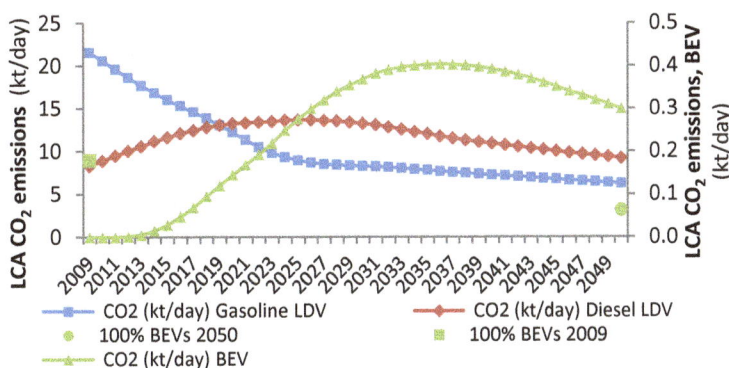

Figure 12. Evolution scenario of LCA CO_2 emissions (kton/day) regarding energy production efficiency and electricity mix, regarding energy production efficiency, LDV fleet evolution and technology improvements to 2050

Note that the evolution shown in Figures 11 to 13 accounts with the influence of the vehicle technology improvements (Figure 9), the electricity generation evolution (Figure 10), and the Portuguese vehicle fleet evolution, including the BEV penetration. In Figures 11 to 13 two extreme scenarios are highlighted in single data points: "100% BEVs 2009" and "100% BEVs 2050" concerning to 100% of the Portuguese LDV fleet represented by BEVs in 2009 or in 2050 respectively. This means the total LDV fleet to be composed by BEVs in those cases.

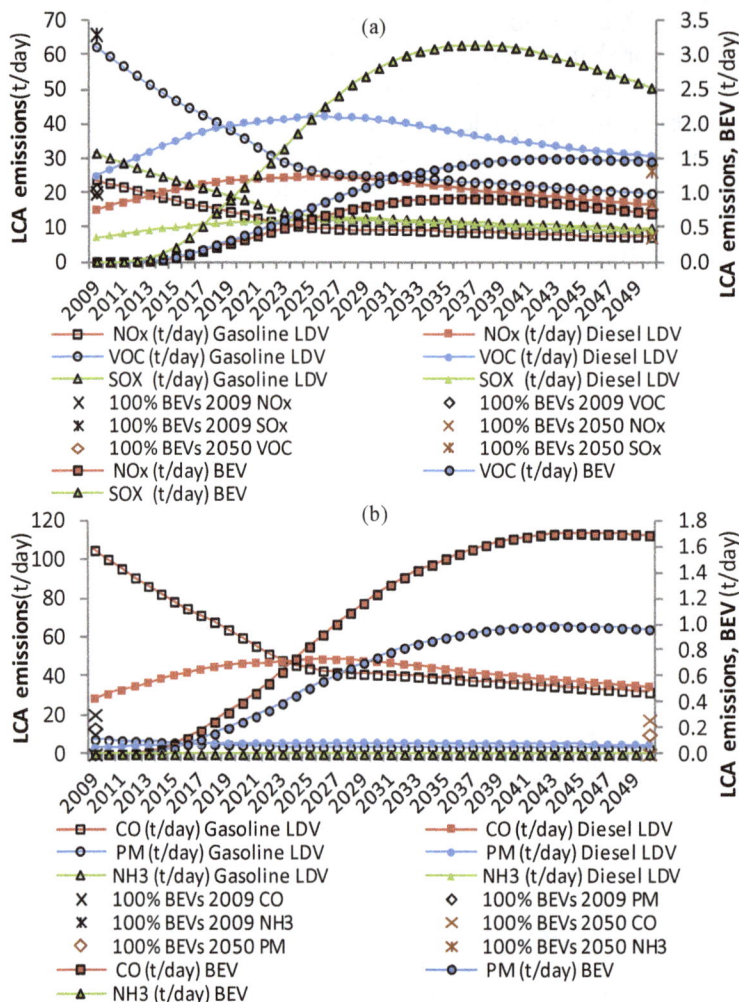

Figure 13. Evolution scenario of LCA NO_x, SO_x and VOC emissions (ton/day) (a) and CO, PM and NH_3 emissions (ton/day) (b), regarding energy production efficiency, LDV fleet evolution and technology improvements to 2050

As the number of the gasoline vehicles decreases and its efficiency gets better, the gasoline demand diminishes. In the other hand, the diesel demand is expected to grow as long as the number of diesel vehicles increases (Figure 9), however, the efficiency improvement of the vehicles overcomes this tendency and the diesel demand should decrease to values near the 2009 (Figures 11 and 12). Regarding the BEV introduction, the increase of the vehicle number lead directly to the increase of the energy and emissions associated to BEV energy use, production, and vehicle fabrication. Nevertheless that increase is overcome by the technology and efficiency improvements which invert the growing tendency.

In resume, the electrification of the LDV sector in Portugal, when followed by technology and energy production evolution, has the tendency to reduce the energy consumption and emissions in the life cycle of the road transport sector. The technology and electricity generation efficiency improvement evolution are clearly an important issue. If 100% BEV's scenario was introduced in nowadays a large amount of electricity would be required to supply the entire BEV fleet, and then the energy consumption and emissions due to the electricity production should rapidly increase. Nevertheless, the efficiency of the electric vehicle (TTW) is still a great advantage relatively to conventional vehicles; and emissions and energy consumption regarding the LCA maintain lower than gasoline and diesel vehicles.

Figures 14 and 15 show the evaluation of the energy and emissions variation for the total LDV Portuguese fleet in respective year, regarding the considered scenario of BEVs market penetration. A reduction of around 44% and 47% of energy consumption and CO_2 emissions respectively, of the LCA associated to the LDV sector can be achieved by 2050, and around 40%-52% the air for pollutant emissions (17% for PM emissions). In these figures, it can be seen the two extreme scenarios of 100% BEVs fleet share highlighted by single data points.

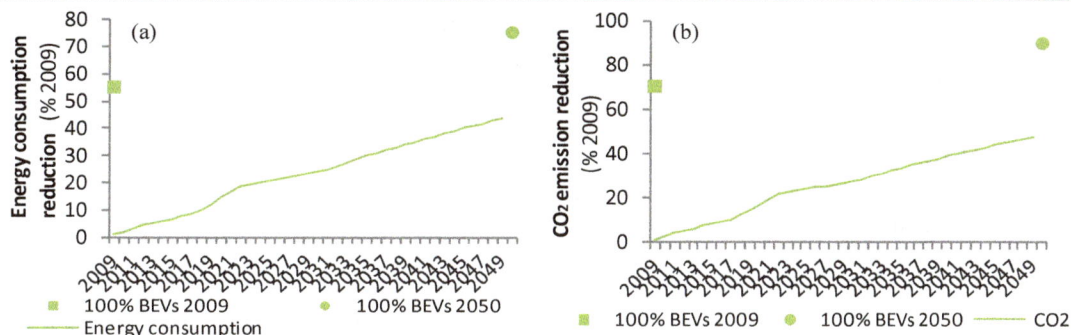

Figure 14. Percentage of energy consumption (a) and CO_2 emissions (b) decrease, for the total LDV fleet in comparison to 2009 values

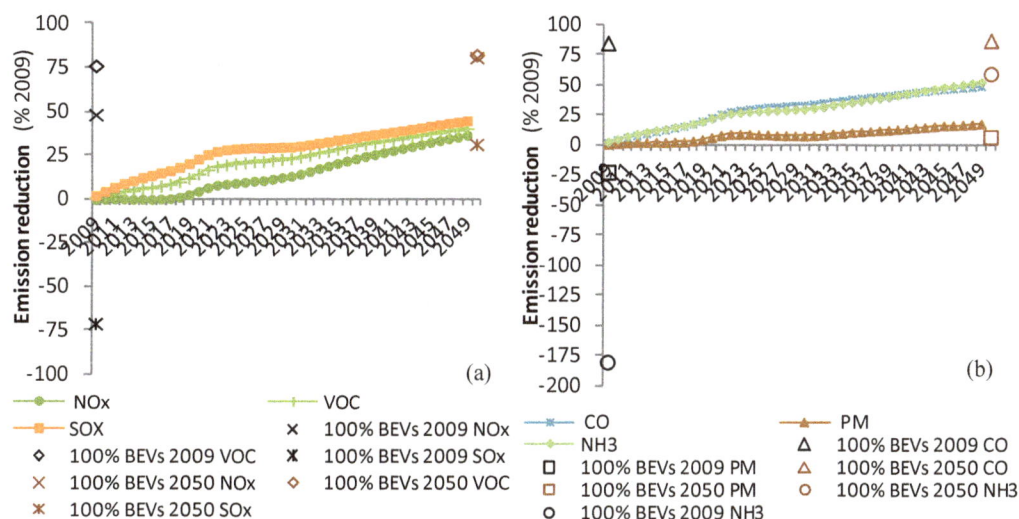

Figure 15. Percentage of NO_x, SO_x, VOC emissions (a) and NH_3, CO, PM emissions (b) decrease, for the total LDV fleet in comparison to 2009 values

The increase of the electricity demand due to the electric vehicle increase in the Portuguese fleet will allocate the resulting energy and emissions from the fuels sector to the electricity generation sector. The Tables B.1 to B.4 show that the electric vehicle is more efficient and less pollutant in the TTW stage, however the inverse occurs in WTT.

As in Figure 14, it can be seen in Figure 15 that emissions are decreasing due to the electrification of LDV sector. However if 100% BEV were introduced in Portuguese LDV fleet in 2009 some emissions, such as PM, NH_3 and SO_x, would not decrease, but would suffer an increase. This result was expected because in 2009 the electricity production is still much dependent of thermal power plants, responsible for those emissions that result from combustion processes. On the other hand, if the total Portuguese fleet was replaced by BEVs in 2050, it would lead to a decrease in energy consumption and emissions with no exceptions.

If no electricity generation improvements evolution were accounted the reduction of energy would have lower values, of around 37%. The CO_2 and pollutant emissions would have lower reduction tendencies also, of around 42% and 20%-47% respectively (6% for PM emissions). If additionally, the vehicle technology improvements evolution were none, the energy consumption reduction would be around 10%, the CO_2 emissions reduction 15%, and the reduction of SO_x and VOCs would be 4% and 25%, respectively. The NO_x, PM, and NH_3 emissions tendency would be inverted and increase 3%, 3%, and 16% respectively.

In Figure 16 the share of the different LCA stages (TTW, WTT, and CTG) is highlighted, and the influence of the efficiency of the vehicles (TTW) can be observed, especially regarding the energy consumption, CO2, CO and NO_x emissions. On the other hand, the other pollutant emissions, VOC, SO_x, NH_3, are associated to the energy production efficiency. The CTG stage has a major influence in the PM emissions. Although the increase of the BEV share lead to the increase of electricity demand, and to the increase of the vehicle battery fabrication impact (which has a major influence in the CTG), the evolution of the WTT and CTG factors cease a possible energy and emissions growth. The energy consumption has its major influence in TTW stage, since the fleet is most composed by internal combustion engine vehicles, but the process of diesel and gasoline production (WTT) is more efficient than the use of the fuel itself in the vehicle (TTW stage). Following the same idea, due to the energy consumption share and the energy production characteristics the CO_2 and CO emissions are most relevant in TTW. NO_x emissions appear to be divided between TTW and WTT stages although with a little more expression in TTW. Since the diesel internal combustion engine is responsible for a large quantity of NO_x and PM, it can be seen in Figure 16 that these emissions increase along the number of vehicles until technology improvements invert this tendency. The production of energy (WTT), especially electricity, is responsible for the largest share of SO_x, VOC, NH_3 and PM emissions. Although the electrification of the road sector should lower the fossil fuel importation the electricity demand should increase, depending on the power plants evolution.

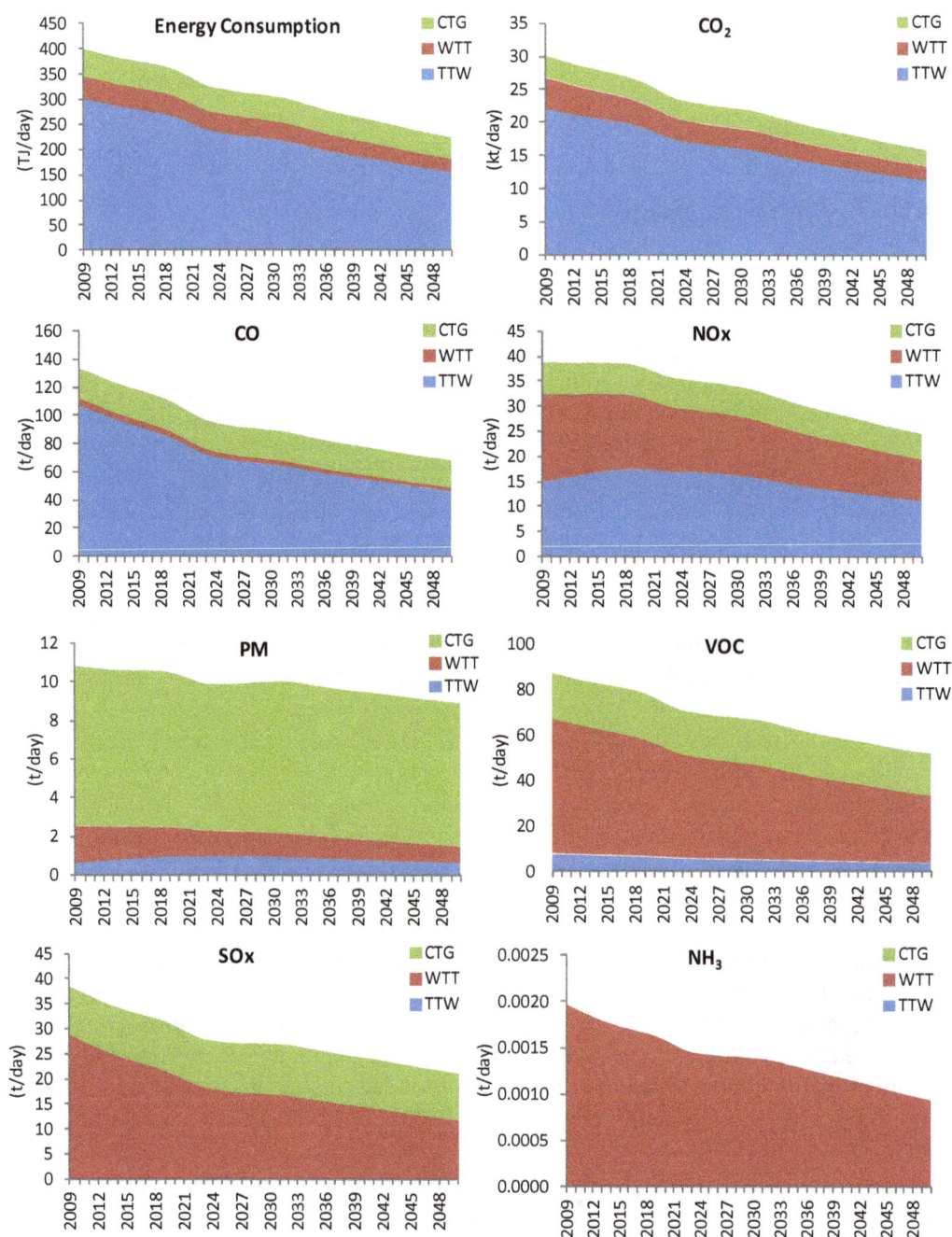

Figure 16. Energy production (WTT), energy use (TTW) and vehicle materials (CTG) energy consumption and emissions share, considering the total LDV fleet. Estimations from 2009 to 2050

6.2 Welfare cost of emission

The total air emissions in Portugal, accounted a total cost of 5,640 M€ (regarding the price per ton of Table 6), and the LDV sector is responsible for around 3% of the total air pollutants emissions in Portugal [25]. The price of pollutants will be taken static to 2050; however the cost of emissions of LDV sector will take into account the evolutions already estimated in previous sections (Figure 17). The cost of pollutants in Table 6 represents the welfare cost of emission associated to the GNNI (Section 2.3). Figure 17 shows the evolution of the damage cost of air emissions (in 2010 euro) regarding the electric vehicle penetration scenario. The cost of emissions is decreasing, however if 100% of electric vehicle substitute the entire fleet in nowadays only PM, VOCs and NO_x emissions would be lower.

Table 6. Damage costs of air pollutants emissions (in 2010 euro) regarding the LDV sector, and value of welfare cost of LDV emissions in 2010

	(€2010/ton)	Emissions LDV sector (Gg)	e.E (€)
SO_2	6900	0.32766	2260854
NH_3	7400	1.27029	9400146
NO_x	2200	32.16328	70759216
VOC	1200	10.52491	12629892
$PM_{2.5}$	44000	2.15363	94,759,720
Total			189,809,828

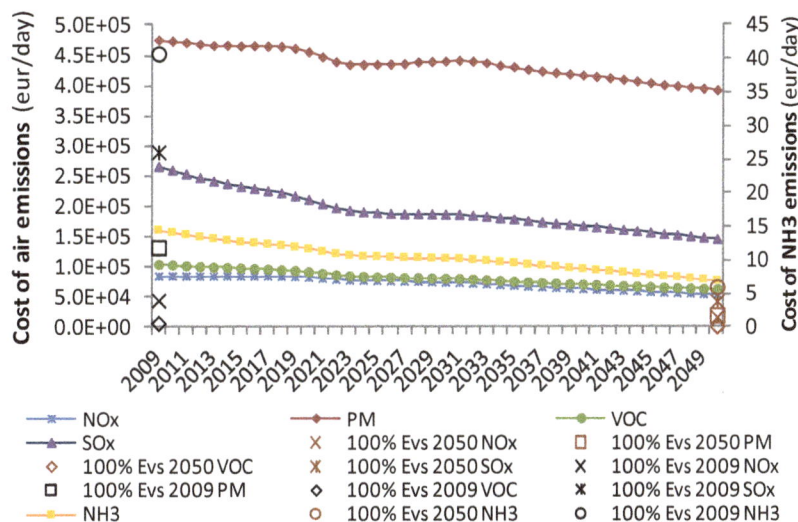

Figure 17. Evolution of the damage cost of air emissions (in 2010 euro)

6.3 Costs to the consumer

Nowadays, the cost of the vehicle sustenance is very important. The energy prices, such as gasoline, diesel and electricity, belong to a very dynamic market, and they are far to be static. Figure 18 resumes the calculated energy cost per distance traveled regarding energy costs and vehicles energy consumption evolution. In Section 5 it was shown that the price of the energy should increase. However, in Figure 18, the costs per kilometer increase but do not follow the same rate of increase. This is due to the technology improvements that slow down this tendency since a better efficiency of the vehicle reduces the energy consumption. If this scenario maintains, and if the BEV infrastructure becomes successful, the lower cost per kilometer of the BEVs may accelerate the BEV purchase and the fossil fuels turnover.

7. Final considerations

Regarding the evolution of energy consumption and emissions as function of the Portuguese GDP evolution, presented in results, a clear tendency can be seen. This tendency may have a similar shape of a portion of a Kuznets curve. This portion is exemplified in Figure 19, where along the GDP increase a variable X takes an approximated polynomial tendency. Along the GDP growth the variable X is decreasing. The variable X can take the form of the main results of this study. The continuation of the curve, in Figure 19 (direction (a)), can be seen as the evolution of the cost of emissions, energy consumption, and emissions per GDP in Figures 20 and 21. Then, although in some variables the variation with the GDP is not very wide, it can be accepted that either price of emissions, energy consumption, and emissions are decreasing along the GDP growth, suggesting the possible Kuznets effect.

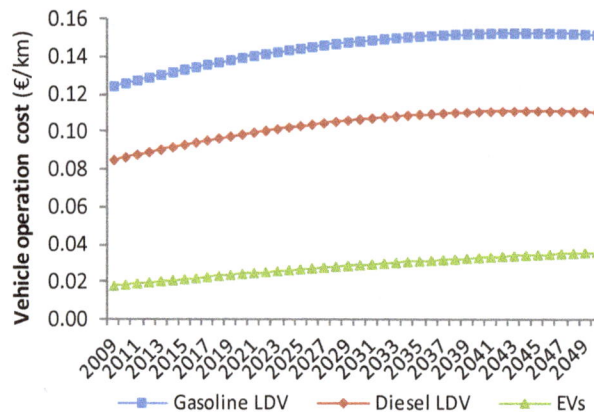

Figure 18. The price of the energy per kilometer to the consumer

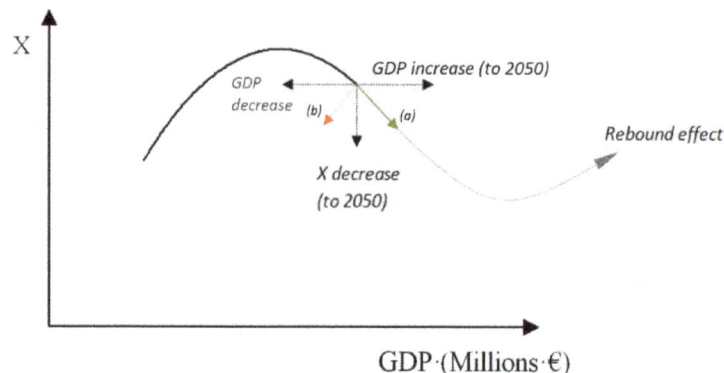

Figure 19. Example of a possible Kuznets effect regarding variable X and GDP. Conservative and fénix scenario (a) and pessimist scenario (b)

In Figure 19 a possible rebound effect is shown. This event occurs when the tendency of the Kuznets effect is inverted, and can be associated to indirect factors. The GDP from literature data is expected to be growing continuously, and maintaining that tendency, a possible rebound effect in the future can occur, if hypothetically:

- The electricity production efficiency improvements lead to cheaper electricity. Thus, since it becomes cheaper to travel per km (regarding BEV market), it is possible that the distance travelled per year increases, then consuming more electricity. The same effect can be achieved if the LDV fleet increases more than the expected.
- The BEV market increases but the electricity production efficiency doesn't meet improvements. The higher demand for electricity in power plants can lead to an unexpected growth in energy consumption and emissions. This can be seen in Figure 15 for PM, NH_3 and VOC emissions in the scenario of 100% BEVs 2009.
- The vehicle technology evolution achieves no or very little improvements.

Although the most reliable data researched was used at the time of this work, it is known that is very difficult to gather consistent data covering all the topics in this study. Additionally, these data estimations can easily vary accordingly to the economic evolution and political targets of a country. The more accurate the estimations and scenarios used in the input data in the methodology, more accurate results are produced. The estimations in this study can suffer some variations due to that variance of the inputs however the tendencies should remain. Nevertheless, it is expected that in long term the estimations of the GDP growth, and the decrease of the energy use, and emissions, as also the increase the electrification of the road vehicle sector, are expected to remain. In nowadays it's possible to see that the objectives regarding CO_2 emissions and energy use are slowly being accomplished.

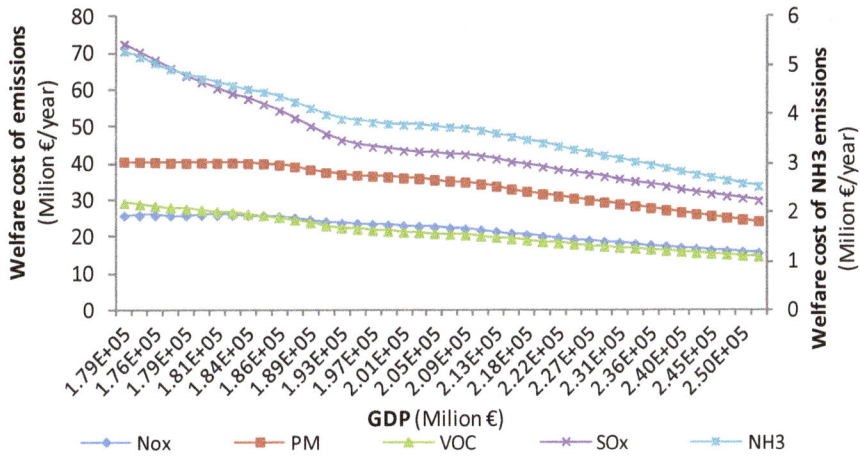

Figure 20. Welfare cost of pollutant emissions per GDP estimation from 2009-2050

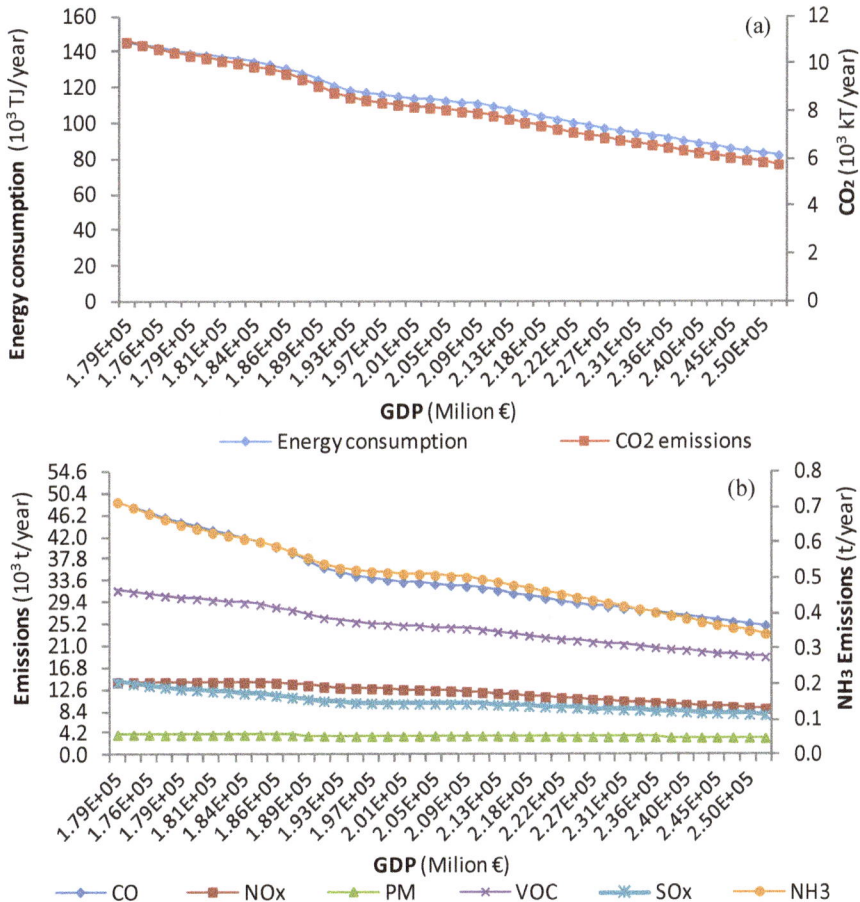

Figure 21. Energy and CO_2 emissions (a), and pollutant emissions (b), regarding GDP estimation from 2009-2050

A brief sensitivity analysis to the GDP growth rate is presented in Appendix C (Figures C1 and C2). From this analysis it can be seen that the fénix scenario (as described in Section 2 (see Figure 2)) has a tendency similar to that observed in (a) in Figure 19, which resulted from the conservative scenario assumed in the beginning of the study. Although welfare cost, the energy and emissions evolution maintain the same throughout the years, the GDP evolution has a higher growth. Therefore the Kuznets effect is more relaxed as expected to wealthier regions where the GDP growth continuous to increase (discussed in Section 1). On the other hand, if a pessimist scenario of an annual decay of 1% for the Portuguese GDP was assumed, the results tendency should take the form observed in (b) in Figure 19. The decrease of the welfare cost, the energy and emissions is accompanied by the decrease of the Portuguese GDP. Despite the technology continues to evolve the economic growth is not verified. This case indicates that the environmental effect regarding the road transport sector is disaggregated from the economic growth of the country.

8. Conclusions

From the scenarios discussed in this study by 2050, it is possible that 64% of gasoline and diesel for LDV sector can be reduced. This means that a large portion of oil imports would be also reduced. Since Portugal is entirely dependent on oil importation, the dependence on foreign energy would decrease. Therefore, the GDP may become more independent of energy imports and then be freer to grow. The electric vehicle penetration in Portuguese LDV fleet is expected to reduce the air pollutant emissions especially local emissions. However, the allocation of energy demand from the fossil fuel sector to the electric energy sector can impute to the power plants a larger share of emissions and energy demand. Nevertheless the evolution of electricity generation mix and power plant efficiency, and the evolution of the vehicle technology, should lead a decreasing tendency of the energy and emissions. Based on a Kuznets effect hypothesis, the impacts regarding cost of air emissions, energy consumption, and emissions from the road transport sector, were possible to co-relate with the gross domestic product evolution in Portugal. Note that a specific sector (light duty vehicle sector) is considered, and the Kuznets effect estimations are only regarding the studied sector, and therefore the energy consumption and emissions conclusions cannot be extrapolated to Portugal global energy and emissions evolution.

Glossary

LDV - Light duty vehicle
BEV – Battery electric vehicles
CO – Carbon oxide
CO_2 – Carbon dioxide
CTG – Cradle-To-Grave
GDP – Gross domestic product
GNNI – Green net national income
HC – Hydrocarbons
LCA – Life cycle analysis
NH_3 - Ammonia

NMVOC – Non methane volatile organic compounds
NO_x – Nitrous oxides
PM – Particulate matter
SO_x – Sulfur oxides
toe – Metric tons of oil equivalent
ton – Metric ton
TTW – Tank-To-Wheel
VOC – Volatile organic compounds
WTT – Well-To-Tank
WTW – Well-To-Wheel

Appendix A. Vehicles characteristics and LCA data

Table A1. Conventional and electric vehicle characteristics considered in this study

		Gasoline ICE	Diesel ICE	BEV
Traction Power/Weight (kW/kg)		0.061	0.059	0.062
Maximum Speed (km/h)		183.4	186.5	156.7
time 0-100 km/h (s)		11.7	10.4	8.7
Weight (kg)		1215	1239	1088
Traction Electric Motor	Nominal Power (kW)	-	-	68
	Torque (N.m)@ rpm /max rpm	-	-	260 @ 2500/8500
Battery (Li-ion), Energy capacity (kWh) / Max. power (kW)				16.05 / 110
SI Engine	Nominal Power (kW)@rpm	75 @ 6000	77 @ 4000	-
	Torque (N.m)@ rpm / maximum rpm	140 @ 3500/6500	248 @ 2000/4400	-
Wheelbase (mm)		2578	2578	2550

Table A2. Electricity generation mix regarding the Portuguese electrical grid system (data per 1 MJ of electricity generated). (power plant location: PT - Portugal, ES - Spain)

		Energy (MJ/MJ)	CO_2 (g/MJ)
Ordinary regime. production		-	-
hydraulic	ES+PT	0.314	-
thermic	ES+PT	-	-
coal	ES+PT	0.377	61.33
Natural Gas	ES+PT	0.480	23.03
Fuel	ES+PT	0.004	2.56
Nuclear	ES	0.037	-
Combine cycle	ES	0.013	-
Especial regime production		-	-
hydraulic	ES+PT	0.028	-
thermic	ES+PT	0.569	0.27
wind	ES+PT	0.194	-
photovoltaic/solar	ES+PT	0.006	-
Total		2.021	87.19
Net Total		1.021*	

*MJ used to produce fuel/MJ final fuel used

Table A3. CO_2 and pollutant emissions factor regarding gasoline, diesel, and electricity production in Portugal 2004 (EcoInvent database [34])

	CO_2	CH_4	CO	VOC	NO_x	PM	SO_x	NH_3
Gasoline (kg/ton)	786.3	4.76	0.69	8.68	2.71	0.321	5.06	0.00032
Diesel (kg/ton)	403.7	4.25	0.61	7.98	1.95	0.163	1.92	0.00019
Electricity (kg/TJ)	*174940*	267.6	50.2	189.1	455.2	109	1524	0.2000

Table A4. Energy consumption and CO_2 emissions values associated to each process of gasoline and diesel production, WTT stage (data per 1 MJ of fuel produced)

	Energy consumed (MJ/MJ)			CO_2 (g/MJ)		
	Best est.	Min.	Max.	Best est.	Min.	Max.
Crude oil to gasoline						
Crude Extraction & Processing	0.03	0.01	0.04	3.6	-	-
Crude Transport	0.01			0.9	-	-
Refining	0.08	0.06	0.10	7.0	-	-
Distribution and dispensing	0.02			1.0	-	-
Net Total	0.14	0.12	0.17	12.5	-	-
Crude oil to diesel					-	-
Crude Extraction & Processing	0.03	0.01	0.04	3.7	-	-
Crude Transport	0.01			0.9	-	-
Refining	0.10	0.08	0.12	8.6	-	-
Distribution and dispensing	0.02			1.0	-	-
Net Total	0.16	0.14	0.18	14.2	-	-

Table A5. Average number of replacement of the main consumable elements in the vehicle (accounted for CTG stage). [39, 42, 43]

Maintenance/replacements	
Tires	2
Battery	2 (Pb, ICEV)
	1 (Lithium, BEV)
Engine Oil (ICEV only)	30
Brake Fluid	3
Transmission Fluid (ATF)	1
Powertrain Coolant	2
Winshield Fluid	19

Appendix B. TTW, WTT, and CTG detailed results

Table B1. TTW energy consumption and CO_2 emissions evolution scenario 2009-2050, per vehicle type. (Gas - Gasoline LDV, Di - Diesel LDV, BEV - Battery electric vehicle)

TTW					
year	Energy Consumption (MJ/km)			emissions CO_2 (g/km)	
	Gas	Di	BEV	Gas	Di
2009	2.46	2.10	0.43	179.2	156.5
2010	2.44	2.08	0.43	177.7	155.0
2020	2.22	1.89	0.40	162.1	140.7
2030	2.01	1.70	0.37	146.5	126.4
2040	1.80	1.50	0.35	130.9	112.1
2050	1.58	1.31	0.32	115.1	97.8

Table B2. TTW pollutant emissions per vehicle evolution (scenario 2009-2050). (Gas - Gasoline LDV, Di - Diesel LDV, BEV - Battery electric vehicle)

	TTW emissions											
year	CO (g/km)			NO_x (g/km)			PM (g/km)			VOC (g/km)		
	Gas	Di	BEV	Gas	Di	BEV	Gas	Di	BEV	Gas	Di	BEV
2009	*1.00*	*0.50*	--	*0.07*	*0.22*	--	--	*0.02*	--	*0.07*	*0.05*	--
2010	0.99	0.50	--	0.07	0.21	--	--	0.01	--	0.07	0.05	--
2020	0.90	0.45	--	0.06	0.19	--	--	0.01	--	0.06	0.04	--
2030	0.82	0.40	--	0.06	0.17	--	--	0.01	--	0.06	0.04	--
2040	0.73	0.36	--	0.05	0.15	--	--	0.01	--	0.05	0.04	--
2050	0.64	0.31	--	0.05	0.13	--	--	0.01	--	0.04	0.03	--

Table B3. WTT energy production efficiency and CO_2 emissions evolution for a scenario 2009-2050. (Gas - Gasoline LDV, Di - Diesel LDV, BEV - Battery electric vehicle)

	WTT					
Year	Energy Consumption (MJ/MJ)			emissions CO_2 (g/MJ)		
	Gas	Di	BEV	Gas	Di	BEV
2009	0.140	0.160	1.021	18.076	9.432	87.192
2010	0.140	0.160	1.021	18.076	9.432	87.192
2020	0.140	0.160	0.836	18.076	9.432	69.918
2030	0.140	0.160	0.651	18.076	9.432	52.644
2040	0.140	0.160	0.466	18.076	9.432	35.370
2050	0.140	0.160	0.281	18.076	9.432	18.096

Table B4. WTT pollutant emissions evolution for a scenario 2009-2050. (Gas - Gasoline LDV, Di - Diesel LDV, BEV - Battery electric vehicle)

Year	WTT emissions																	
	CO (g/MJ)			NOx (g/MJ)			PM (g/MJ)			VOC (g/MJ)			SOx (g/MJ)			NH3 (g/MJ)		
	Gas	Di	BEV	Gas	Di	BEV	Gas	Di	BEV	Gas	Di	BEV	Gas	Di	BEV	Gas	Di	BEV
2009	0.016	0.014	0.025	0.062	0.046	0.227	0.007	0.004	0.054	0.200	0.186	0.094	0.116	0.045	0.760	7.36E-06	4.44E-06	9.97E-05
2010	0.016	0.014	0.025	0.062	0.046	0.227	0.007	0.004	0.054	0.200	0.186	0.094	0.116	0.045	0.760	7.36E-06	4.44E-06	9.97E-05
2020	0.016	0.014	0.020	0.062	0.046	0.182	0.007	0.004	0.044	0.200	0.186	0.076	0.116	0.045	0.609	7.36E-06	4.44E-06	7.99E-05
2030	0.016	0.014	0.015	0.062	0.046	0.137	0.007	0.004	0.033	0.200	0.186	0.057	0.116	0.045	0.459	7.36E-06	4.44E-06	6.02E-05
2040	0.016	0.014	0.010	0.062	0.046	0.092	0.007	0.004	0.022	0.200	0.186	0.038	0.116	0.045	0.308	7.36E-06	4.44E-06	4.04E-05
2050	0.016	0.014	0.005	0.062	0.046	0.047	0.007	0.004	0.011	0.200	0.186	0.020	0.116	0.045	0.158	7.36E-06	4.44E-06	2.07E-05

Table B5. CTG energy consumption and CO_2 emissions evolution for a scenario 2009-2050 , per distance (km) travelled during the vehicle life time. (Gas - Gasoline LDV, Di - Diesel LDV, BEV - Battery electric vehicle)

	CTG					
Year	Energy Consumption (MJ/km life)			emissions CO_2 (g/km life)		
	Gas	Di	BEV	Gas	Di	BEV
2009	0.416	0.420	0.531	24.3	25.1	32.0
2010	0.414	0.419	0.528	24.2	24.9	31.7
2020	0.397	0.401	0.500	22.9	23.6	28.8
2030	0.380	0.384	0.472	21.6	22.4	25.9
2040	0.363	0.367	0.443	20.4	21.1	23.1
2050	0.346	0.350	0.415	19.1	19.8	20.2

Table B6. CTG pollutant emissions evolution for a scenario 2009-2050, per distance (km) travelled during the vehicle life time. (Gas - Gasoline LDV, Di - Diesel LDV, BEV - Battery electric vehicle)

Year	CTG emissions																	
	CO (g/km life)			NOx (g/km life)			PM (g/km life)			VOC (g/km life)			SOx (g/km life)			NH3 (g/km life)		
	Gas	Di	BEV	Gas	Di	BEV	Gas	Di	BEV	Gas	Di	BEV	Gas	Di	BEV	Gas	Di	BEV
2009	0.167	0.163	0.149	0.050	0.049	0.058	0.069	0.055	0.080	0.156	0.156	0.120	0.076	0.076	0.187	0	0	0
2010	0.167	0.163	0.149	0.050	0.049	0.058	0.069	0.055	0.080	0.156	0.156	0.120	0.076	0.076	0.186	0	0	0
2020	0.166	0.162	0.148	0.048	0.047	0.055	0.069	0.055	0.080	0.156	0.156	0.120	0.074	0.074	0.180	0	0	0
2030	0.165	0.161	0.147	0.046	0.045	0.052	0.069	0.054	0.080	0.156	0.156	0.120	0.072	0.072	0.175	0	0	0
2040	0.164	0.160	0.146	0.044	0.043	0.049	0.069	0.054	0.080	0.156	0.155	0.120	0.070	0.070	0.171	0	0	0
2050	0.163	0.160	0.145	0.042	0.041	0.046	0.069	0.054	0.080	0.155	0.155	0.120	0.067	0.068	0.168	0	0	0

Appendix C.

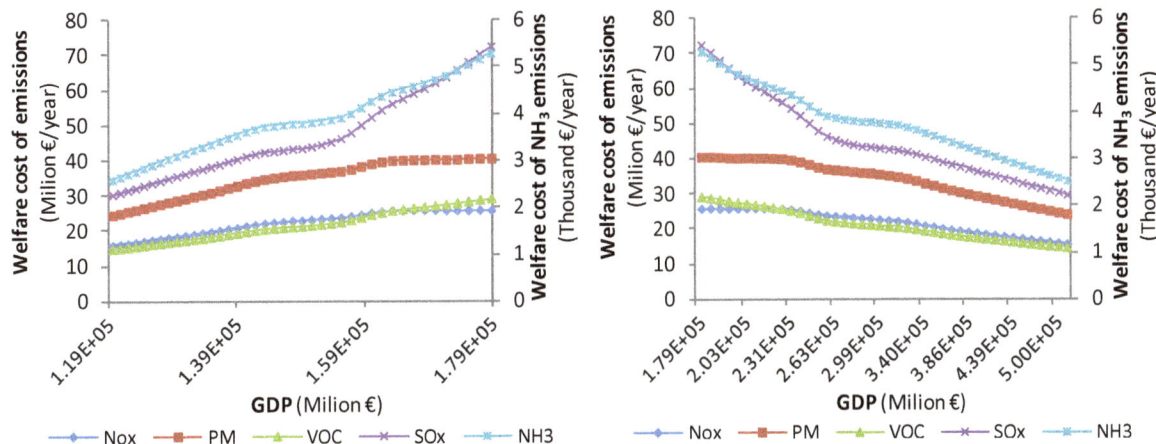

Figure C1. Welfare cost of pollutant emissions regarding GDP estimation from 2009-2050 considering a pessimist scenario (1% annual decrease in GDP) (a) and fénix scenario (2.6% annual increase in GDP) (b)

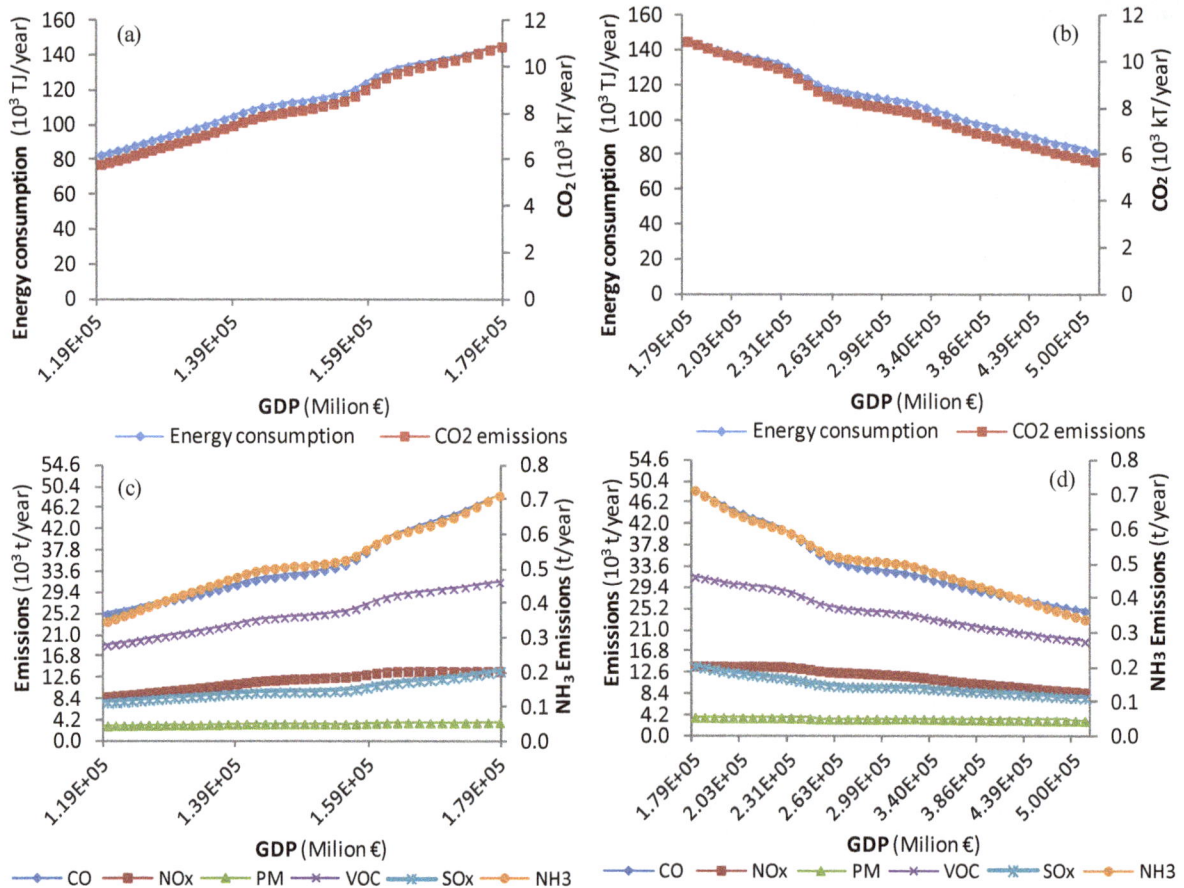

Figure C2. Energy and CO_2 emissions regarding GDP estimation from 2009-2050 considering a pessimist scenario (1% annual decrease in GDP) (a) and fénix scenario (2.6% annual increase in GDP) (b); and air pollutant emissions (c) and (d) for the same respective scenarios

Acknowledgments
Thanks are due to FCT - Fundação para a Ciência e Tecnologia (National Science and Technology Foundation) for the PhD financial support SFRH/ BD/ 68569/ 2010, and Post-Doctoral financial support SFRH/BPD/95098/2013.

References
[1]   Eurostat, European Commission. Portugal, 2012, see http://epp.eurostat.ec.europa.eu/ (accessed 26.02.2014).
[2]   DGEG. Direcção Geral de Energia e Geologia. Portugal, 2012, see http://www.dgeg.pt/ (accessed 26.02.2014).
[3]   Directive 2003/30/EC of the European Parliament and of the Council of 8 May 2003.
[4]   C. Silva, M. Ross, T. Farias, Energy Convers. Manage. 50, 215 (2009).
[5]   J. Ribau, C. Silva, F. P. Brito, J. Martins, Energy Convers. and Manage. 58, 120 (2012).
[6]   F. Ferreira, A. C. Marques, A. P. Batista, P. A. S. S. Marques, L. Gouveia, C. M. Silva, Int. J. Hydrogen Energy 37, 179 (2012).
[7]   F. Ferreira, J. Ortigueira, L. Alves, L. Gouveia, P. Moura, C. M. Silva, Biomass and Bioenergy 49, 249 (2013).
[8]   F. Ferreira, J. Ribau, C. M. Silva, Int. J. Hydrogen Energy 36, 13547 (2013).
[9]   F. Ferreira, L. Ribeiro, A. P. Batista, P. A. S. S. Marques, B. P. Nobre, A.M.F. Palavra, P. P. Silva, L. Gouveia, C. M. Silva, Bioresour. Technol. 138, 235 (2013).
[10]   N. S. Djomo, D. Blumberga, Bioresour. Technol. 102, 2684 (2011).
[11]   P. Campbell, T. Beer, D. Batten, Bioresour. Technol. 102, 50 (2010).
[12]   L. Lardon, A. Helias, B. Sialve, J. P. Steyer, O. Bernard, Environ. Sci. Technol. 43, 6475 (2009).

[13]   Demirbas, Biohydrogen – For Future Engine Fuel Demands (1st ed. Springer, London, U.K., 2009).

[14]   P. Baptista, Ph.D thesis, Technical University of Lisbon, Instituto Superior Técnico, Lisbon, 2011, see https://fenix.ist.utl.pt/homepage/ist151313/phd-thesis (accessed 20.01.2014).

[15]   P. Baptista, J. Ribau, J. Bravo, C. Silva, P. Adcock, A. Kells, Energy Policy 39, 4683 (2011).

[16]   Van Vliet, T. Kruithof, W.C. Turkenburg, A.P.C. Faaij, J. Power Sources 195, 6570 (2010).

[17]   M. A. Kromer and J. B. Heywood, Electric Powertrains: Opportunities and Challenges in the U.S. Light-Duty Vehicle Fleet (Sloan Automotive Laboratory, Laboratory for Energy and the Environment, Cambridge, USA, 2007).

[18]   R. Mota, Ph.D thesis, Technical University of Lisbon, Instituto Superior Técnico, Lisbon, 2011, see   http://bibliotecas.utl.pt/cgi-bin/koha/opac-detail.pl?biblionumber=447787   (accessed 26.02.2014).

[19]   Q. Zhang, B.C. Mclellan, T. Tezuka, K.N. Ishihara, Energy 61, 118 (2013).

[20]   P.K. Narayan and S.Narayan, Energy Policy 38, 661 (2010).

[21]   CAFE, Clean Air for Europe (2008), see http://ec.europa.eu/environment/archives/cafe/.

[22]   European Commission (2011), see http://ec.europa.eu/energy/energy2020/roadmap/index_en.htm.

[23]   J. Seixas, P. Fortes, J. Gouveia, L. Dias, S. Simões, B. Alves (2011), see http://climate.cense.fct.unl.pt/docs/Cen_socioeconomicos_2050.pdf (accessed 20.02.2014).

[24]   World Bank (2012), see http://databank.worldbank.org/ddp/home.do.

[25]   Agência   Portuguesa   do   Ambiente   (2011),   see http://www.apambiente.pt/_zdata/DPAAC/INERPA/NIR_20130517.pdf (accessed 22.02.2014).

[26]   R.P. Mota, T. Domigos, V. Martins, Ecological Economics 69, 1934 (2010).

[27]   S. Smulders, Green National Accounting. The New Palgrave Dictionary of Economics.(Palgrave MacMillan, Economics Online, 2008).

[28]   MOBI.E (2012), see http://www.mobie.pt/ (accessed 07.02.2014).

[29]   McKinsey & Company (2009), see http://www.fch-ju.eu/sites/default/files/documents/ Power_trains_for_Europe.pdf (accessed 23.02.2014).

[30]   CONCAWE (2011), see http://ies.jrc.ec.europa.eu/uploads/media/WTW_Report_010307.pdf (accessed 6.02.2014).

[31]   ISO 14040-14044, 2006. Environmental management - Life cycle assessment - Principles and framework, International Organisation for Standardisation, Geneva (2006).

[32]   T. Patterson, S. Esteves, R. Dinsdale, A. Guwy, J. Maddy, Bioresour. Technol. 131, 235 (2013).

[33]   K. Wipke, M. Cuddy, S. Burch, IEEE Transactions on Vehicular Technology 48, 1751 (1999).

[34]   Dieselnet. Emission Standards, 2012, see http://www.dieselnet.com/standards/.

[35]   M. Goedkoop, A. Schryver, M. Oele, S. Durksz, D. Roest (2008), see http://www.pre-sustainability.com.

[36]   EDP, Energia de Portugal S.A., 2011, see http://www.edp.pt (accessed 26.03.2014).

[37]   REN, National electricity grid, 2012, see http://www.ren.pt/ (accessed 26.03.2014).

[38]   Burnham, M. Wang, Y. Wu. U.S. Department of Energy's FreedomCar and Vehicle Technologies Program (2006).

[39]   Directive 2009/33/CE of the European Parliament and of the Council of 23 Apr 2009.

[40]   M. L. Valente, Master's thesis, Technical University of Lisbon, Instituto Superior Técnico, Lisbon, 2011, see http://bibliotecas.utl.pt/cgi-bin/koha/opac-detail.pl?biblionumber=454058 (accessed 6.02.2014).

[41]   Tandberg, (2006), see http://internetinnovation.org/files/special-reports/TAN_WhtPpr_Green_ FINAL.pdf (accessed 26.03.2014).

[42]   FORD. FORD Owners Guide, see https://owner.ford.com/servlet/ContentServer?pagename= Owner/Page/YearMakeModelSelectorPage&BackToLogin=Owner/Page/OwnerGuidePage (accessed 2.01.2014).

[43]   Green Car Congress, 2010, see http://www.greencarcongress.com/2010/07/chevrolet-volt-msrp-starts-at-41000-33500-net-of-full-federal-credit-3year-lease-program-with-option.html#more (accessed 3.01.2014).

2

An experimental study of temperatures in cloud from release of flashing liquid CO_2 in 3m long channel

Amrit Adhikari, André V. Gaathaug, Dag Bjerketvedt, Knut Vaagsaether

Telemark University College, Porsgrunn, Norway.

Abstract

Flashing of the liquid CO_2 is an accidental hazards that may occurs in many industrial sector such as, process industries, carbon capture and storage projects, crude oil extraction process etc. Sometimes the accidental release of liquid CO_2 that causes the health hazards which may costs loss of lives and properties. In order to alleviate the aforementioned probable hazards, the experiment will be highly beneficial. This research activity is conducted through the temperature measurement in the cloud of flashing liquid CO_2 confirming the formation of dry ice and measuring frontal velocity of the cloud as well as its height formed from the vapour CO_2 dispersion. The liquid CO_2 was released in the 3m long channel from the cylinder through two nozzles of diameter 0.5mm and 1.0mm. This leads the formation of dry ice measuring $-73^{\circ}C$ and $-71^{\circ}C$ from the nozzle sized 0.5mm and 1.0mm respectively. 0.5mm nozzle and 1.0mm nozzle having mass flow rate of liquid CO_2 as 0.0089 kg/s and 0.029 kg/s, overall frontal velocity of 0.52 m/s and 1.51 m/s thus formed cloud height measuring 0.05m and 0.1m respectively. Frontal velocity of the cloud was found to be highest at distance of 0.5m from nozzle with 1.49 m/s and 5.5 m/s for both nozzles 0.5mm and 1.0mm diameter respectively. Upon the increasing distance from the nozzle, the temperature of the formed cloud was seen to be in increasing order.

Keywords: CCS; Carbon dioxide; Dispersion; Flashing liquid; Temperature measurement.

1. Introduction

Accidental release of the pressurized liquid gases has caused numerous accidents all over the world. The release of pressurized gas generates flammable and toxic gases clouds. The hazard of CO_2 is a natural events, in Cameroon, West Africa, sudden release of Carbon Dioxide(CO_2) gas from Lake Nyos caused the deaths of around 1700 people on 21 August 1986 [1].

Carbon Capture and Storage(CCS) are known as globally mitigating technology as it can capture and store large amount of CO_2 and preventing it to reach the atmosphere. In CCS projects, large quantity of CO_2 is compressed to high pressure that can be transported to the storage site. Accidental release of pressurized CO_2 while transporting and storing cause flashing of liquid and a huge amount of gas is escaped which can cause frostbite and asphyxiation [2].

With the increase of the respiration problem, if CO_2 concentration is above 5% approximately it causes other health symptoms, such as headache, palpitation, breathing difficulty, weakness and dizziness. 20% of CO_2 gas is considered to be instantaneously fatal for the human health [3].

Lisbona et al. [4] wrote, "Currently, the source term for a CO_2 release is not well understood because of its complex thermodynamic properties and its tendency to form solid particles under specific pressure and temperature conditions. This is a key knowledge gap and any subsequent dispersion modelling, particularly when including topography, may be affected by the accuracy of the source term". It is very important to know and predict the behavior of flashing liquid so that an accidental release could be prevented and mitigated. The temperature measurement gives knowledge of understanding two phase dense gas dispersion. This paper presents the result from the experiments to investigate flashing scenarios of liquid CO_2. The major goal of this paper is to ensure and include:

- Released liquid CO_2 includes phase transition into mixture of gas and solid with formation of dry ice.
- Froude Scaling can be used to validate high pressure CO_2 dispersion.

CO_2 is a heavy-gas and under normal pressure and temperature it is a colorless and odorless gas. It is denser than air because of different specific weight of the air and CO_2 which leads to form a gravity current flow [5]. When liquid CO_2 is emptying from a horizontal channel with constant cross-section, it changes its phase to dry ice and vapour [6]. The dry ice is formed and deposited in the horizontal channel base and the vapour mixes with air to form cloud and moves forward with frontal velocity, u. The mass of the heavy gas cloud increased due to air entrainment. Figure 1 describes the flow design of the liquid CO_2 flowing in a horizontal channel with constant cross section area.

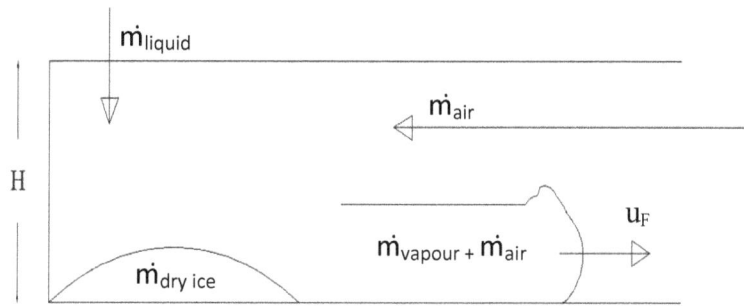

Figure 1. Flow model for liquid CO_2 released in the rectangular channel

The mass balance of the model from Figure 1 can be described as below:

$$\dot{m}_{liquid} + \dot{m}_{air} = \dot{m}_{dryice} + \dot{m}_{vapour} + \dot{m}_{air} \tag{1}$$

where $\dot{m}_{liquid}$ is mass flow rate of liquid CO_2 (kg/s), $\dot{m}_{dry\ ice}$ is mass flow rate of solid CO_2 (kg/s) and $\dot{m}_{air}$ is mass flow rate of air (kg/s).

2. Experiment setup

The experiment setup is shown in Figure 2 and Figure 3. The experiment was conducted on long channel tube comprising of 3m length, 0.1m width and 0.1m height. The system was made up of transparent polycarbonate and steel. The designing of tube was closed on one end and open at the other. The system comprised of five type K thermocouple with different time responding units. The thermocouple T#1 was placed at the side glass wall at height of 0.05m, thermocouple T#2 at distance 0.5m, thermocouple T#3 and T#4 at 1.5m, and thermocouple T#5 at 2.5m from the side wall of the tube. T#1 and T#4 has response time of 0.08 s, T#3 and T#5 with a respond time 1 μs and T#2 is slow responding thermocouple. The layout of the experimental setup of the channel has been shown in Figure 2.

Liquid CO_2 is released from the top of the channel with the nozzle of 0.5mm and 1.0mm diameter. The nozzle is placed 0.1m from the closed end of the channel.

The experimental setup for measuring temperature in the cloud from flashing of liquid CO_2 is illustrated in Figure 3.

The liquid CO_2 gas cylinder is connected to the pneumatic valve which is further connected to the nozzle. The liquid CO_2 was supplied from the cylinder. The mass flow of liquid CO_2 released was measured by standard 'HBM RSCA 100 kg' load cell device, and the cylinder was suspended in the load cell device. The initial and final weight of the cylinder during the release of liquid CO_2 with respect to time was noted down, which in fact, helps to calculate the mass flow rate of liquid CO_2 from the cylinder.

Signal from the thermocouples and load cell were connected to the 'QuantumX MX410' amplifier via which data were logged. Sampling rate of data was used as 2400 Hz.

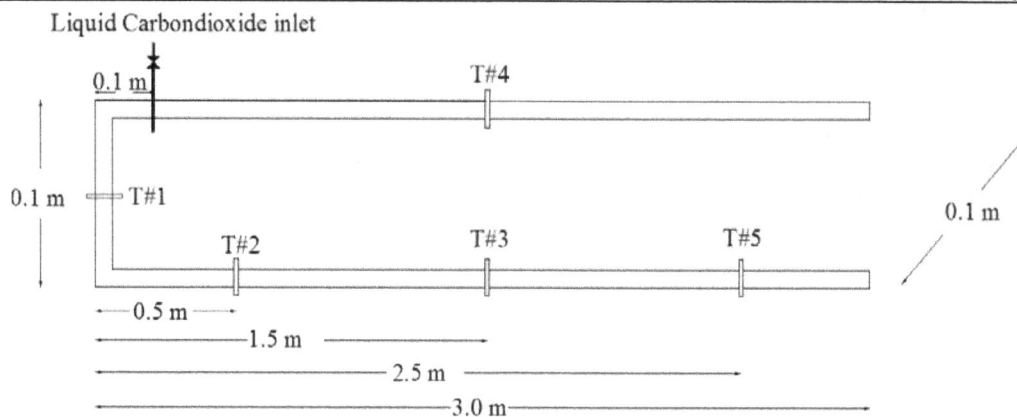

Figure 2. Schematic setup showing temperature measurement location and liquid CO_2 inlet

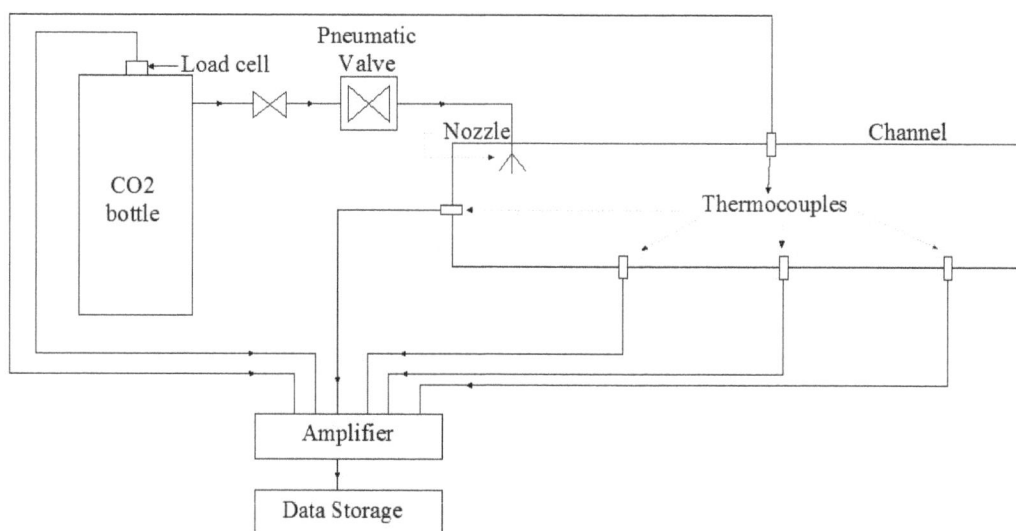

Figure 3. Experimental setup for the liquid CO_2 release system

3. Results and discussion

3.1 Frontal velocity and cloud height

Frontal velocity in this experiment is the velocity of CO_2-air cloud which is defined as, u_F (m/s) and which was found from HD video recording. As liquid CO_2 was released from the 0.5mm nozzle diameter, the overall frontal velocity of the cloud was found to be 0.52 m/s as shown in Figure 4.

Flow from the 1.0mm diameter nozzle gives the overall frontal velocity of 1.51 m/s which is shown in Figure 5.

Figure 4. Photo of carbon dioxide propagation with the 0.5 mm nozzle

Figure 5. Photo of carbon dioxide propagation with the 1.0 mm nozzle

The experimental results has shown two different regimes of the flow of vapour CO_2 inside the channel. The flow regimes are shown in Figures 4 and 5. Gravity current flow was observed from the release of 0.5mm nozzle, whereas plug flow was observed when release from 1.0mm nozzle diameter.

The time taken for CO_2 cloud to reach the various distance in the channel is shown in Figure 6 which gives the information about the initial and later frontal velocity of the cloud. Frontal velocity of the cloud was found to be increased rapidly until the cloud distance of 0.5m as shown in Figure 6.

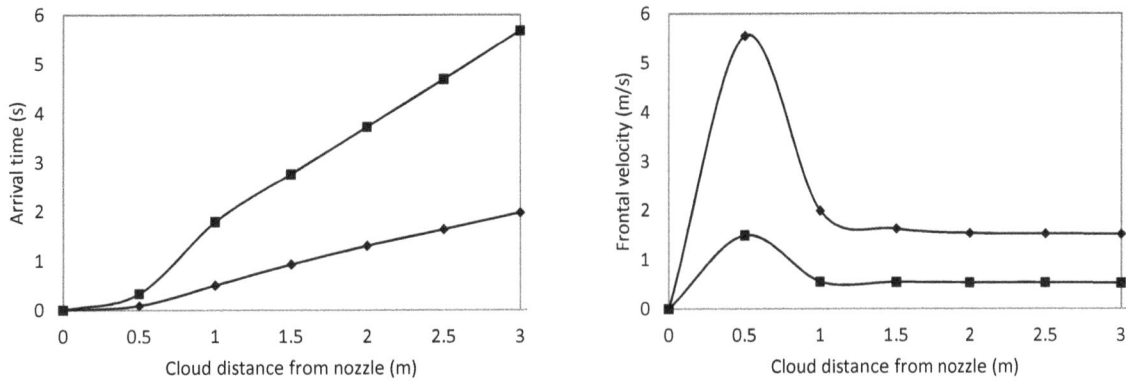

Figure 6. Distance variation with time and frontal velocity, symbols refers to nozzle diameter:
→◆ 1.0mm nozzle; →■ 0.5mm nozzle

At distance 0.5m form the nozzle, the frontal velocity of cloud when released from 0.5mm and 1.0mm nozzle diameter were found to be 1.49 m/s and 5.5 m/s respectively. Later after 1m of release distance, frontal velocity were approximately constant.

Using the video, the height of the cloud was estimated. As it can be noticed from Figure 5, there is a plug flow in the channel after liquid CO_2 is released from 1.0mm nozzle diameter. The height of the cloud is same as the height of the channel i.e. 0.1m .

Height of the cloud after it is release from 0.5mm nozzle diameter was found to be half of the height of the channel. From the video, it can be concluded that, the height of the cloud was approximately 0.05m as shown in Figure 4.

Nozzle diameter affects the height and frontal velocity of the cloud. Higher the nozzle diameter for the release of liquid CO_2 in the atmosphere, higher will be the cloud height as well as frontal velocity of the cloud.

3.2 Froude scaling

With knowledge of the frontal velocity, Froude scaling with this type of experiments can be performed. Froude number is defined as the ratio between the momentum and gravity force acting in the fluid flow which is given as:

$$Fr = \frac{u}{\sqrt{gh}}$$

(2)

where u is a velocity (m/s), g is the acceleration of gravity (m/s^2), h is height of cloud (m).
Experiment from 0.5mm nozzle shows frontal velocity of 0.52 m/s with Froude number of 0.74. Based on the observed Froude number, it suggests to be in a range of Froude scaling which was done in same type of experiment but with hydrogen-air [7]. Gravity current effect was observed from the release of 0.5mm diameter nozzle as seen in Figure 4 and with smaller nozzle than 0.5mm gravity current effects can be studied and used for the Froude scaling.

3.3 Mass flow of liquid CO_2

The mass flow of liquid CO_2 in 1.0mm and 0.5mm diameter nozzles were 0.029 kg/s and 0.0089 kg/s respectively as shown in Figure 7 verifying that the mass flow of liquid CO_2 is dependent on nozzle diameter. Wider the diameter of the nozzle, higher will be the mass flow of liquid CO_2.

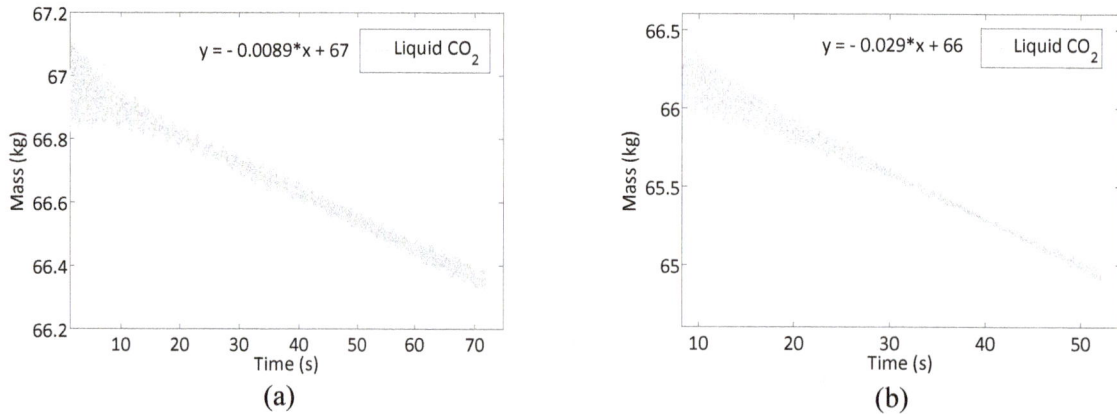

(a)   (b)

Figure 7. Mass flow of liquid CO_2 from (a) 0.5mm nozzle, (b) 1.0mm nozzle

3.4 Temperature measurements

In this experiment, the temperature of the cloud after flashing of liquid CO_2 was considered . The temperature reading of CO_2 (solid and vapour) when released from 0.5mm diameter nozzle as shown in Figure 8 below. Initially, all the thermocouple were at room temperature. With the release of liquid CO_2, the significant reduction in temperature was noticed. Whereas, air temperature by T#3 into the channel, remained almost constant throughout the flow.

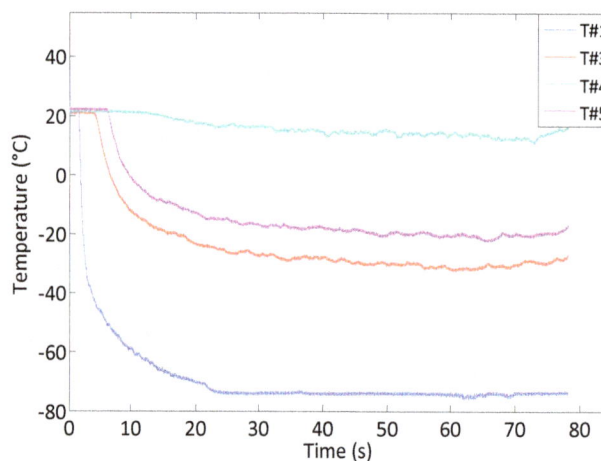

Figure 8. Temperature of CO_2 (solid and vapour) when released from 0.5mm diameter nozzle

T#1 suggests the formation of the dry ice whose temperature is about -73°C at 1 bar which is almost accurate as we find in literatures that a dry ice temperature at 1 bar is -78.4°C or 194.75 K [8].
Released of liquid CO_2 in the atmosphere from cylinder of high pressure changes its phase to solid and vapour. The temperature of the CO_2 cloud and dry ice is constant after it get stabilized at specific point

over time as shown by thermocouple T#1, T#3 and T#5 from both Figure 8 and Figure 9. The temperature reading of CO_2(solid and vapour), when released from 1.0mm diameter nozzle, is shown in Figure 9.

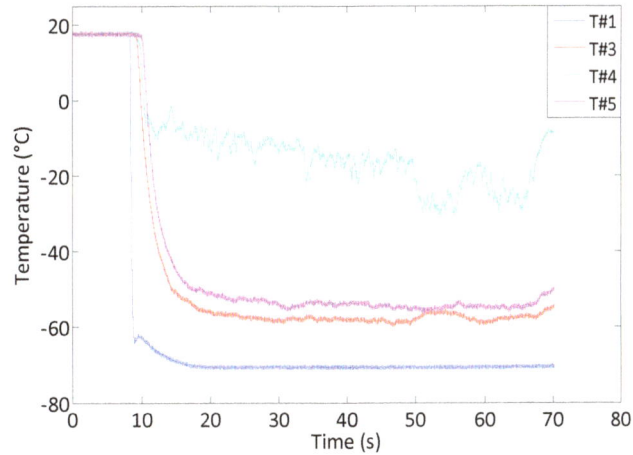

Figure 9. Temperature of CO_2 (solid and vapour) when released from 1.0 mm diameter nozzle

The air temperature(T#4) is not constant which is due to heat transfer from cloud CO_2 to air, and it is a plug flow as shown in Figure 5. Dry ice is formed in T#1 and T#2 which are at minimum temperature of -71°C and -63.5°C. Slow thermocouple T#2 takes relatively long time to reach the stabilized temperature and its graph is not shown in both Figure 8 and Figure 9. The image of the formation of dry ice on the tip of thermocouple T#1 and T#2 and on channel base can be seen in the Figure 10. Dry ice stays in the channel for the long time and vaporize into gas in atmospheric pressure.

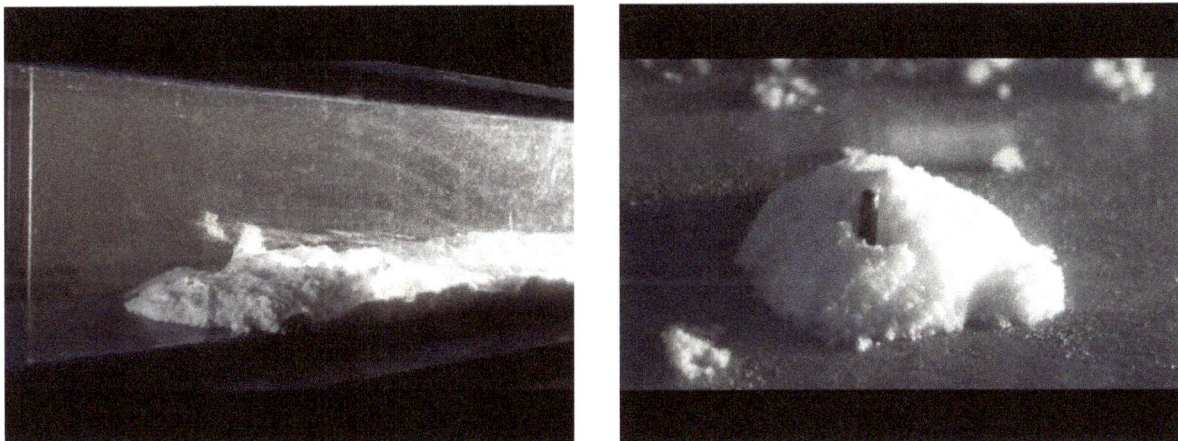

Figure 10. Picture of formation of dry ice

Temperature of CO_2 cloud increases with increasing distance of release from the nozzle. The reason behind this is due to the heat transfer into the system from the surrounding. Figure 11 shows that there is a continuous increase in temperature of the cloud as cloud hits the thermocouples down the channel. The stabilized temperature data were used to find out the relation between the opening area of release and temperature at various distance.

However, from both nozzles we can find the difference in the temperature at different distance. When the nozzle diameter is increased to twice(from 0.5mm to 1.0mm) then there is a decrement of CO_2 cloud temperature by approximately -32°C and -37°C at T#3 and T#5 distance respectively.

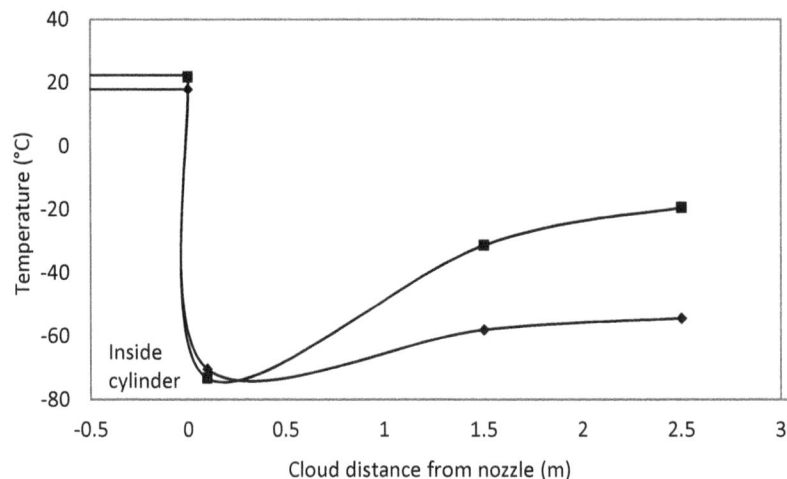

Figure 11.Stablized temperature of CO_2 cloud at different distance, symbols refers to nozzle diameter:
——◆——, 1.0mm nozzle;——■——, 0.5mm nozzle

4. Conclusion

The temperature measurement of cloud of the flashing liquid CO_2 was investigated by conducting experiments in a 3m long square cross section channel. The liquid CO_2 was released from two different nozzles of diameter 0.5mm and 1.0mm. Release of liquid CO_2 form 0.5mm and 1.0mm nozzle diameter gives the constant frontal velocity of 0.52 m/s and 1.51 m/s respectively over 3m long channel. Sudden release from the nozzle has higher frontal velocity of 1.49 m/s and 5.5 m/s for both nozzles of 0.5mm and 1.0mm diameter respectively. The cloud was found to be 0.05m and 0.1m when liquid CO_2 was released from 0.5mm and 1.0mm diameter nozzle respectively. The mass flow of liquid CO_2 in 0.5mm and 1.0mm diameter nozzle was found to be 0.0089 kg/s and 0.029 kg/s respectively. Using of small diameter nozzles to perform Froude scaling was suggested. The nozzle diameter suggests that wider the nozzle diameter higher will be the frontal velocity, cloud height and mass flow rate of liquid CO_2.

Dry ice and vapour CO_2 was formed from flashing of pressurized CO_2. Dry ice stays for the long time that influence the dispersion. Temperature of cloud was found to be in increasing with increase of the distance from the nozzle. The temperature of the cloud and dry ice is found to be constant at specific point and over time which also helped understanding the behavior of slow and fast responding thermocouple.

Further work is to study the heat transfer into the system. The heat transfer is important parameter in determining the temperatures and how it affects to vapourize dry ice into CO_2 gas. Further work will help to develop Froude scaling and models that can predict the concentration and dispersion phenomena of accidental release of liquid CO_2.

Acknowledgements
The authors acknowledge the support from the Statoil ASA.

References

[1]   Kling, G.W., et al., The 1986 Lake Nyos Gas Disaster in Cameroon, West Africa. Science, 1987. 236(4798): p. 169-175.

[2]   Scott, J.L., D.G. Kraemer, and R.J. Keller, Occupational hazards of carbon dioxide exposure. Journal of Chemical Health and Safety, 2009. 16(2): p. 18-22.

[3]   Kruse, H. and M. Tekiela, Calculating the consequences of a CO2-pipeline rupture. Energy Conversion and Management, 1996. 37(6-8): p. 1013-1018.

[4]   Lisbona, D., et al., Risk assessment methodology for high-pressure CO2 pipelines incorporating topography. Process Safety and Environmental Protection, 2014. 92(1): p. 27-35.

[5]   Simpson, J.E., Gravity Currents: In the Environment and the Laboratory. 2 ed. 1997, Cambridge University Press.

[6]   Molag, M. and C. Dam, Modelling of accidental releases from a high pressure CO2 pipelines. Energy Procedia, 2011. 4(0): p. 2301-2307.

[7]   Sommersel, O.K., et al., Experiments with release and ignition of hydrogen gas in a 3 m long channel. International Journal of Hydrogen Energy, 2009. 34(14): p. 5869-5874.

[8]   Witlox, H.W.M., M. Harper, and A. Oke, Modelling of discharge and atmospheric dispersion for carbon dioxide releases. Journal of Loss Prevention in the Process Industries, 2009. 22(6): p. 795-802.

3

Decomposition analysis of industry sector CO_2 emissions from fossil fuel combustion in Kazakhstan

Almaz Akhmetov[1,2]

[1] ENCA Management Ltd., 7 Pobedy Str., Esik, 040400, Kazakhstan.
[2] Orizon Consulting, 6481 Elm Str., Suite 161, McLean, VA, USA.

Abstract

The changes in industrial structure of Kazakhstan resulted in significant transformation on its CO_2 emissions profile. Understanding the driving factors in CO_2 emissions profile is essential given the emissions reduction targets committed by Kazakhstan. The study applies Index Decomposition Analysis to identify factors affecting industrial CO_2 emissions caused by fossil fuel combustion for the period 1990-2011. The results of the analysis indicated that the main factor affecting increase in total industrial emissions was the change in the industrial activity, while improvements in energy intensity helped to reduce the emissions. Analysis of six subsectors was used to define the main reasons underlying changes in CO_2 emissions.

The study underlines policy contradictions between national plans on expansion of carbon intense commodity based industries and Kazakhstan's international commitments on CO_2 reduction. Furthermore, the changes in structure of industrial output towards overreliance on commodity based industries and decline of manufacturing could indicate that Kazakhstan is vulnerable to resource curse.

Keywords: Kazakhstan; Index decomposition analysis; Industry; Decomposition; Fossil fuel.

1. Introduction

The Republic of Kazakhstan is a landlocked country located in the center of the Eurasian continent. Kazakhstan is the ninth largest in the world and it represents around 0.2% of the world's population, 0.3% of the world's GDP and 0.7% of world total CO_2 emissions [1]. The breakup of the USSR in 1991 has resulted in a sharp contraction of the economy. That led the Government of Kazakhstan to undertake reforms to establish a market economy, improve economic freedom and extensively develop its oil sector. As a result since 2000 the economy of Kazakhstan has been steadily growing mainly due to increased prices of oil on the world market. The economy of Kazakhstan is among the most energy and carbon intense in the world.

Kazakhstan's industry is primarily based on the extraction and export of the natural resources, primarily crude oil that country possesses in enormous amounts. Share of industrial output in GDP has increased from 20.5% in 1990 till 31.6% in 2011 [2]. The industry of Kazakhstan has undergone a significant structural transformation since the Soviet period as seen in Figure 1. Kazakhstan has transformed from diverse economy with a dominant share of processing industries into mostly oil export-dependent economy. Hence, the economy of Kazakhstan could be vulnerable to oil price volatility.

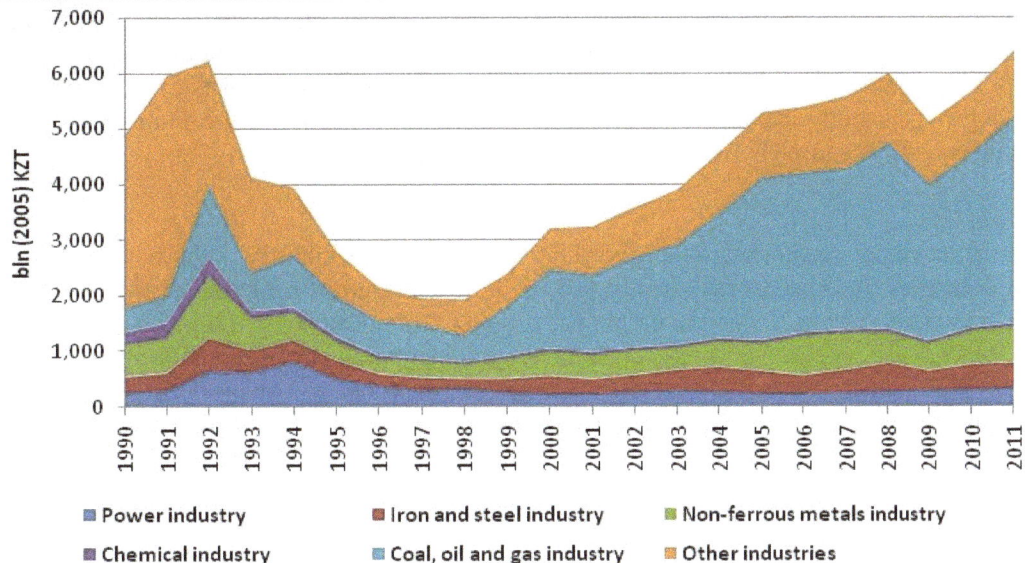

Figure 1. Structure of industrial output in Kazakhstan

1.1 Power industry

Power production in Kazakhstan mainly relies on thermal power plant. About 80% of all heat and electricity is produced by coal combustion. Traditionally, most of electricity generation comes from coal-fired power plants mainly built in the Soviet time. Due to harsh climatic conditions in winter, a significant amount of energy is utilized for district heating purposes. The heat is provided by cogeneration plants and boiler stations. Existing power plants and distribution infrastructure are often highly deteriorated and ineffective, what results in significant energy losses. The power generation sector is responsible for a majority of CO_2 emissions from industrial production and it is the most energy intense among the sectors.

1.2 Iron and steel industry

Although productions of steel and pig iron have decreased by 1.8 and 1.7 times respectively since 1990, the industry remains one of the most developed in the country. The industrial output has been on rise since 2000 following the decline caused by demand disruption after the collapse of the USSR in 1990s. The peak of industrial production was in 1992, and the output has not reached that value yet. Historically, coal has been the fuel of choice for the industry due to its abundance and cheap mining and transportation costs as the iron and steel production plants are located near the major coal mines.

1.3 Non-ferrous metals industry

Copper, lead and zinc has long been produced in the country. However, more recently production of aluminum, titanium, magnesium and other metals have become the focus. The industrial output has already surpassed 1990 level and the further growth is expected. The industry has increased its coal consumption by almost eight times since 1990. The metals are mainly exported.

1.4 Chemical industry

The chemical industry is based on the utilization of phosphate and various salt reserves and petrochemical industry. The industrial output has reduced 3.6 times since 1990 and the share of the industry in the total industrial output has shrunk from 4% till just 0.1% for the same period. However, the industry has been on the recovery path with the average annual growth of 5% since 1998, when the size of the industry has shrunk almost six times.

1.5 Coal, oil and gas industry

The industry is significantly important for the economy of Kazakhstan due to large reserves of hydrocarbons. The reserve-to-production (R/P) ratios of oil, natural gas and coal are 46, 82.5 and 293 years respectively [3]. Kazakhstan has strategic plans to increase production of coal, oil and gas. While oil and gas present a valuable export commodity, domestic consumption of coal is expected to increase by 12% by 2020 [4]. The share of coal, oil and gas industries in total industrial output has increased from just fewer than 10% in 1990 till almost 60% in 2011, while the shares of all other industries have reduced almost twice for the same period. This may indicate that the economy of Kazakhstan maybe vulnerable to the oil curse [5]. This study includes CO_2 emissions related to production and refining activities and does not include emissions caused by flaring associated petroleum gas.

1.6 Other industries

The industries included are machinery, food processing, pulp and paper industry, light industry and other non-specified industries. In other words, mostly processing and manufacturing industries. The share of the industries in the total national industrial output has dropped from 64% in 1990 till 19% in 2011.

Due to heavy dependence on cheap domestic coal, the environmental impact of the industry is significant in Kazakhstan. The industry causes almost 60% of the total national CO_2 emissions [6]. The power industry has had the biggest contribution to CO_2 emissions due to industrial activities as seen in Figure 2. Total CO_2 emissions caused by coal, oil and gas industry, non-ferrous metals industry and other industries increased comparing to 1990 level, while CO_2 emissions from iron and steel and chemical industries have reduced. The main driving factor for CO_2 emissions increase is rise in coal consumption.

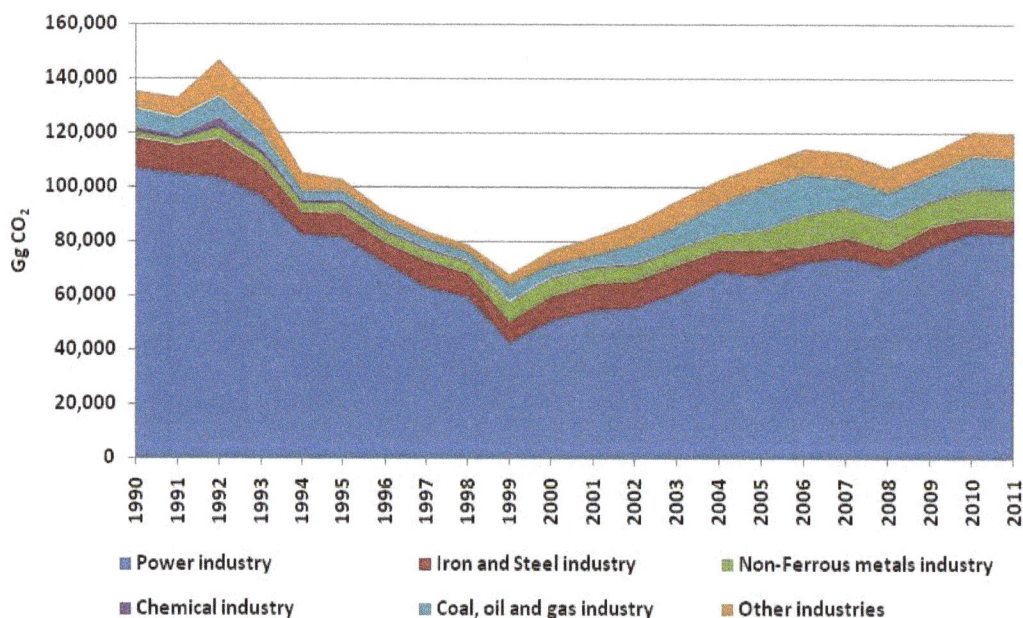

Figure 2. The total amount of CO_2 emissions related to fossil fuel combustion in six different industries

Kazakhstan is the non-Annex I party to the United Nations Framework Convention on Climate Change (UNFCCC). Quantified Emission Limitation or Reduction Objectives (QELROs) committed by Kazakhstan is 95% of 1990 base level by 2020 in Annex B of the Kyoto Protocol for the second commitment period [7]. The Government has developed the Concept of transition to green economy, where it has set the targets to reduce by 50% from the 2008 level, cut CO_2 emissions caused by electricity production by 40% and increase the share of renewable energy in electricity production by 50% by 2050 [8]. On top of that, the national Emission Trading Scheme with cap and trade approach has been launched. Hence, it is important to determine the factors affecting the growth of industry sector CO_2 emissions related to fossil fuel combustion.

In 2005, Karakaya and Ozcag [9] used decomposition analysis to define the driving forces of CO_2 emissions in Central Asia from fossil fuel combustion. The study distinguished between total primary energy supply and total final energy consumption. The factors investigated were: emission factor, energy intensity, fossil fuel intensity, conversion efficiency, economic output per capita and population for the period 1992-2001. The study revealed that the main driving force for CO2 emissions reduction in Kazakhstan due to reduction of economic activities following the collapse of the Soviet Union. Kojima and Bacon (2009) performed a multi-country decomposition analysis of CO_2 emissions from energy use for several time periods from 1994 till 2006 [10]. The methodologies used are five-factor decomposition (carbon intensity, fossil fuel share, energy intensity, GDP per capita and population effect) and six-factor decomposition that used methodology similar to Karakaya and Ozcag (2005) study. The study indicated that primarily economic activity was the main driving force for CO_2 emissions change in Kazakhstan.

2. Methodology

Index Decomposition Analysis (IDA) has been identified as the preferred methodology in energy and environmental studies to investigate the factors influencing energy consumption and its environmental impact [11]. Among the existing IDA methodologies the Logarithmic Mean Divisia Index (LMDI) method has become popular due to its theoretical robustness, adaptability, and ability to provide perfect decomposition [12].

This study aims to conduct a year-to-year decomposition analysis of the factors affecting industrial CO_2 emissions from 1990 to 2011 in Kazakhstan. The existing studies suggest decomposition of CO_2 emissions into five explanatory effects as follows [11-13]:

$$C = \sum_{i,j} C_{ij} = \sum_{i,j} Q \frac{Q_i}{Q} \frac{E_i}{Q_i} \frac{E_{ij}}{E_i} \frac{C_{ij}}{E_{ij}} = \sum_{i,j} Q S_i I_i M_{ij} U_{ij} \tag{1}$$

where C is the total CO_2 emissions (kt), C_{ij} are CO_2 emissions caused by consumption of fuel j by i industry, Q is total industrial output (billion Kazakhstani Tenge (bln KZT)), Q_i is the output of i industry (bln KZT), E_i is the use of fossil fuel by i industry (PJ), E_{ij} is the fossil fuel consumption of j type by i industry (PJ), S_i is the share of i industry in total industrial output, I_i is the energy intensity of i industry, M_{ij} is the energy mix of i industry, U_{ij} is the CO_2 emission factor of j fuel consumed by i industry.

Total changes in CO_2 emissions between target year T and base year (1990) could be expressed as follows:

$$\Delta C_{tot} = C^T - C^{1990} = \Delta C_{act} + \Delta C_{str} + \Delta C_{int} + \Delta C_{mix} + \Delta C_{emf} \tag{2}$$

where ΔC_{act} is the changed in CO_2 emissions caused by changes in activity, ΔC_{str} is the changes in CO2 emissions caused by industrial output structure, ΔC_{int} is the changes in CO_2 emissions caused by energy intensity, ΔC_{mix} is the changes in CO_2 emissions caused by fuel mix, ΔC_{str} is the changes in CO_2 emissions caused by emission factor. Where:

$$\Delta C_{act} = \sum_{ij} w_{ij} \ln \left(\frac{Q^T}{Q^{1990}} \right) \tag{3}$$

$$\Delta C_{str} = \sum_{ij} w_{ij} \ln \left(\frac{S_i^T}{S_i^{1990}} \right) \tag{4}$$

$$\Delta C_{int} = \sum_{ij} w_{ij} \ln \left(\frac{I_i^T}{I_i^{1990}} \right) \tag{5}$$

$$\Delta C_{mix} = \sum_{ij} w_{ij} \ln \left(\frac{M_{ij}^T}{M_{ij}^{1990}} \right) \tag{6}$$

$$\Delta C_{emf} = \sum_{ij} w_{ij} \ln \left(\frac{U_{ij}^T}{U_{ij}^{1990}} \right) \tag{7}$$

where w_{ij} is the logarithmic mean of industrial CO_2 emissions in year T and base year (1990) and expressed as follows:

$$w_{ij} = \frac{C_{ij}^T - C_{ij}^{1990}}{\ln C_{ij}^T - \ln C_{ij}^{1990}} \tag{8}$$

3. Data

The time interval under investigation ranged from 1990 till 2011 (Table 1). The industrial outputs in current prices for each sector were obtained from the Agency of the Committee on Statistics of the

Ministry of National Economy of the Republic of Kazakhstan [2], and adjusted to the constant prices of FY2005 using price deflator the United Nations Statistics Division [14].

Data on fossil fuel consumption, CO_2 emissions and implied CO_2 emission factors were acquired from Kazakhstan's national GHG inventory submitted to United Nations Framework Convention on Climate Change [6]. Biomass combustion and related CO_2 emissions are excluded from analysis as they are carbon-neutral.

Table 1. Fuel mix by industries in Petajoule (PJ)

Industry	1990					2011				
	Oil	Coal	Gas	Other	Total	Oil	Coal	Gas	Other	Total
Power	190.4	871.6	193.7	2.5	1258.3	12.8	797.7	145.8	0.0	956.3
Iron and steel	27.2	59.5	14.5	0.0	101.1	13.2	51.2	10.6	0.0	75.0
Non-ferrous metals	13.6	13.9	1.9	0.1	29.5	11.2	107.8	0.1	0.0	119.1
Chemical	2.0	1.9	23.9	2.7	30.6	0.0	0.5	11.9	0.0	12.4
Coal, oil and gas	47.4	4.3	42.1	5.9	99.7	60.7	3.0	112.8	0.0	176.6
Other	25.1	36.1	22.3	0.2	83.7	9.5	70.8	34.3	0.0	114.6
Total	305.7	987.3	298.4	11.4	1602.9	107.5	1031.0	315.6	0.0	1454.0

4. Results and discussion

Table 2 represents the results of decomposition analysis of CO_2 emissions in Gg. The results indicate that total CO_2 emissions from industrial activities in Kazakhstan have reduced by 15,455.1 Gg or 11% from 1990, while the total fossil fuel consumption have reduced by 9% for the same period. Coal and gas combustion have increased by 4% and 6% respectively, while oil consumption have dropped by 65% for the period 1990-2011. The activity effect indicates that CO_2 emissions would have grown by 24% if other effects had stayed constant. Improved energy intensity was the main factor for total CO_2 emissions reduction.

Table 2. Results of decomposition analysis 1990-2011 (Gg of CO2)

	ΔC_{tot}	ΔC_{act}	ΔC_{str}	ΔC_{int}	ΔC_{mix}	ΔC_{emf}
Power industry	-24,175.5	24,665.7	-6.485.7	-43,451.1	2,269.1	-1,173.5
Iron and steel industry	-5,720.5	2,170.1	1,949.4	-6,541.9	274.8	-3,572.9
Non-ferrous metals industry	8,419.9	1,399.7	-902.6	6,802.6	1,126.8	-6.6
Chemical industry	-1,138.4	290.2	-1,666.3	401.0	-41.7	-121.5
Coal, oil and gas industry	4,319.3	2,153.0	15,238.5	-12,795.7	-38.1	-238.3
Other industries	2,840.2	1,992.1	-9,084.9	9,426.1	342.4	164.5
Total	-15,455.1	32,670.9	-951.8	-46,159.0	3,933.3	-4,948.5

The results of the analysis are presented in the form of indexed time-series charts. The results indicate that coal, oil and gas industry, non-ferrous metals industry and other industries surpassed CO_2 emissions level of 1990, while power industry, iron and steel industry and chemical industry are still below that level.

The total CO_2 emissions from the power industry have reduced by 23% since 1990. However, the industry remains the biggest cause of CO_2 emissions in Kazakhstan. The main driving factor affecting CO_2 emissions changes caused by power industry is the industrial activity of the sector as seen in Figure 3. Energy intensity had the biggest contribution to the emissions reduction. The share of oil consumption in the fuel mix of the industry has dropped by 93% from 1990. Despite being the main cause of CO_2 emissions from total industry in Kazakhstan, the share of the power industry never exceeded 25% of the total industrial output.

CO_2 emissions related to the iron and steel industry have reduced by 51% since 1990. The decline in CO_2 emissions in 1990s was caused by output contraction, while improvements in energy intensity and emission factor due to fuel switching were the main causes of CO_2 emissions reduction in 2000s as in

Figure 4. For the whole period from 1990 to 2000, the main factors affecting emissions increase were the industrial activity and the output structure, while the energy intensity drove down the emissions by 58%.

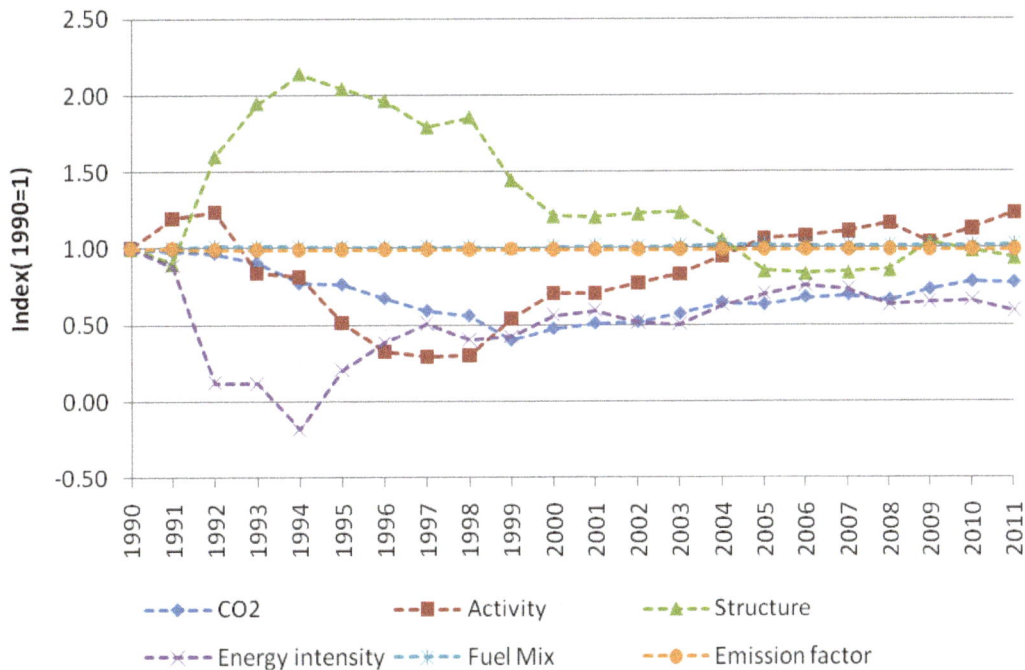

Figure 3. Results of decomposition analysis for power industry

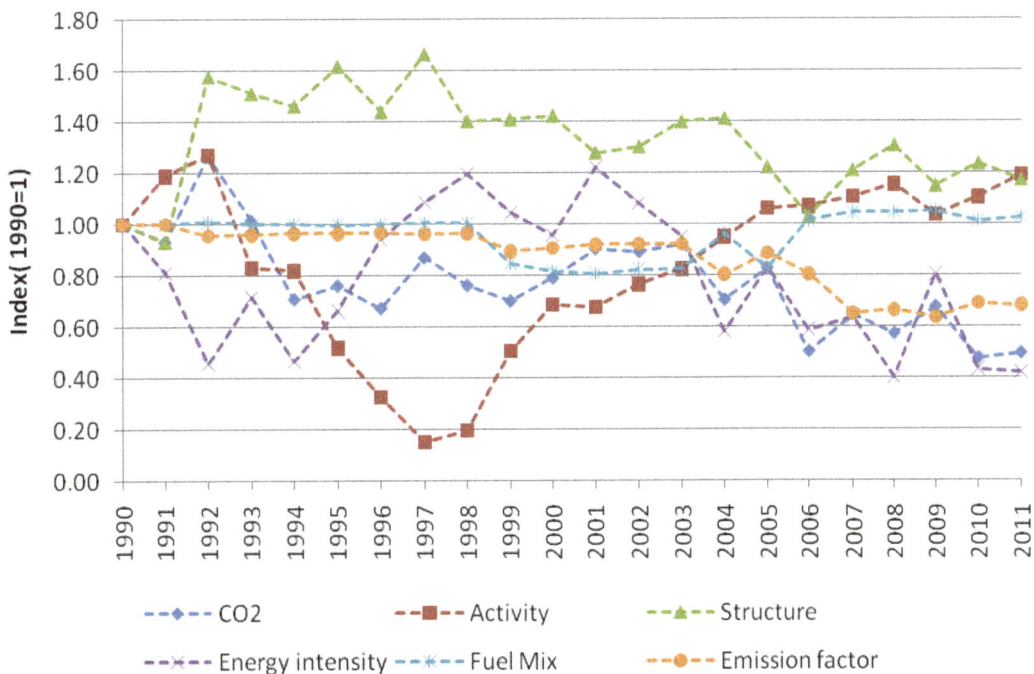

Figure 4. Results of decomposition analysis for iron and steel industry

The total CO2 emissions caused by the non-ferrous industry have increased by 347% for the whole period. The main reason behind CO_2 emissions increase is the energy intensity rise by 3.7 times since 1990 as displayed in Figure 5. Furthermore, the share of coal has reached 90% from 47% in 1990. This combination caused significant boost in CO_2 emissions caused by the industry.

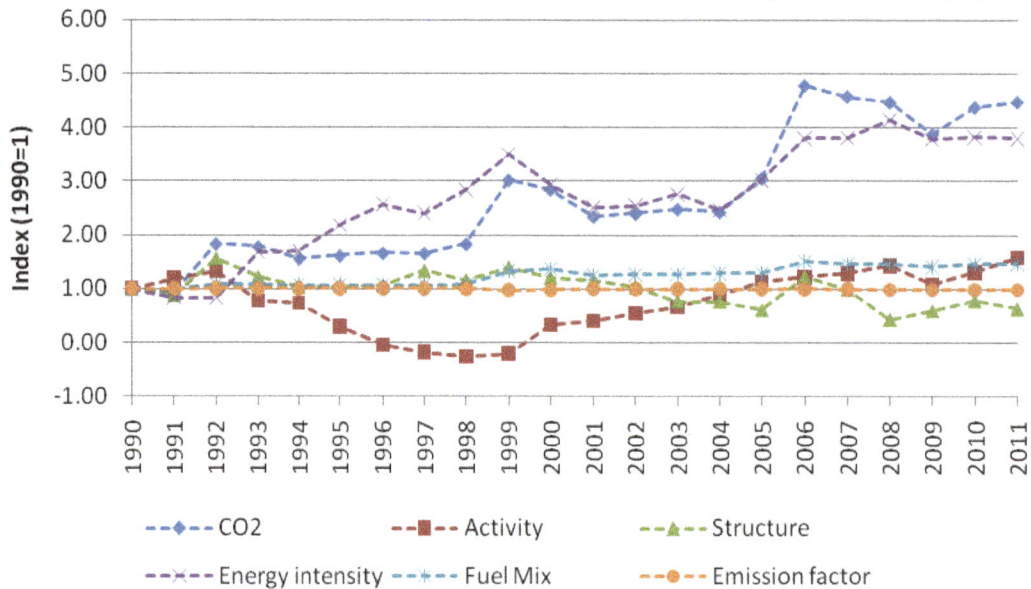

Figure 5. Results of decomposition analysis for non-ferrous metals industry

The total CO_2 emissions from chemical industry have declined by 62% since 1990. The main driving factor behind the decrease was the industrial structure effects that caused over 90% decline as seen in Figure 6. The industry is the only sector where gas is the dominant fuel in the mix and consisted 96% of the total fuel mix in 2011.

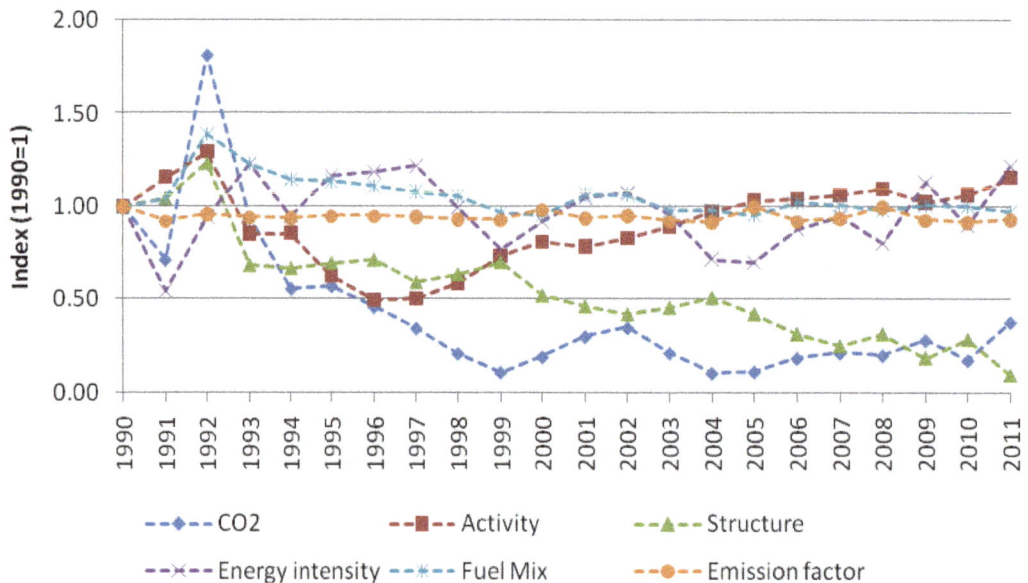

Figure 6. Results of decomposition analysis for chemical industry

The industrial output of coal, oil and gas industry has been increasing at the average rate of 14% annually since 1999 following the decline from 1992. Total CO_2 emissions caused by the industry have increased by 67% for analysis period. The main factor affecting the dynamics of CO_2 emissions from coal, oil and gas sector is the structure of industrial output in spite of improvements in energy intensity as seen in Figure 7. The industrial structure effect caused 237% increase in CO_2 emissions, while energy intensity factor pushed down emissions by almost 200% for the whole period. Furthermore, the results of the study most likely indicate that the industrial output of the coal, oil and gas sector highly depends on oil price fluctuations on the world market. This possibly explains energy intensity improvements of the sector despite increased fossil fuel consumption by 77% since 1990.

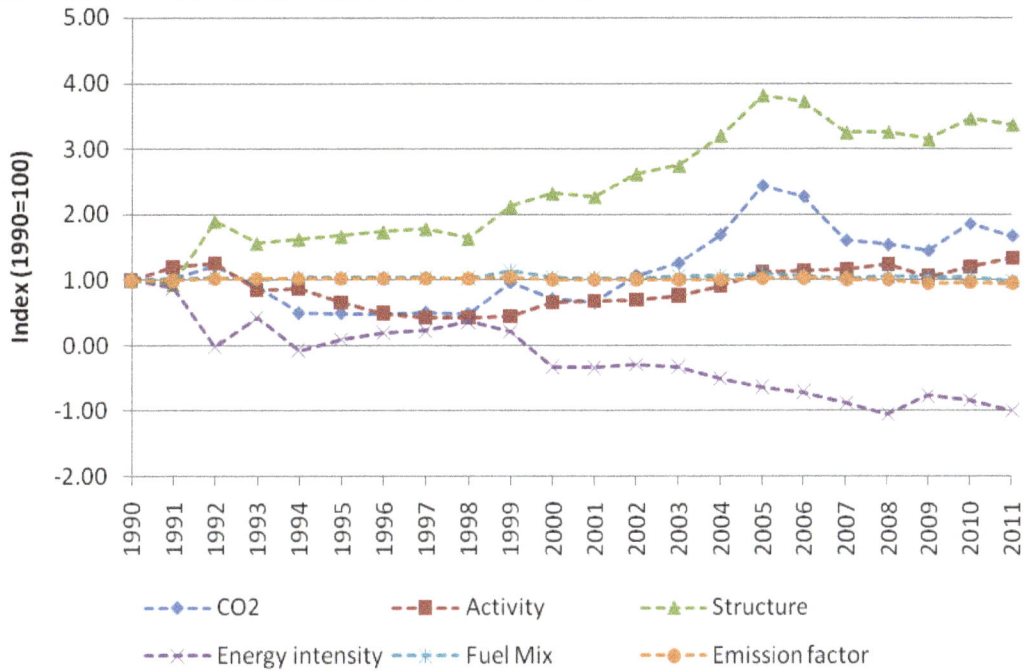

Figure 7. Results of decomposition analysis for coal, oil and gas industries

CO_2 emissions from the other sectors have increased by 45% since 1990. However, the emissions are below 1992 level when the industrial output and consequent environmental impact were at the peak as displayed in Figure 8. The biggest cause of CO2 emissions increase from the industry was energy intensity factor, while structure effect was the main driving force for reduction. The industry has increased coal consumption almost twice since 1990.

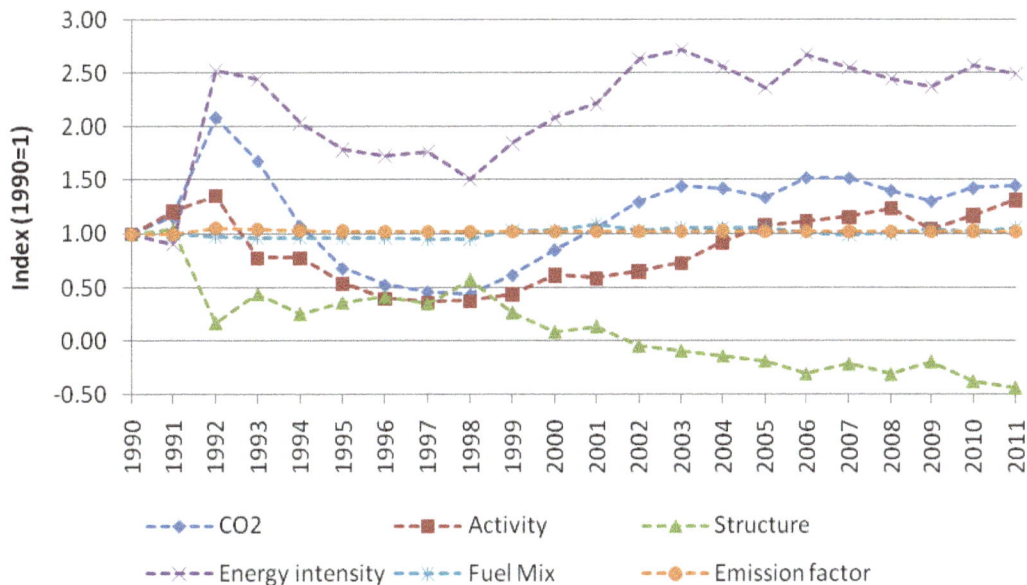

Figure 8. Results of decomposition analysis for other industries

5. Conclusion

From 1990 to 2011, CO_2 emissions related to fuel combustion by industry have increased by 11%. By applying LMDI methodology it was identified that changes in industrial activity pushed up total CO_2 emissions from industry by 24% followed by fuel mix with 3% increase, while changes in emission intensity, emission factor and structure of industrial output pushed down emissions by 34%, 4% and 1% respectively. Analysis of industries revealed that the relative CO_2 emissions reduction was achieved in chemical and iron and steel and power industries by 62%, 51% and 23% respectively since 1990.

Meanwhile, CO_2 emissions caused by non-ferrous metals, coal, oil and gas and other industries have increased by 347%, 67% and 45% respectively. Furthermore, it was identified that changes in industrial activity was the main driving force in emissions increase in power and iron and steel industries; energy intensity in non-ferrous metals, other and chemical industries; and the structure effect has significantly pushed up CO_2 emissions in coal, oil and gas industry. The energy intensity was the main factor to push down CO_2 emissions from coal, oil and gas, iron and steel and power industries, while changes in industrial output structure pushed down emissions in non-ferrous metals, chemical and other industries.

Although, Kazakhstan has achieved 31% increase in total industrial output since 1990, the growth occurred in power, iron and steel, non-ferrous metals and coal, oil and gas industries by 22%, 66%, 10% and impressive 770% respectively. On the other hand, chemical and other industries have dropped in size by 72% and 61% respectively. The transformation of industrial output towards over-reliance on natural resources export, crude oil in particular, may indicate that Kazakhstan is vulnerable to the phenomenon of the oil curse. The Government of Kazakhstan has been trying to diverse its industry from heavy dependence on export of hydrocarbons by development of non-energy intensive industries, measures for energy efficiency and energy saving improvement. However, a number of national industrial diversification programs have not succeeded. This is most likely due to greater corruption that often hits countries that undergo the oil curse [5].

Another important discovery from the analysis is the increase in coal consumption and reduction of oil presence in the fuel mix of the industry. In other words, coal, a fuel with a bigger environmental impact but cheaper cost, have become a main fuel for domestic industry, while oil and gas, major export commodities, have been sold on the world market. The fuel switch raises the questions of environmental justice and social equity in Kazakhstan.

A general policy conclusion on the basis of the study is that national strategy on increasing domestic coal consumption [4] and development of energy- and carbon-intense commodity based industries [15] contradicts Kazakhstan's international commitments on CO_2. This could create the incompatibility between national plans on transition to green economy and economic development of the country.

References

[1]   The World Bank Data, 2014. Kazakhstan. [online] Available at:<http://data.worldbank.org/country/kazakhstan> [Accessed 22 October 2014].

[2]   Ministry of National Economy of the Republic of Kazakhstan Committee on Statistics, 2014. Operational data. [online] Available at: < http://www.stat.gov.kz> [Accessed 22 October 2014].

[3]   BP, 2014. BP statistical review of world energy June 2014. 63rd ed. [pdf]. London. Available at: <http://www.bp.com/statisticalreview> [Accessed 22 October 2014].

[4]   Concept of development of fuel-energy complex of the Republic of Kazakhstan until 2030. 2014 SI 724. [in Russian] Astana.

[5]   Ross L. M. The oil curse: How petroleum wealth shapes the development of nations. Princeton University Press, 2012.

[6]   UNFCCC, 2014. National inventory submissions 2013. [online] Available at: <http://unfccc.int/national_reports/annex_i_ghg_inventories/national_inventories_submissions/items/7383.php>.

[7]   Sergazina G., Khakimzhanova B. Kazakhstan: Status of ETS development and need for support. Marrakech, Morocco, 23 October 2013. Marrakech: Partnership for market readiness.

[8]   Conception of Kazakhstan on transition to green economy. 2013 SI 577. [in Russian] Astana.

[9]   Karakaya E., Ozcag M. Driving forces in Central Asia: A decomposition analysis of air pollution from fossil fuel combustion. Arid Ecosystems Journal 2005, 11(26-27), 49-57.

[10]   Kojima M., Bacon R. Changes in CO2 emissions from energy use: A multicountry decomposition analysis. The World Bank, 2009.

[11]   Ang B.W., Zhang F.Q. A survey of index decomposition analysis in energy and environmental studies. Energy, 2000, 25, 1149-1176.

[12]   Ang B.W. Decomposition analysis for policymaking in energy: which is preferred method? Energy Policy, 2004, 32, 1131-1139.

[13]   Ang B.W., Xu X.Y. Tracking industrial energy efficiency trends using index decomposition analysis. Energy Economics, 2013, 40, 1014-1021.

[14]   The United Nations Statistics Division, 2014. National accounts main aggregates database. Implicit price deflators in national currencies and US dollars. [online] Available at: <http://unstats.un.org/unsd/snaama/dnllist.asp>. [Accessed 22 October 2014].

[15] Social-economic development forecast of the Republic of Kazakhstan 2014-2018. 2013 SI 33. [in Russian] Astana: Ministry of Economy and Budget Planning of the Republic of Kazakhstan.

4

Use of sugarcane straw ash for zeolite synthesis

Denise Alves Fungaro, Thais Vitória da Silva Reis

Instituto de Pesquisas Energéticas e Nucleares, IPEN–CNEN/SP- Av. Prof. Lineu Prestes, 2242, Cidade Universitária, CEP 05508-000 São Paulo SP, Brasil.

Abstract

The amount of biomass combustion residue is growing nowadays due to constant increasing demands of biomass utilization. The biomass ash produced currently is disposed on agricultural fields. The presence of metals, chlorine, sulphur and other species may have significant impacts on soils and the recycling of soil nutrient. The main challenge is related to the increase of possible applications of this byproduct. Sugarcane straw ash (SCSA) was used in a study on synthesis of zeolitic material by alkaline conventional hydrothermal treatment. Different experimental conditions, such as, reaction time, alkali hydroxide concentration and liquid/solid ratio were studied. Raw ash material and synthesis products were characterized by X-Ray Fluorescence, Fourier transform infrared spectroscopy powder, X-ray diffraction, cation exchange capacity and scanning electron microscopic. The presence of zeolite hydroxysodalite confirms successful conversion of native SCSA into zeolitic material. Sugarcane straw ash utilization minimizes the environmental impact of disposal problems and further appears as an alternative for the future sustainable large-scale management of biomass ash.

Keywords: Sugarcane straw; Biomass ash; Zeolite; Hydrothermal treatment.

1. Introduction

In recent years, there has been an increasing trend towards more efficient utilization of lignocellulosic agro-industrial residues and among them, sugarcane straw and bagasse, wastes generated by the sugar and alcohol production processes.

The sugar industry is one of the most important agro-based industries in Brazil, India, and others developing countries. Brazil is the largest producer of sugarcane of the world generating 400 million tons per year, on average [1].

Sugarcane straw is the material that is removed before the cane is crushed and comprises the dried/fresh leaves and the top of the plant [2, 3]. A typical sugarcane harvest in Brazil will generate at least 90 million tonnes of straw, of which 3-5% will be turned into ashes [4].

Inappropriate dumping method of sugarcane straw ash (SCSA) creates soil pollution and also causes air pollution with an allergic problem to human being [5-7].

It has been reported that sugarcane straw ash obtained from heaps of open-air burnt straw in the vicinity of a sugar factory showed a high pozzolanic activity [8].

Villar-Cociña et al. [9] studied the pozzolanic behaviour of a mixture of sugarcane straw ash with 20 and 30% clay burned at 800 and 1000 $^{\circ}$C and calcium hydroxide and proposed a kinetic–diffusive model for describing the pozzolanic reaction kinetics.

Other possible utilization of SCSA is the conversion into higher level product zeolitic material. Sugarcane straw ash contains aluminosilicates glass, mullite and quartz, which are the required building blocks of zeolite formation [10-12]. Zeolitic materials have high potential to be developed as efficient low cost adsorbents with application in environmental problems, particularly in liquid effluents treatment through the process of ion exchange.

Some authors have reported the conversion of bagasse fly ash into zeolites [13-19]. To the best of our knowledge, are no reports on the conversion of sugarcane straw ash into zeolites.

The objective of this study is to investigate the conversion of sugarcane straw ash, a sugar industry solid waste into zeolites by conventional hydrothermal treatment. During the hydrothermal synthesis procedure, several variables were evaluated: concentration of the activator; reaction time and solution/SCSA ratio.

2. Experimental

2.1 Materials

All the regents used were of analytical grade. The sugarcane straw was collected in Piracicaba city located at state of São Paulo, Brazil. This waste was gathered as dry straw, it was cut in order to obtain fibres between 20 and 30mm long and the material was ground by oscillating mill. Sugarcane straw ashes (SCSA) from the combustion of sugarcane straw were obtained in a muffle furnace at $700^{\circ}C$ during 20 min.

2.2 Zeolite synthesis

The SCSA was converted to zeolite using the conventional hydrothermal treatment. SCSA was mixed with aqueous NaOH solution in a Teflon vessel in a pre-determined ratio. This mixture was heated to $100^{\circ}C$ in oven. After hydrothermal treatment, the reaction mixtures were filtered and washed with distilled water until the washing water had pH ~ 11 and the synthesis products were oven dried at $50^{\circ}C$ for 12 h. The synthesis conditions used with different samples such as, reaction time, alkali hydroxide concentration and NaOH solution volume/mass of SCSA ratio (L/S) are presented in Table 1.

Table 1. Synthesis conditions for alkaline activation of SCSA by using conventional hydrothermal treatment

Reaction time (h)	[NaOH] (mol L^{-1})	L/S (mL/g)	Sample
45	2.0	8	TA-1
45	3.5	8	TA-2
72	2.0	10	TA-3
72	3.0	10	TA-4

2.3 Characterization of materials

The chemical compositions of materials in the form of oxides were analyzed by energy dispersive XRFX-ray fluorescence spectrometry (RX Axios Advanced, PANalytical, Phillips spectrometer). Phase and crystallinity were confirmed by powder XRD (Rigaku Multiflex) with nickel filter CuKα radiation (λ = 1.54060 Å) scanning from 5 to 80° at a rate of 1°/min with current 40 kV and 20 mA. The Fourier transform infrared spectroscopy (FTIR) spectra were recorded on Nexus 670 Thermo Nicolet using KBr pellet method. Scanning Electron micrograph was obtained by using XL-30 Philips scanning electron microscope (SEM). The particle size of the materials was measured using a laser based particle size analyzer, namely a Malvern MSS Mastersizer 2000 Ver. 5.54. For the cation exchange capacity (CEC) measurements, the samples were saturated with sodium acetate solution (1 mol L^{-1}), washed with 1L of distilled water and then mixed with ammonium acetate solution (1 mol L^{-1}) [20]. The sodium ion concentration of the resulting solution was determined by optical emission spectrometry with inductively coupled plasma - ICP-OES (Spedtroflame - M120).

3. Results and discussion

3.1 Sugarcane straw ash characterization

The composition of biomass ash is dependent on the plant species, materials that the plant absorbed from the water or the soil during its growth, growth conditions and ash fraction. The chemical compositions of SCSA in the form of oxides are shown in Table 2. The major component was SiO_2 along with small amounts of CaO, K$_2$O, Al$_2$O$_3$ and Fe$_2$O$_3$. Loss on ignition (LOI) implied in high weight loss of about 60%, and is mainly attributed to the presence of organic matter in the waste sample.

Table 2. Chemical composition of the sugarcane straw ash

Oxides	% wt	Oxides	% wt
SiO$_2$	25.3	P$_2$O$_5$	0.44
Al$_2$O$_3$	0.68	MnO	0.38
Na$_2$O	0.05	TiO$_2$	0.06
K$_2$O	3.78	BaO	0.14
CaO	3.98	Cr$_2$O$_7$	0.08
SO$_3$	0.92	SrO$_2$	0.01
Fe$_2$O$_3$	1.19	NiO	0.001
MgO	0.96	ZnO	0.01
Cl	0.41	LOI	61.6

Silica is absorbed from the soil through the roots of sugarcane. Accumulated silica between the plant's cuticle and cell walls acts as a physical barrier against the penetration of pathogenic fungi and reduces water loss through transpiration [21].

The percentages of the main oxides present in sugarcane straw ash are lower than values found in the literature [10, 11]. This could be related to the different calcining temperature used and controlled or uncontrolled calcining temperatures conditions. Furthermore, the silica content is probably related to the soils where sugarcane grows, as well as to other factors, such as fertilization methods and soil management.

Generally, the ash from biomass combustion content include Ca, Si, Al, Ti, Fe, Mg, Na, K, S and P. The composition of ash affects its behavior under high temperatures of combustion and gasification reactors and is responsible for technological and environmental problems during biomass processing. Problems like clogged ash-removal caused by slagging ash, sintering, deposition, erosion, corrosion and pollutant emissions are mainly created by the presence of alkali metals, alkaline earth metals, silicon, chlorine and sulphur in the ashes. Ash weight content (in dry basis) of different herbaceous biomass types have reported values ranging from less than 2% up to 8-12 % [22].

An extended overview of the phase–mineral and chemical composition and classification of biomass ash was conducted by Vassilev et al. [23].

The powder XRD pattern of SCSA is shown in Figure 1. SCSA shows a very low crystallinity and wide band is observed between 10 and 30 (2-theta), which implies the presence of vitreous matter. This fact is attributed to glass forming constituents since the organic constituents would have been removed during combustion. It has been reported that at burning temperatures up to 800°C silica was in amorphous form and silica crystals grew with time and temperature of incineration [24].

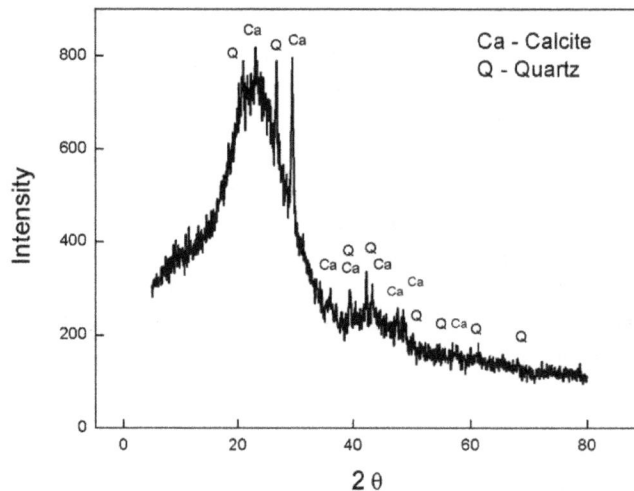

Figure 1. XRD patterns of sugarcane straw ash

Moreover, the XRD patterns suggest the presence of calcite ($CaCO_3$, ICDD/JCPDS 01-071-3699) as the main crystalline compound with quartz (SiO_2, ICDD/JCPDS 001-0649) as minor compounds. The amorphous aluminosilicate nature of SCSA makes its chemical composition difficult to characterize, but also very versatile, since the glassy phase both react and also goes into solution before the crystalline phase.

SEM micrograph of SCSA (Figure 2) seems like fibrous material containing large shallow pores with strands in each fold. The SCSA contain noncrystalline glass phase, with a loose structure and possess smooth surface particle because the surface is covered by an aluminosilicate glass phase. SCSA also possesses irregular shapes and surfaces.

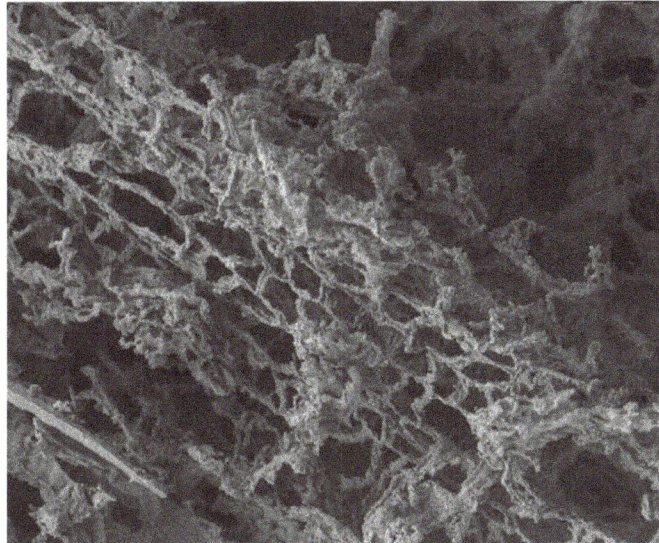

Figure 2. Scanning electron microscope micrographs of SCSA

The particle size distributions are important for predicting the size of ash particles and the tendency to impact surfaces. Figure 3 shown the absolute and cumulative, respectively, particle size distributions for the sugarcane straw ash (SCSA).

Relevant parameters are again listed in Table 3. These distributions specify that the majority of particles (90%) lie below 246.239 μm. The surface area determined by wet method was $0.0894 m^2 g^{-1}$.

It must be noted that fine-grained materials are more effective, when utilized for heavy metals retaining, a procedure that is also strongly dependent on the nature of the contaminated.

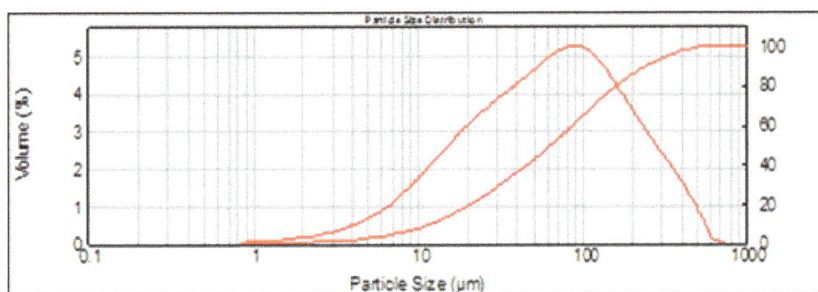

Figure 3. Particle size distribution of the sugarcane straw ash

Table 3. Parameters characteristics of size distribution of sugarcane straw ash

	Particle diameter
D_{10} (μm)	11.293
D_{50} (μm)	63.888
D_{90} (μm)	246.239
D ([3,2])	26.665

Classic alkaline conversion of ash is based on the combination of different activation solution/ash ratios, with temperature, pressure, and reaction time to obtain different zeolite types. The applied technique mainly aims at the dissolution of Al-Si bearing phases of the ash and the subsequent precipitation of the zeolitic materials.

The ash activation is usually carried out varying the activation agent (mainly KOH and NaOH), temperature ($20-250^{\circ}$C), conversion time (2 h a 7 days), solution concentration (0.1– 8 mol L^{-1}), pressure (the vapor pressure at the temperature selected), and solution/sample ratio (1-20 mL g^{-1}) [25-27].

In the present study, experimental conditions shown in Table 1 were selected on the basis of conditions from previous reports in the literature [25]. NaOH was selected as an activation solution, since its solutions present higher conversion efficiency than the respective KOH under the same conditions. The NaOH concentrations and temperature are typical for a pure alkaline activation, taking place at low temperatures and intermediate activation periods.

As the chemical analysis results indicated (Table 2), the SCSA ash contains very little Al_2O_3, so longer reaction times (45-72 h) are required to obtain zeolitic material without adding of aluminum or preliminary fusion with alkali.

Samples TA-2 and TA-4 synthesized using more concentrated NaOH solutions (NaOH > 2 mol L^{-1}) showed colloidal properties and the solid-liquid separation by filtration was very difficult. Therefore this condition was excluded as a parameter of synthesis.

The XRD patterns of the two samples TA-1 and TA-3 appeared in Figure 4. The common crystalline phases identified in TA-1 and TA-3 from the d-spacing values are quartz and calcite, both of them are present in the unmodified sugarcane straw ash as major content (Figure 1). This significantly demonstrates that under the given experimental conditions, these phases have not been digested completely.

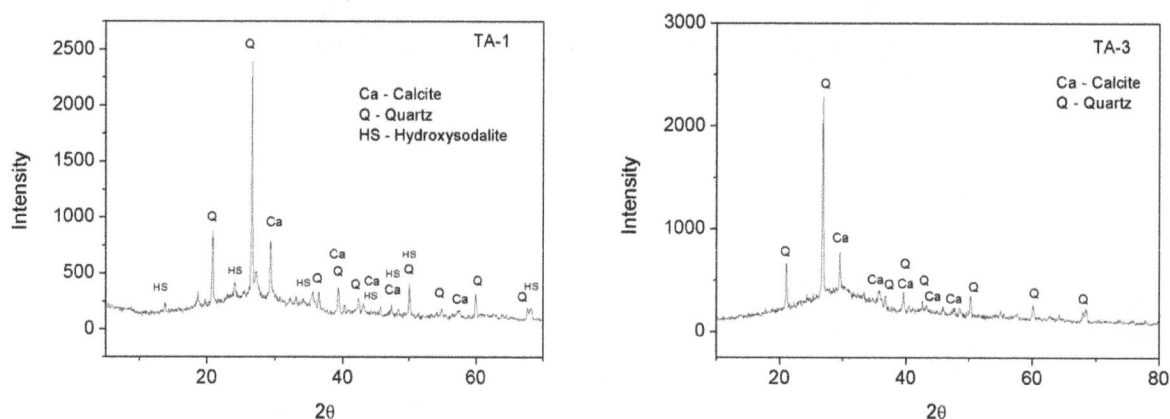

Figure 4. XRD patterns of the synthesis products obtained by hydrothermal treatment of sugarcane straw ash: TA-1 and TA-3

The high quantity of amorphous material in the SCSA (Figure 1) was found to reduce signals of crystalline material present and this can be confirmed by the increase of intensity of calcite and quartz peaks after the hydrothermal treatment (Figure 4).

During the ageing process, the amorphous phase undergoes dissolution in the alkaline media before new crystalline phases are formed and this can be confirmed by the disappearance of the broad diffraction hump that had appeared in SCSA the XRD pattern (Figure 1).

Several new sharp diffraction peaks obtained in TA-1 which were not observed in PXRD of TA-3. The observed newly peaks with d-spacing values of 14, 20, 24.7, 35, 43.4, 45.7 and 50° are ascribed to formation of zeolitic units in TA-1. The zeolite hydroxysodalite (JCPDS 011-0401) was positively identified in TA-1. Thus, under treatment conditions of sample TA-1, quartz phases and some parts of mullite get digested and consequently these phases involve in the conversion of ash into zeolite.

The activation of the studied SCSA with treatment conditions of sample TA-3 did not produce noticeable quantities of crystalline zeolites. This fact indicated that when higher crystallization time and/or higher solution/ash ratio are applied, impurities of other compounds may effect on the formation of zeolite.

Hydroxysodalite (HS) is crystalline aluminosilicate with a three dimensional channel network, belonging to clathrasil group. It exhibits similar structure of sodalite consisting of six-member oxygen aperture where framework charges are balanced with hydroxide anions.

HS has a pore size of 2.65 Å, which provides access to small molecules, e.g., helium (~2.6 Å), water (~2.7 Å), ammonia (~2.5 Å), making it a potential membrane material in separation of small molecules. Further, it is attractive as a functional material for a wide range of applications, such as removal of heavy metals and dyes from water, desalinating seawater and other salt solutions, optical materials, hydrogen storage, hydrogen separation, catalysts, and pigment occlusion [27-31].

The cation exchange capacity (CEC) and selectivity are definitely the most important characteristics of zeolites, because a substantial part of water treatment processes are based on the selective adsorption of metal cations, ammonium or polar organic compounds. In addition, cation exchange properties of zeolites can be exploited to modify their surface chemistry such that other classes of compounds can also be retained [29, 31].

The CEC rose from 0.245 meq g^{-1} for the unmodified Sugarcane straw ash to 0.602 meq g^{-1} for the modified product TA-1. The alkaline hydrothermal treatment increased CEC value of the ash due to particle modification from smooth surface to plate-and rod-shape crystals, which was indicated by the XRD analysis as hydroxysodalite zeolite. Moreover, most of raw biomass ash has low metal adsorption capacity because they do not contain suitable functional group for effective adsorption.

FTIR spectroscopy has been widely performed as an effective tool to explain the alteration of chemical structure of lignocellulosic substrates. Figure 5 shows the spectrum of treated (TA-1) and untreated sugarcane straw ash (SCSA).

The spectra of the untreated and thermally treated sugarcane straw ashes are very similar. The stretching at 3433-3415 cm^{-1} is attributed to both the silanol groups (Si-OH) and adsorbed water on the surface in SCSA and TA-1.

The weak band at 1626 cm^{-1} is attributed to the bending mode of H_2O molecules. The band at 1082 cm^{-1} (SCSA) was shifted to 1008 cm^{-1} (TA-1), which confirms the tetrahedral coordination of aluminum in the zeolite framework. The peaks at 779-796 cm^{-1} are indicative of quartz and amorphous silica. Bands in the region 465-460 cm^{-1} are mostly attributed to internal tetrahedron vibrations of Si-O and Al-O of the materials. The lack of absorption bands in the range 3600-3700 cm^{-1}, characteristic of Al-OH-Al stretching, suggests that discrete, poorly crystalline Al-rich phases are not present.

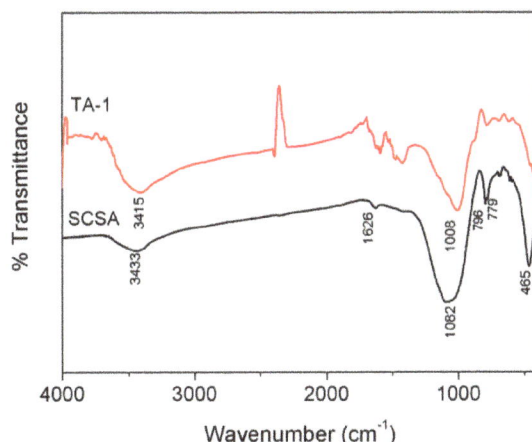

Figure 5. FTIR spectra of untreated (SCSA) and treated sugarcane straw ash (TA-1)

5. Conclusion

The presented studies revealed that the zeolitic material (zeolite hydroxysodalite) was successfully developed from the low cost sugarcane straw ash (SCSA) in the absence of organic templates, without addition of aluminum solution and without fusion prior to hydrothermal treatment.

The formation of the zeolite structure was confirmed by XRD. The efficiency in the conversion of SCSA was affected by the experimental conditions, such as, reaction time, alkali hydroxide concentration and solid/solution ratio. The conversion of SCSA into zeolitic material both contributes to the mitigation of environmental problems and turns this biomass waste of sugar industry into an attractive and useful material, which will be used in the future as an ion exchanger in removing pollutants from wastewater.

Acknowledgements

The authors are grateful to Conselho Nacional de Desenvolvimento Científico e Tecnológico (CNPq) for financial support.

References

[1]   ÚNICA - União da Indústria de Cana-De-Açúcar http://www.unica.com.br/ Accessed 18 Dez 2013, 2011.

[2]   Saad M.B.W., Oliveira, L.R.M., Cândido, R.G., Quintana, G., Rocha, G.J.M., Gonçalves, A.R. Preliminary studies on fungal treatment of sugarcane straw for organosolv pulping Enzyme and Microb. Technol. 2008, 43, 220-225.

[3]   Moriya R.Y., Gonçalves A.R., Duarte M.C. Ethanol/water pulps from sugar cane straw and their biobleaching with xylanase from Bacillus pumilus. Appl. Biochem. Biotechnol. 2007, 137–140, 501-513.

[4]   Costa S.M., Mazzola P.G., Silva J.C.A.R., Pahl R., Pessoa A., Costa S.A. Use of sugarcane Straw as a source of cellulose for textile fiber production. Ind. Crops. Prod. 2013, 42, 189-194.

[5]   Cançado J.E., Saldiva P.H., Pereira L.A., Lara L.B., Artaxo P., Martinelli L.A., Arbex M.A., Zanobetti A., Braga A.L. The Impact of Sugar Cane–Burning Emissions on the Respiratory System of Children and the Elderly. Environ. Health Perspect. 2006, 114, 725-729.

[6]   Ribeiro H. Sugar cane burning in Brazil: respiratory health effects. Rev. Saúde Pública 2008, 42, 370-376.

[7]   Andrade S.J., Cristale J., Soares S.F.S., Zocolo G.J., Marchi M.R.R. Contribution of sugar-cane harvesting season to atmospheric contamination by polycyclic aromatic hydrocarbons (PAHs) in Araraquara city, such showed Southeast Brazil. Atmos. Environ. 2010, 44, 2913-2919.

[8]   Martirena J.F., Middendorf B., Budelman H. Use of wastes of the sugar industry as pozzolan in lime-pozzolan binders: Study of the reaction. Cem. Concr. Res. 1998, 28, 1525-1536.

[9]   Villar-Cociña E., Valencia-Morales E., Gonzalez R.R., Hernandez-Ruiz J. Kinetics of the pozzolanic reaction between lime and sugar cane straw ash by electrical conductivity measurement: A kinetic–diffusive model. Cem. Concr. Res. 2003, 33,517-524.

[10]   Hernandez J.F.M., Middendorf B., Gehrke M., Budelmaun H. Use of wastes of the sugar industry as pozzolana in lime pozzolana binders: Study of the reaction. Cem. Concr. Res. 1998, 28, 1528-1536.

[11]   Frias M., Villar-Cociña E., Valencia-Morales E. Characterisation of sugar cane straw waste as pozzolanic material for construction: calcining temperature and kinetic parameters. Waste Manage. 2007, 27, 533-538.

[12]   Morales E.V., Villar –Cociña E., Fría, M., Santos S.F., Savastano Jr. H. Effects of calcining conditions on the microstructure of sugar cane waste ashes (SCWA): Influence in the pozzolanic activation. Cem. Concr. Compos. 2009, 31, 22-28.

[13]   Shah B., Tailor R., Shah A. Adaptation of bagasse fly ash, a sugar industry solid waste into zeolitic material for the uptake of phenol. Environ. Prog. Sustainable Energy 2011a, 30:358–367.

[14]   Shah B.A., Patel H.D., Shah A.V. Equilibrium and kinetic studies of the adsorption of basic dye from aqueous solutions by zeolite synthesized from bagasse fly ash. Environ. Prog. Sustainable Energy 2011b, 30, 549-557.

[15]   Shah B., Tailor R., Shah A. Sorptive sequestration of 2- chlorophenol by zeolitic materials derived from bagasse fly ash. J. Chem. Technol. Biotechnol. 2011c, 86, 1265–1275.

[16]   Shah B., Shah A.V., Mistry C.B., Tailor R.V., Patel H.D. Surface modified bagasse fly ash zeolites for removal of Reactive Black 5. J. Dispersion Sci. Technol. 2011d, 32, 1247-1255.

[17]   Shah B., Tailor R., Shah A. Zeolitic bagasse fly ash as a low-cost sorbent for the sequestration of p-nitrophenol: equilibrium, kinetics, and column studies. Environ. Sci. Pollut. Res. Int., 2012a, 19, 1171-86.

[18]   Shah B., Tailor R., Shah A. Equilibrium, Kinetics, and Breakthrough Curve of Phenol Sorption on Zeolitic Material Derived from BFA. J. Dispersion Sci. Technol. 2012b, 33, 41-51.

[19]   Worathanakul P., Kittipalarak S., Anusarn K. Utilization Biomass from Bagasse Ash for Phillipsite Zeolite Synthesis. Adv. Mater. Res. 2011, 383-390, 4038-4042.

[20]   Scott J., Guang D., Naeramitmarnsuk K., Thabuot M.J. Zeolite synthesis from coal fly ash for the removal of lead ions from aqueous solution, Chem. Technol. Biotechnol. 2002, 77, 63-69.

[21]   Barboza Filho M.P., Prabhu A.S. Aplicação de silicato de cálcio na cultura do arroz (Application of calcium silicate in rice culture) – Circular Técnica 51, Santo Antônio de Goiás: EMBRAPA (in portuguese), 2002.

[22]   James A.K., Thring R.W., Helle S., Ghuman H.S. Ash management review. Applications of biomass bottom ash. Energies 2012, 5, 3856-3873.

[23]   Vassilev S., Baxter D., Andersen L., Vassileva C. An overview of the composition and application of biomass ash. 1. Phase-mineral and chemical composition and classification. Fuel, 2013, 105, 40-76.

[24]   Le Blond J.S., Horwell C.J., Williamson B.J., Oppenheimer C. Generation of crystalline silica from sugarcane burning. J Environ Monit. 2010, 12, 1459-70.

[25]   Querol X., Moreno N., Umaña J.C., Alastuey A., Hernández E., López-Soler A., Plana, F. Synthesis of zeolites from fly ash: an overview. Int. J. Coal Geol. 2002, 50, 413-423.

[26]   Jha B., Singh D.N. A review on synthesis, characterization and industrial applications of fly ash zeolites. J. Mater. Educ. 2011, 33, 65-132.

[27]   Shoumkova A. Zeolites for water and wastewater treatment: An overview. Research Bulletin of the Australian Institute of High Energetic Materials. Special Issue on Global Fresh Water Shortage 2011, 2, 10.

[28]   Khajavi S. Separation of Process Water using Hydroxy Sodalite Membranes. PhD Thesis, Delft University of Technology, The Netherlands, 2010.

[29]   Naskar M.K., Kundu D., Chatterjee M. Effect of process parameters on surfactant-based synthesis of hydroxy sodalite particles. Mater. Lett. 2011, 65, 436-438.

[30]   Izidoro J.C., Fungaro D.A., Wang S.B. Zeolite synthesis from Brazilian coal fly ash for removal of. Zn^{2+} and Cd^{2+} from water. Adv. Mater. Res. 2012, 356-360, 1900-1908.

[31]   Fungaro D.A., Magdalena C.P. Counterion Effects on the adsorption of Acid Orange 8 from aqueous solution onto HDTMA-modified nanozeolite from fly ash. Environ. Ecol. Res. 2014, 2, 97-106.

5

CO$_2$ emission optimization for a blast furnace considering plastic injection

Xiong Liu[1,2,3], Xiaoyong Qin[1,2,3], Lingen Chen[1,2,3], Fengrui Sun[1,2,3]

[1] Institute of Thermal Science and Power Engineering, Naval University of Engineering, Wuhan 430033, P. R. China.
[2] Military Key Laboratory for Naval Ship Power Engineering, Naval University of Engineering, Wuhan 430033, P. R. China.
[3] College of Power Engineering, Naval University of Engineering, Wuhan 430033, P. R. China.

Abstract

An optimization model based on mass balance and energy balance for a blast furnace process is established by using a nonlinear programming method. The model takes the minimum CO$_2$ emission of a blast furnace as optimization objective function, and takes plastic injection or pulverized coal injection into account. The model includes sixteen optimal design variables, six linear equality constraints, one linear inequality constraint, six nonlinear equality constraints, one nonlinear inequality constraint, and thirteen upper and lower bound constraints of optimal design variables. The optimization results are obtained by using the Sequential Quadratic Programming (SQP) method. Comparative analyses for the effects of plastic injection and pulverized coal injection on the CO$_2$ emission of a blast furnace are performed.

Keywords: Blast furnace; CO$_2$ emission; Iron-making; Plastic injection; Optimization.

1. Introduction

The iron and steel industry is one of the higher industrial CO$_2$ emission sources and energy consumers. Around the world, between 4% and 7% of the anthropogenic CO$_2$ emissions originate from this industry [1-3]. Blast furnace iron-making is a vital process in integrated iron and steel works. The technical improvement and process optimization of blast furnace iron-making is a key step to the development of the iron and steel industry, energy conservation and CO$_2$ emission reductions [4, 5]. A blast furnace, however, is a rector containing many very complex physical and chemical processes. Mathematical modeling is an efficient way to obtain further understanding of blast furnace process, and can achieve further improvements of the operations. Currently, some scholars have established different kinds of models for blast furnaces. The models for blast furnace may approximately be divided into three classes: Statistical models [6, 7], kinetic models [8-10] and mass and energy balance models [11-19]. The mass and energy balance model, which is based on thermodynamic theory and takes the characteristics of blast furnace into account, is an effective method to conduct macro analyses and calculations for blast furnace performance. Rasul *et al* [11] established an model for a blast furnace based on mass and energy balances, and analyzed the influences of blast temperature, silicon content in hot metal and ash content in coke on the blast furnace performance. Emre *et al* [12] established a model for a blast furnace based on the first law of thermodynamics, and analyzed the energy balance of Erdemir No.1 blast furnace. Ziebik *et al* [13, 14] established exergy analysis models for a blast furnace based on mass and energy balances, and analyzed the effects of the operation parameters such as blast temperature and oxygen enrichment degree on exergy and exergy loss of the blast furnace.

In addition, based on mass and energy balances, some optimization models for blast furnace iron-making have been established by using mathematical programming method. Helle *et al* [15] established an optimization model of iron-making process using a linear programming method with biomass as an auxiliary reductant in the blast furnace, and investigated the economy of biomass injection and its dependence on the price structure of materials and emissions. Helle *et al* [16] established a blast furnace iron-making optimization model using nonlinear programming method by taking production as objective function on the basis of the given production rate of hot metal, and analyzed the optimum performance of iron-making system including a blast furnace. Yang *et al* [17] established an optimization model for a blast furnace using linear programming method by taking coke rate as objective function, and proposed some guidelines for the operation of a blast furnace after comparing the optimization result with production reality. Zhang *et al* [18] established a multi-objective optimization model of blast furnace iron-making system using linear programming method by taking energy consumption, cost and CO_2 emissions as objective functions, and analyzed the effects of coke rate, coal rate, blast temperature and sinter ore grade on the energy consumption and cost of production.

The plastic is mainly composed of carbon and hydrogen, and its composition is similar to heavy oil. Thus, the application value of plastic for blast furnace smelting is obvious. To a certain extent, the technology of injecting plastic into a blast furnace can solve environmental problem caused by the extensive use of plastic. Hence, the industrial application value and environmental protection value of plastic injection in blast furnace have been noted by researchers [19-21]. Minoru *et al* [19] described the development of waste plastics injection for blast furnaces. Dongsu *et al* [20] conducted an experiment on plastic injection for blast furnaces and discovered that the combustion efficiency of plastic in tuyere zone could be improved by improving blast temperature and oxygen enrichment degree, and reducing plastic particle size. Minor *et al* [21] conducted experiments on plastic injection in blast furnaces and found that the combustion performance of plastic in a blast furnace is equivalent to pulverized coal when a plastic particle is less than 1.44 mm.

Based on the studies mentioned above, a blast furnace optimization model, in which CO_2 emissions of the blast furnace is taken as an objective function, is established, and the plastic injection and pulverized coal injection are considered. Then, the model is solved by using the Sequential Quadratic Programming (SQP) method from MATLAB optimization toolbox. In addition, the effects of plastic injection and pulverized coal injection on the CO_2 emissions of a blast furnace are analyzed and contrasted. The conclusions obtained herein can provide some guidelines for the design and operation of blast furnaces.

2. The CO_2 emission optimization model for a blast furnace
2.1 Physical model
As shown in Figure 1, a physical model of a blast furnace is considered based on the temperature characteristics inside the blast furnace and some division methods proposed in Refs. [22, 23]. The blast furnace is divided into three zones along its height: the upper preparation zone (PZ), the middle reserve zone (RZ) and the bottom elaboration zone (EZ). The inputs of material flows include sinter ore, pellet ore, lump ore, coke, blast and fuel injected into tuyere area. The outputs of material flows include hot metal, slag and blast furnace gas. The limit temperature of the bottom elaboration zone is set as $950\,°C$; the middle reserve zone is considered as an isothermal region of $950\,°C$, and the upper preparation zone is a lumpish zone while its temperature is lower than $950\,°C$. Furthermore, the following assumptions are considered: (1) All the high valence iron oxides in the preparation zone are reduced into wustite; (2) The gasification of carbon only takes place in the elaboration zone; (3) Behaviors in a blast furnace are described according to the theory of Rist operation; (4) The combustion efficiency of fuel in blast furnace is 100%; (5) Both free water and crystal water in raw material and fuel are evaporated or separated in the preparation zone.

The chemical reaction relations exist in the elaboration zone are listed in Table 1.

The main chemical reactions present in the middle reserve zone are: indirect reduction of wustite ($FeO+CO=Fe+CO_2$) and water gas shift reaction ($CO+H_2O=CO_2+H_2$).

The main chemical reactions present in the preparation zone are: decomposition of carbonate (excluding flux); both the free water and crystal water of raw material and fuel are evaporated or separated; carbon deposition ($2CO = CO_2 + C$); hematite and magnetite are completely reduced to wustite.

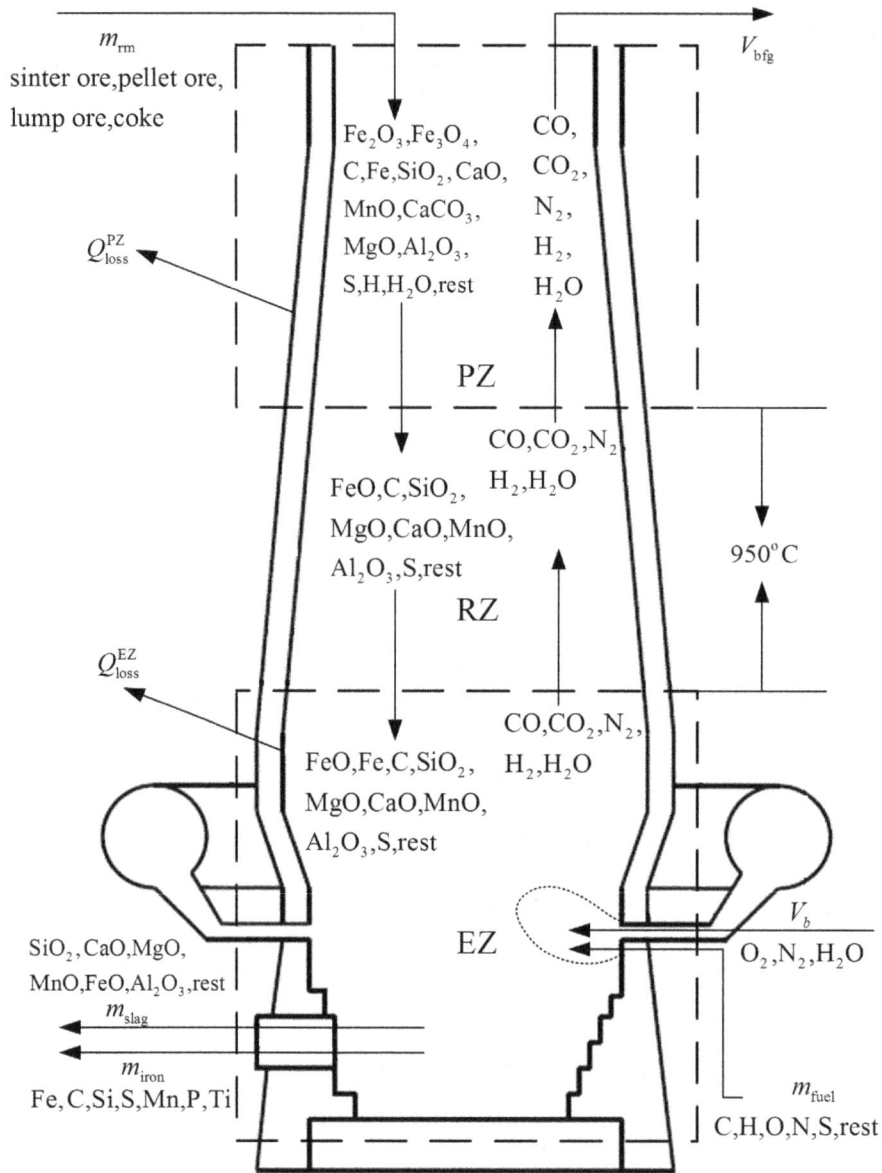

Figure 1. Physical model of a blast furnace

Table 1. Chemical reactions and their introductions in the elaboration zone

chemical reaction	introduction
$FeO+C=Fe+CO$	direct reduction of wustite
$SiO_2+2C=Si+2CO$	direct reduction of SiO_2
$MnO+C=Mn+CO$	direct reduction of MnO
$P_2O_5+5C=2P+5CO$	direct reduction of P_2O_5
$FeS+CaO+C=CaS+Fe+CO$	desulfurization
$C+O_2=2CO$	combustion of carbon
$CO_2+C=2CO\,(>1000\,°C)$	reduction of CO_2
$C+H_2O=CO+H_2\,(>1000\,°C)$	reduction of water in blast

2.2 Optimal design variables

The performance of a blast furnace is affected by many factors. These factors include three classes: (1) raw material and fuel parameters, (2) process parameters and (3) product quality parameters. The raw material parameters refer to the dosage of iron ore and flux. The fuel parameters refer to the coke rate and injected fuel rate. The process parameters refer to the direct reduction degree of iron, blast parameters (including volume, temperature, humidity and oxygen enrichment degree), slag basicity, volume of blast furnace gas and coke load. The product quality parameters refer to the content of each ingredient in hot metal.

Some main techno-economic indexes of iron-making process are often influenced by these parameters. Thus, as listed in Table 2, sixteen parameters are chosen from these three kinds of parameters as optimal design variables.

Table 2. Optimal design variables and introductions

parameter categories	variables	symbols	units	introductions
raw material parameters	x_1	m_{sinter}	kg/t	sinter ore rate
	x_2	m_{pellet}	kg/t	pellet ore rate
	x_3	m_{lump}	kg/t	lump ore rate
	x_4	m_{ls}	kg/t	flux rate
fuel parameters	x_5	$m_{fuel,injected}$	kg/t	injected fuel rate
	x_6	m_{coke}	kg/t	coke rate
technological parameters	x_7	r_d	-	direct reduction degree of iron
	x_8	V_b	Nm^3/t	blast volume
	x_9	T_b	°C	blast temperature
	x_{10}	φ	%	blast humidity
	x_{11}	f	%	blast oxygen enrichment degree
quality parameters of production	x_{12}	[Fe]	%	Fe content in hot metal
	x_{13}	[C]	%	C content in hot metal
	x_{14}	[P]	%	P content in hot metal
	x_{15}	[Mn]	%	Mn content in hot metal
	x_{16}	[S]	%	S content in hot metal

2.3 Objective function

In fact, there are various carbon gases in the blast furnace gas. Thus, the CO_2 emissions value should be the mass of all the CO_2 when the carbon gases are converted to CO_2 [24]. According to this method of calculation on CO_2 emissions, and the carbon gas in blast furnace is composed of CO and CO_2, the CO_2 emission objective function is expressed as

$$F = \frac{44 V_{bfg} \cdot (\omega_{CO_2,bfg} + \omega_{CO,bfg})}{2.24} \ (\text{kg/t}) \tag{1}$$

where V_{bfg} is the blast furnace gas volume (Nm^3/t), $\omega_{CO,bfg}$ is the volume content of CO within blast furnace gas (%), and $\omega_{CO_2,bfg}$ is the volume content of CO_2 within blast furnace gas (%).

2.4 Constraint conditions

The process of blast furnace iron-making must obey the laws of mass and energy balances, and also needs to conform to a certain process system and some material conditions. Thus, all the constraint conditions are classified into mass and energy balance constraints, process constraints, and upper and lower bound constraints of the optimal design variables.

2.4.1 Mass and energy balance constraints

Mass and energy balance constraints include hot metal composition balance constraint, ferrum element balance constraint, manganese element balance constraint, phosphorus element balance constraint, sulfur element balance constraint, dissolved carbon balance constraint, heat balance constraint for the elaboration zone, and carbon and oxygen balance constraints for the elaboration zone.

The hot metal composition balance constraint for blast furnace means that the sum of the contents of each kind of element in hot metal is 100%, so its constraint function is

$$\sum[j] = 100 \tag{2}$$

where [j] is the content of each kind of element in hot metal (%).

The balance constraints of ferrum element, manganese element, phosphorus element and sulfur element mean that the inputs of each kind element within a blast furnace should be equal to the outputs of it. Thus, the constraint function is

$$\sum(m_i \cdot \omega_{i,j} / 100) = 10[j] \, (\text{kg/t}) \tag{3}$$

where m_i is the dosage of each kind of raw material and fuel (kg/t), and $\omega_{i,j}$ is the content of element j (Fe, P, Mn, S) in each kind of raw material and fuel (%).

The dissolved carbon balance constraint means that the carbon content of hot metal has a relationship with the other element content within the hot metal. As it is hard to control the content of carbon in hot metal, the corrected formula is adopted in this model according to Ref. [25]:

$$[C] = 4.3 - 0.27[Si] - 0.32[P] - 0.032[S] + 0.03[Mn] \, (\%) \tag{4}$$

The heat balance constraint in the elaboration zone means that the heat inputs should be equal to the heat outputs in the elaboration zone [26]. Thus, its constraint function is

$$Q_c + Q_b + Q_{fuel} = Q_{df} + Q_{dr} + Q_{dcar} + Q_{bfg} + Q_{iron} + Q_{slag} + Q_{loss}^{EZ} \, (\text{kJ/t}) \tag{5}$$

where Q_c, Q_b and Q_{fuel} are, respectively, heat release of carbon combustion, physical heat of blast (excluding decomposition heat of water in blast) and physical heat of injected fuel (kJ/kg); Q_{df}, Q_{dr}, Q_{dcar}, Q_{bfg}, Q_{iron}, Q_{slag} and Q_{loss}^{EZ} are, respectively, decomposition heat of injected fuel, demanded heat of direct reduction of ferrum element and other alloying elements, decomposition heat of carbonate, physical heat of blast furnace gas, physical heat of hot metal, physical heat of slag, and heat loss of the elaboration zone (kJ/kg).

When the blast furnace iron-making process is in equilibrium state, the coke rate from calculation is the lowest coke rate, namely theoretical coke rate [25]. Actually, because the blast furnace iron-making process is always in a non-equilibrium state, the constraint function of carbon oxygen balance for the elaboration zone is

$$10[Fe]/56 - \eta_{H_2} \cdot V_{H_2,r} / 0.0224 - (m_{C,b} + m_{C,da} + m_{C,dFe} - 10[C])/12/3.237 \leq m_{C,dFe}/12 \tag{6}$$

where η_{H_2} is the hydrogen utilization ratio, $V_{H_2,r}$ is the volume of hydrogen involved in reduction reaction, $m_{C,b}$, $m_{C,da}$ and $m_{C,dFe}$ are, respectively, the mass of carbon burning in raceway, the mass of carbon involved in direct reduction for alloying elements (including the mass of carbon involved in solution loss reaction and desulfurization), and the mass of carbon involved in direct reduction for iron.

2.4.2 Process constraints

Process constraints include constraint of slag basicity, constraint of the content of MgO in slag, constraint of the content of Al_2O_3 in slag, constraint of coke load, constraint of sulfur load, constraint of blast temperature, constraint of oxygen enrichment degree, constraint of blast humidity, and constraint of the relationship between hydrogen utilization ratio and carbon monoxide utilization ratio. These constraints are listed in Table 3.

Table 3. Process constraints and constraint functions

process constraints	constraint functions
constraint of slag basicity (R)	$R_{min} \leq R \leq R_{max}$
content constraint of MgO in slag ($\omega_{MgO,slag}$)	$\omega_{MgO,slag} = \sum \omega_{MgO,i} \cdot m_i / m_{slag}$
content constraint of Al$_2$O$_3$ in slag ($\omega_{Al_2O_3,slag}$)	$\omega_{Al_2O_3,slag} \leq \omega_{Al_2O_3,slag,max}$
constraint of coke load (L_{coke})	$L_{coke,min} \leq L_{coke} \leq L_{coke,max}$
constraint of sulfur load (L_S)	$L_S \leq L_{S,max}$
constraint of blast temperature (t_b)	$t_{b,min} \leq t_b \leq t_{b,max}$
constraint of oxygen enrichment degree (f)	$f_{min} \leq f \leq f_{max}$
constraint of blast humidity (φ)	$\varphi_{min} \leq \varphi \leq \varphi_{max}$
constraint of the relationship between hydrogen utilization ratio and carbon monoxide utilization ratio (η_{H_2})	$\eta_{H_2} = 0.88 \times \omega_{CO_2,bfg} \cdot (\omega_{CO,bfg} + \omega_{CO_2,bfg}) + 0.1$

2.4.3 Upper and lower bound constraints for optimal design variables

All of the optimal design variables in the model come from raw material parameters, fuel parameters, process parameters and product quality parameters. These optimal design variables should be within the allowable ranges. In addition, as blast temperature, oxygen enrichment degree of blast and blast humidity have been contained in process constraints, the upper and lower bounds of the other thirteen optimal design variables needed to be given. The constraint functions of upper and lower bound of the optimal design variables can be written as

$$lb_i \leq x_i \leq ub_i \tag{7}$$

where x_i is optimal design variable, lb_i and ub_i are, respectively, upper and lower bounds of optimal design variables.

3. Description of the optimization problem and its solution

3.1 Description of the optimization problem

The optimization problem in this model is a nonlinear programming problem with multivariable and multi-dimensional constraints [27]. Its mathematical description can be expressed as follows:

$$\begin{cases} \min & f(x) \\ \text{s.t.} & c(x) \leq 0 \\ & c_{eq}(x) = 0 \\ & Ax \leq b \\ & A_{eq}x = b_{eq} \\ & lb \leq x \leq ub \end{cases} \tag{8}$$

where $f(x)$ is objective function, x, b, b_{eq} and lb are, respectively, n dimension column vector, m_1 dimension column vector, and m_2 dimension column vector. $c(x)$ and $c_{eq}(x)$ are, respectively, nonlinear functions of return vectors, ub and lb are, respectively, upper and lower bounds of optimal design variables, while both ub and lb have the same dimension with x.

3.2 Solutions of constraint conditions and objective function

In order to obtain the values of constraint conditions and objective function, the results of material balance calculation and heat balance calculation should be substituted into constraint conditions and objective function, when the initial values of the optimal design variables are given. Thus, at first, it is necessary to calculate the material and heat balances [26].

3.2.1 Material balance calculation

The material balance calculation includes calculation of slag mass and its composition contents, blast volume, blast furnace gas volume and its composition contents.

The calculation methods of slag mass and its composition contents are listed in Table 4.

The blast volume V_b is

$$V_b = \frac{22.4 m_b}{24 \varphi_{O_2,b}} (\text{Nm}^3/\text{t}) \tag{9}$$

where m_b is the mass of carbon burned in the raceway (kg/t), and $\varphi_{O_2,b}$ is the content of oxygen in the blast air.

Blast furnace gas is composed of H_2, CO_2, CO and N_2. The calculation methods of blast furnace gas volume and its composition contents are listed in Table 5.

Table 4. Calculation of slag mass and its composition content*

symbol	introduction	unit	calculation method
$m_{SiO_2,slag}$	SiO₂ mass in slag	kg/t	$m_{SiO_2,slag} = \sum \omega_{SiO_2,i} \cdot m_i / 100 - 10[\text{Si}] \times 30 / 28$
$m_{CaO,slag}$	CaO mass in slag	kg/t	$m_{CaO,slag} = \sum \omega_{CaO,i} \cdot m_i / 100$
$m_{MgO,slag}$	MgO mass in slag	kg/t	$m_{MgO,slag} = \sum \omega_{MgO,i} \cdot m_i$
$m_{Al_2O_3,slag}$	Al₂O₃ mass in slag	kg/t	$m_{Al_2O_3,slag} = \sum \omega_{Al_2O_3,i} \cdot m_i$
$m_{FeO,slag}$	FeO mass in slag	kg/t	$m_{FeO,slag} = \sum (\omega_{TFe,i} \cdot m_i \cdot \eta_{Fe,slag}) \times 72 / 56 / 100$
$m_{Mn,slag}$	Mg mass in slag	kg/t	$m_{Mn,slag} = \sum (\omega_{Mn,i} \cdot m_i \cdot \eta_{Mn,slag}) \times 71 / 55 / 100$
$m_{S,slag}$	S mass in slag	kg/t	$m_{S,slag} = 0.5 \sum (\omega_{S,i} \cdot m_i \cdot \eta_{S,slag}) \times 32 / 100$
m_{slag}	slag mass	kg/t	$m_{slag} = m_{SiO_2,slag} + m_{CaO,slag} + m_{MgO,slag} + m_{Al_2O_3,slag}$ $+ m_{FeO,slag} + m_{Mn,slag} + m_{S,slag}$

* $\omega_{SiO_2,i}$, $\omega_{CaO,i}$, $\omega_{MgO,i}$, $\omega_{TFe,i}$, $\omega_{Mn,i}$ and $\omega_{S,i}$ are, respectively, the contents of SiO₂, CaO, MgO, TFe, Mn and S in each kind of raw material (%), i is each kind of raw material, $\eta_{Fe,slag}$, $\eta_{Mn,slag}$ and $\eta_{S,slag}$ respectively are the distribution rate of Fe, Mn and S in slag.

Table 5. Calculation of blast furnace gas volume and its composition content*

symbol	introduction	unit	calculation method
$V_{H_2,bfg}$	volume of H₂ in blast furnace gas	Nm³/t	$V_{H_2,bfg} = (1 - \eta_{H_2}) \cdot (V_{H_2,b} + V_{H_2,fuel})$
$V_{CO,bfg}$	volume of CO in blast furnace gas	Nm³/t	$V_{CO,bfg} = V_{CO,b} + V_{CO,d} - V_{CO,id}$
$V_{CO_2,bfg}$	volume of CO₂ in blast furnace gas	Nm³/t	$V_{CO_2,bfg} = V_{CO_2,r} + \sum V_{CO_2,i}$
$V_{N_2,bfg}$	volume of N₂ in blast furnace gas	Nm³/t	$V_{N_2,bfg} = V_{N_2,b} + V_{N_2,fuel}$
V_{bfg}	blast furnace gas volume	Nm³/t	$V_{bfg} = V_{H_2,bfg} + V_{CO,bfg} + V_{CO_2,bfg} + V_{N_2,bfg}$

* η_{H_2} is hydrogen utilization rate, $V_{H_2,b}$ is the volume of water in blast (Nm³/t), $V_{H_2,fuel}$ is the volume of H_2 within injected fuel (Nm³/t), $V_{CO,b}$ is the volume of CO produced by the combustion of carbon in raceway (Nm³/t), $V_{CO,d}$ is the volume of CO produced by the direction reduction of iron and other alloying elements (Nm³/t), $V_{CO,id}$ is the volume of CO used by the indirect reduction (Nm³/t), $V_{CO_2,r}$ is the volume of CO₂ produced in reduction reaction (Nm³/t), $V_{CO_2,i}$ is the volume of CO₂ in each kind of raw material (Nm³/t), $V_{N_2,b}$ is the volume of N₂ in blast (Nm³/t), $V_{N_2,fuel}$ is the volume of N₂ in injected fuel (Nm³/t).

3.2.2 Heat balance calculation

Heat inputs of a blast furnace include heat released by combustion of carbon in raceway and physical heat of the hot blast air. Heat outputs of blast furnace include heat demand of reduction reaction, heat demand of desulfurization, heat demand of carbonate decomposition, physical heat of slag, physical heat of hot metal, physical heat of blast furnace gas, heat demand of evaporation of water in raw materials and heat carried by cooling water and heat loss. The calculation methods of those are listed in Table 6.

Table 6. Calculation of each kind of heat*

	symbol	introduction	unit	calculation method
heat input	$Q_{C,b}$	heat released by combustion of carbon in raceway	kJ/t	$Q_{C,b} = 9781.2 m_{C,b} - q_{dm} \cdot m_{fuel}$
	Q_b	physical heat of hot-blast air	kJ/t	$Q_b = V_b \cdot \overline{C}_{p,t_b} \cdot t_b - 10806(1-f) \cdot \varphi$
heat output	Q_d	heat demand for reduction reaction	kJ/t	$Q_d = 2890 \times 10[Fe] \cdot r_d + 22960 \times 10[Si]$ $+4880 \times 10[Mn] + 26520 \times 10[P]$
	Q_S	heat demand for desulfurization	kJ/t	$Q_S = 4650 \omega_{S,slag} \cdot m_{slag}$
	Q_{carb}	heat demand for carbonate decomposition	kJ/t	$Q_{carb} = \Sigma q_{d,i} \cdot m_{carb,i}$
	Q_{slag}	physical heat of slag	kJ/t	$Q_{slag} = m_{slag} \cdot h_{slag,out}$
	Q_{iron}	physical heat of hot metal	kJ/t	$Q_{iron} = 1000 h_{iron,out}$
	Q_{bfg}	physical heat of blast furnace gas	kJ/t	$Q_{bfg} = V_{bfg} \cdot C_{bfg} \cdot t_d + V_{H_2O,r} \cdot C_{H_2O} \cdot t_d$
	Q_{H_2O}	heat demand for evaporation of water in raw materials and heat carried out by cooling water	kJ/t	$Q_{H_2O} = 2450 \Sigma (\omega_{H_2O,i} \cdot m_i / 100)$
	Q_{loss}	heat loss	kJ/t	$Q_{loss} = 10 Z_0 \cdot \omega_{C,coke} / \eta_V$

* $m_{C,b}$ is the quantity of carbon burned in raceway (kg/t), q_{dm} is heat demanded for injected fuel

decomposition (kg/t), $\overline{C}_{p,t_b}$ is the specific heat capacity of blast (kJ/(m$^3 \cdot$ °C)), f and φ respectively are oxygen enrichment degree and humidity of blast, $m_{carb,i}$ is quantity of carbon within each kind of raw material (kg/t), $q_{d,i}$ is heat demanded for decomposition of carbonate within each kind of raw material (kJ/t), $h_{slag,out}$ is specific enthalphy of slag of hot metal (kJ/kg), C_{bfg} is specific heat capacity of blast furnace gas (kJ/(m$^3 \cdot$ °C)), t_d is temperature of blast furnace gas (°C), $V_{H_2O,r}$ is volume of water produced by reduction reaction in which hydrogen involved (Nm^3/t), C_{H_2O} is the specific heat capacity of water vapor (kJ/(m$^3 \cdot$ °C)), $\omega_{H_2O,i}$ is the content of water within each kind of raw material and fuel (%), η_V is productivity (kJ/(m$^3 \cdot$d)), Z_0 is heat loss of one kilogram carbon when smelting intensity is one (kJ/kgC), $\omega_{C,coke}$ is the content of carbon in coke (%).

3.3 Optimization method
The optimization problem in this model is a nonlinear programming problem with multivariable and multi-dimensional constraints. Its objective function is a nonlinear function. Its constraints include nonlinear equality constraints, nonlinear inequality constraints, linear equality constraints and linear inequality constraints. The function of "fmincon" in the optimization toolbox of the MATLAB is used to find the optimization results of nonlinear programming problem with multivariable and multi-dimensional constraints [27]. As SQP algorithm has global and superlinear convergence, it has been one of the most efficient nonlinear programming algorithms in solving nonlinear programming problem with multivariable and multi-dimensional constraints [28]. Then, the function of "fmincon" in the optimization toolbox of the MATLAB is adopted in this model, and its call form is

$$[x, fval] = \text{fmincon}(@objfun, x_0, A, b, Aeq, beq, lb, ub, @confun, options) \tag{10}$$

where x_0 is a initial point, x is optimal solution, and *fval* is the minimum of the objective function.

4. Optimization results and analyses
A designed blast furnace described in Ref. [26] is taken as an example. The contents of plastic and pulverized coal are listed in Table 7.

Table 7. Contents of plastic and pulverized coal (%)

item	C	S	O	H	N	H_2O	FeO	SiO_2	CaO	MgO	Al_2O_3
plastic	85.60	14.40	-	-	-	-	-	-	-	-	-
pulverized coal	85.40	0.550	0.460	0.300	0.310	0.37	0.847	5.950	0.800	0.710	4.373

The upper and lower bounds of the optimal design variables are listed in Table 8. The upper bound of injected fuel is 170 kg/t-hot metal when pulverized coal is injected. The upper bound of injected fuel is 100 kg/t-hot metal when plastic is injected. The upper and lower bounds of the other optimal design variables with pulverized coal injection are the same as those of optimal design variables with plastic injection.

Table 8. Upper and lower bounds of the optimal design variables

variable	unit	upper bound	lower bound	variable	unit	upper bound	lower bound
x_1	kg/t	1500	0	x_9	°C	1250	1050
x_2	kg/t	1000	0	x_{10}	%	2.0	0
x_3	kg/t	158.52	0	x_{11}	%	6.0	0
x_4	kg/t	80	0	x_{12}	%	100	94
x_5	kg/t	100 (plastic injection) 170 (pulverized coal injection)	0	x_{13}	%	4.9	0
x_6	kg/t	500	200	x_{14}	%	0.4	0
x_7	-	1	0.3	x_{15}	%	1.2	0
x_8	Nm^3/t	1800	700	x_{16}	%	0.07	0

4.1 Optimization results

The optimization results and original ones are listed in Table 9. As shown in Table 9, the optimal pulverized coal rate reaches the lower bound (0 kg/t-hot metal) when pulverized coal is injected. In contrast, the optimal plastic rate reaches the upper bound (100 kg/t-hot metal) when plastic is injected.

Table 9. Optimization results and original results

variable	introduction	symbol	unit	optimization results with plastic injection	optimization results with pulverized coal injection	original results
x_1	sinter ore rate	m_{sinter}	kg/t	840.25	998.23	1030.35
x_2	pellet ore rate	m_{pellet}	kg/t	575.13	436.12	396.29
x_3	lump ore rate	m_{lump}	kg/t	158.52	158.52	158.52
x_4	flux rate	m_{ls}	kg/t	0	0	0
x_5	injected fuel rate	m_{fuel}	kg/t	100	0	170
x_6	coke rate	m_{coke}	kg/t	270.86	448.94	325
x_7	direct reduction degree of iron	r_d		0.39	0.56	0.45
x_8	blast volume	V_b	m^3/t	865.32	1005.32	1000.48
x_9	blast temperature	T_b	°C	1250	1250	1250
x_{10}	blast humidity	φ	%	0	0	2.0
x_{11}	blast oxygen enrichment degree	f	%	0	0	3.5
x_{12}	Fe content in hot metal	[Fe]	%	95.09	95.09	94.34
x_{13}	C content in hot metal	[C]	%	4.16	4.16	4.90
x_{14}	P content in hot metal	[P]	%	0.09	0.10	0.10
x_{15}	Mn content in hot metal	[Mn]	%	0.14	0.13	0.15
x_{16}	S content in hot metal	[S]	%	0.03	0.03	0.025
-	minimum CO_2 emissions	-	kg/t	1013.96	1272.44	1344.30

In addition, both blast humidity (φ) and blast oxygen enrichment degree (f) reaches the lower bound whether plastic or pulverized coal is injected. The CO_2 emissions of blast furnace with pulverized coal injection decrease 6.27% after optimization. In fact, the metal oxide content of coal is higher than that of coke, so both heat demand of reduction and carbon dosage with pulverized coal injection are increased. Hence, the mass of pulverized coal reaches 0 kg/t-hot metal when CO_2 emissions of blast furnace reach the minimum. In contrast, the CO_2 emissions of blast furnace are decreased 24.57% with plastic injection. This is due to the fact that plastic contains high hydrogen content and has no metal oxide. Thus, one can conclude that plastic injection will decrease CO_2 emissions of a blast furnace, while pulverized coal injection will increase CO_2 emissions of a blast furnace. While from the perspective of economics, burning coke only is not practical while plastic injection is economical. Thus, plastic injection has significance for both emission reduction and economic considerations.

4.2 Analyses of influence factors
4.2.1 Influence of injected fuel rate on optimization results

Figures 2-5 show the relationships among the minimum CO_2 emission (F_{min}) and the corresponding fuel rate (m_{fuel}), coke rate (m_{coke}), direct reduction degree of iron (r_d) and injected fuel rate ($m_{fuel,injected}$), respectively.

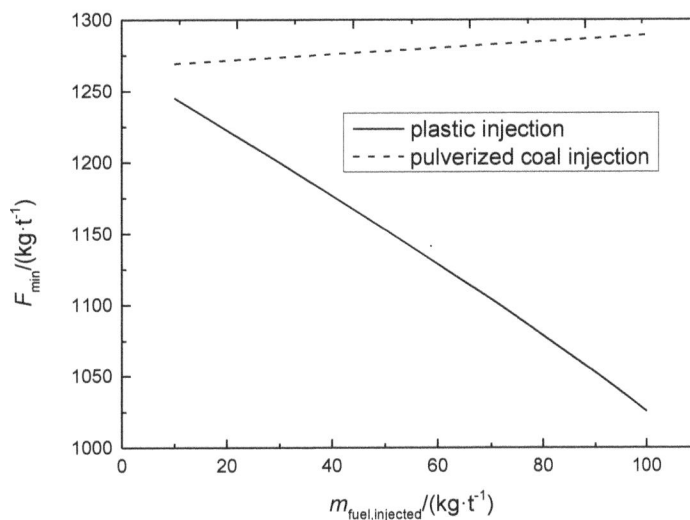

Figure 2. The minimum CO_2 emission (F_{min}) versus injected fuel rate ($m_{fuel,injected}$)

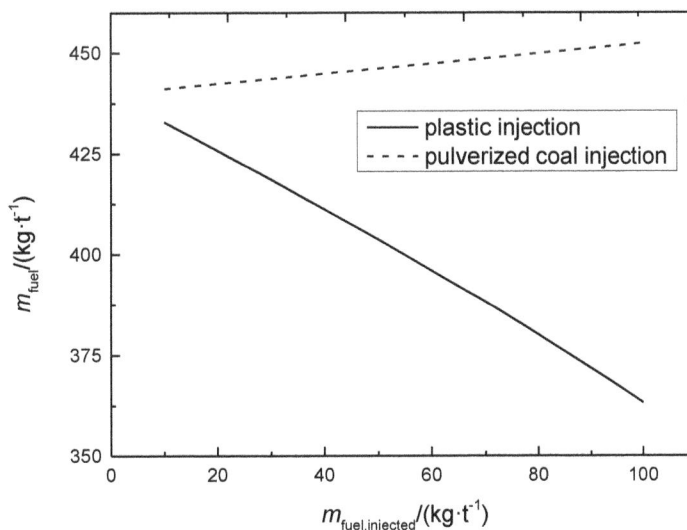

Figure 3. The fuel rate (m_{fuel}) versus injected fuel rate ($m_{fuel,injected}$) corresponding to the minimum CO_2 emission (F_{min})

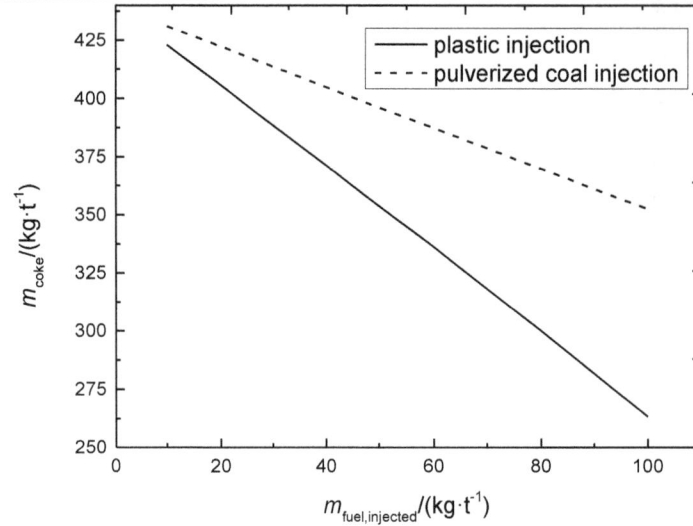

Figure 4. Coke rate (m_{coke}) versus injected fuel rate ($m_{fuel,injected}$)corresponding to the minimum CO_2 emission (F_{min})

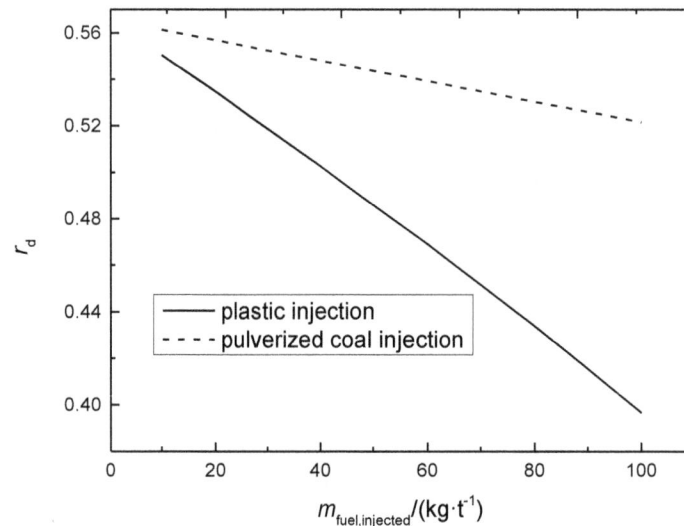

Figure 5. Direct reduction degree of iron (r_d) versus injected fuel rate ($m_{fuel,injected}$) corresponding to the minimum CO_2 emission (F_{min})

From Figures 2 and 3, one can see that the minimum CO_2 emission (F_{min}) and its corresponding fuel rate (m_{fuel}) decrease when the plastic injection rate ($m_{plastic}$) increases. In contrast, the minimum CO_2 emission (F_{min}) and its corresponding injected fuel rate ($m_{fuel,injected}$) increase when pulverized coal rate (m_{coal}) increases. The reason is that the content of hydrogen in plastic is relatively high and the amount of hydrogen takes the place of carbon to take part in reduction, and thus the carbon consumption is decreased. Then, the minimum CO_2 emission (F_{min}) and fuel rate (m_{fuel}) decrease. In contrast, as the content of hydrogen in coal is lower than that in plastic and a certain amount of metal oxide exist in coal, the carbon consumption increases. Then, the minimum CO_2 emission (F_{min}) and fuel rate (m_{fuel}) decrease. From Figures 4 and 5, one can see that the corresponding coke rate (m_{coke}) and direct reduction degree of iron (r_d) decrease when injected fuel rate ($m_{fuel,injected}$) increases. However, the downtrend of both direct reduction degree of iron (r_d) and coke rate (m_{coke}) with plastic injection is more obvious than that with pulverized coal injection. As a certain amount of carbon is replaced by the injected fuel, the coke rate (m_{coke}) with plastic injection or pulverized coal injection decreases. As part of hydrogen in the injected fuel takes part in direct reduction of iron (r_d), the direct reduction degree of iron (r_d) decreases. In addition, as hydrogen content of plastic is higher than that of pulverized coal, the downtrend of direct reduction degree of iron (r_d) with plastic injection is more obvious than that with pulverized coal injection.

From Figures 2-5, one can see that plastic injection is more efficient in both coke conservation and decrease of direct reduction degree of iron (r_d) when the hydrogen content of plastic is higher than that of pulverized coal.

4.2.2 Influence of carbon-hydrogen mass ratio of plastic on optimization results

The carbon-hydrogen mass ratio of plastic ($n_{C/H,plastic}$) means the ratio of the mass of carbon to the mass of hydrogen in plastic. Figures 6 and 7 show the relationships among the minimum CO_2 emission (F_{min}), its corresponding direct reduction degree of iron (r_d), coke rate (m_{coke}) and the carbon-hydrogen mass ratio of plastic ($n_{C/H,plastic}$), respectively.

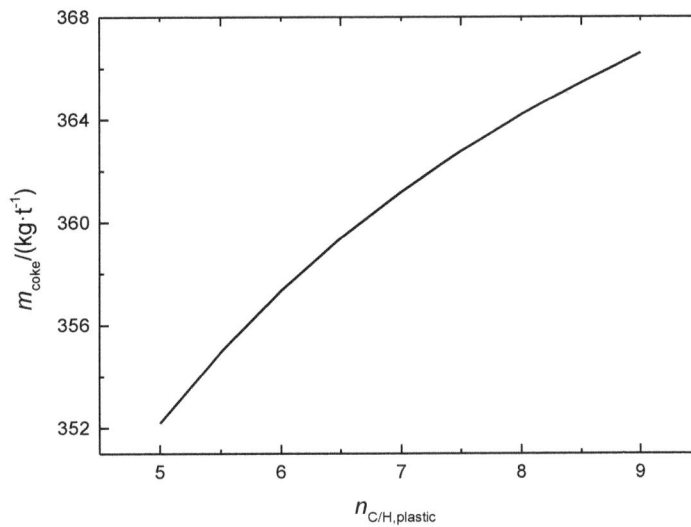

Figure 6. Coke rate (m_{coke}) versus carbon-hydrogen mass ratio of plastic ($n_{C/H,plastic}$)corresponding to the minimum CO_2 emission (F_{min})

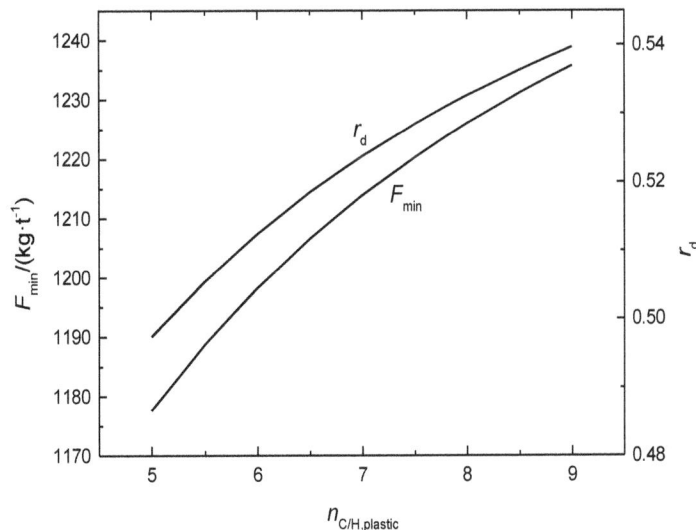

Figure 7. The minimum CO_2 emission (F_{min}) and the corresponding direct reduction degree of iron (r_d) versus carbon-hydrogen mass ratio of plastic ($n_{C/H,plastic}$)

Figure 6 shows that the coke rate (m_{coke}) corresponding to the minimum CO_2 emission (F_{min}) decreases with the decrease of carbon-hydrogen mass ratio of plastic ($n_{C/H,plastic}$). This is due to the fact that the mass of hydrogen getting into blast furnace increases with the decreases of carbon-hydrogen mass ratio of plastic ($n_{C/H,plastic}$), as well as the mass of hydrogen involved in direct reduction of iron. As a result, the mass of carbon involved in direct reduction of iron (r_d) decreases. Thus, the coke rate (m_{coke}) decreases with the decrease of carbon-hydrogen mass ratio of plastic ($n_{C/H,plastic}$).

Figure 7 shows that both the minimum CO_2 emission (F_{min}) and its corresponding direct reduction degree of iron (r_d) decrease with the decrease of carbon-hydrogen mass ratio of plastic ($n_{C/H,plastic}$). As has been noted, coke rate decreases with the decrease of carbon-hydrogen mass ratio of plastic ($n_{C/H,plastic}$). The injected fuel rate ($m_{fuel,injected}$), however, is not changed. Therefore, both fuel rate (m_{fuel}) and carbon consumption decrease, and the minimum CO_2 emission (F_{min}) decreases. As a result of decreasing carbon-hydrogen mass ratio of plastic ($n_{C/H,plastic}$), the mass of hydrogen involved in reduction increases and the level of indirect reduction are improved. Thus, the direct reduction degree of iron (r_d) decreases.

From Figures 6 and 7, one can conclude that injecting plastic with a low carbon-hydrogen mass ratio ($n_{C/H,plastic}$) is more beneficial to coke conservation, emission reduction and strengthening smelting than injecting plastic with a high carbon-hydrogen mass ratio ($n_{C/H,plastic}$).

4.2.3 Influences of blast parameters on optimization results

Figures 8-10 show the relationships among the minimum CO_2 emission (F_{min}) and its corresponding coke rate (m_{coke}), blast temperature (T_b), blast oxygen enrichment degree (f), and blast humidity (φ), respectively.

From Figure 8, one can see that the minimum CO_2 emission (F_{min}) and its corresponding coke rate (m_{coke}) decrease when blast temperature (T_b) increases. The calculations show that the minimum CO_2 emission (F_{min}) and its corresponding coke rate decrease about 3.35 kg/t-hot metal and 1.07 kg/t-hot metal, when blast temperature (T_b) increases about $10\,°C$. Figure 9 shows that both the minimum CO_2 emission (F_{min}) and its corresponding coke rate (m_{coke}) increase when blast oxygen enrichment degree (f) increases. Figure 10 shows that the minimum CO_2 emission (F_{min}) and its corresponding coke rate (m_{coke}) increase when blast humidity (φ) increases.

From Figures 8 and 10, one can conclude that the technology of improving blast temperature (T_b) or dehumidifying blast are beneficial for coke conservation and emission reduction. From Figure 9, one can conclude that blast oxygen enrichment degree (f) should be controlled within a proper range as emission can be increased by a high blast oxygen enrichment degree (f).

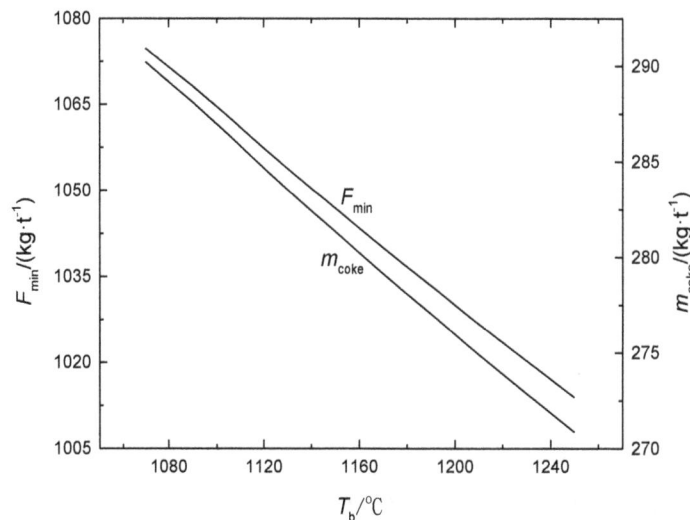

Figure 8. The minimum CO_2 emission (F_{min}) and the corresponding coke rate (m_{coke}) versus blast temperature (T_b)

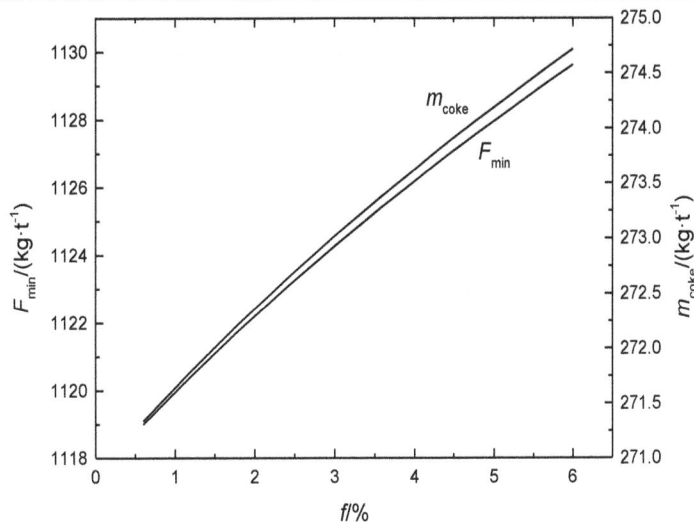

Figure 9. The minimum CO_2 emission (F_{min}) and the corresponding coke rate (m_{coke}) versus blast oxygen enrichment degree (f)

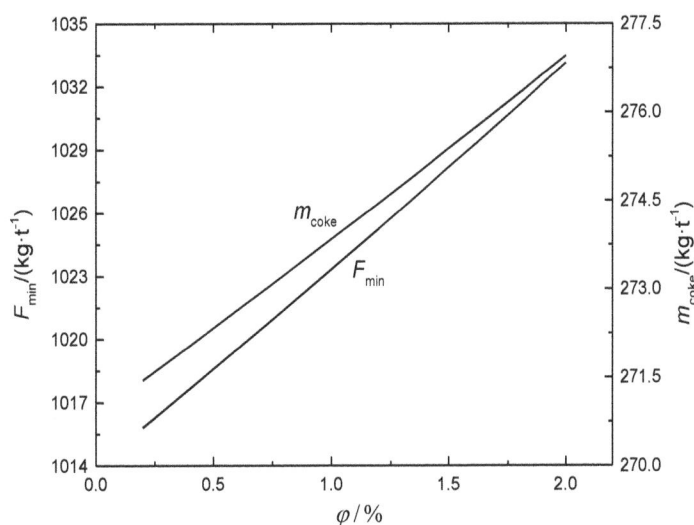

Figure 10. The minimum CO_2 emissions (F_{min}) and the corresponding coke rate (m_{coke}) versus blast humidity (φ)

5. Conclusions

Base on material balance and energy balance of blast furnaces, an optimization model for blast furnace iron-making with the CO_2 emission reduction as optimization objective is established by using nonlinear programming method. The calculation program is compiled on the MATLAB, and the model is solved by using SQP algorithm in the optimization toolbox of the MATLAB. Comparative analyses for the effects of plastic injection and pulverized coal injection on the CO_2 emissions of the blast furnace are performed. The effects of carbon-hydrogen mass ratio of plastic, blast temperature, blast oxygen enrichment degree of blast and blast humidity on coke rate and direct reduction degree of iron are analyzed. The results show that plastic injection is beneficial for decreasing coke rate, fuel rate and direct reduction degree of iron when injecting plastic with a low carbon-hydrogen mass ratio. The CO_2 emission with plastic injection is less than that with pulverized coal injection. Plastic injection with a low carbon-hydrogen mass ratio can do more to decrease coke rate and emission.

Acknowledgments

This paper is supported by the National Key Basic Research and Development Program of China (973) (Project No. 2012CB720405).

References

[1] Sen P K. CO2 accounting and abatement: an approach for iron and steel industry. J. Trans. Indian Inst. Met., 2013, 66(5-6): 711-721.

[2] Sarker T, Coahan M. Energy sources and carbon emissions in the iron and steel industry sector in South Asia. J. IJEEP, 2013, 3(1): 30-42.

[3] Kundak M, Lazic L, Crinko J. CO2 emissions in the steel industry. J. Metalurgija, 2009, 3(48): 193-197.

[4] Hasanbeigi A, Lynn K, Marlene A. Emerging Energy-Efficiency and Carbon Dioxide Emissions-Reduction Technologies for the Iron and Steel Industry. Berkeley: Lawrence Berkeley National Laboratory, 2013.

[5] Ghanbari H, Helle M, Saxen H. Process integration of steelmaking and methanol production for decreasing CO2 emissions - A study of different auxiliary fuels. J. Chem. Eng. Processing, 2012, 61: 58-68.

[6] Ghosh A, Majumdar S K. Modeling blast furnace productivity using support vector machines. J. Int. J. Adv. Manuf. Techno, 2011, 52(9-12): 989-1003.

[7] Gao C H, Zhou Z M, Chen J M. Assessing the predictability for blast furnace system through nonlinear time series analysis. J. Ind. Eng. Chem. Res., 2008, 47(9): 3037-3045.

[8] Hussain M M. Ore Reduction Kinetics and Simulation of a Lead Blast Furnace. The University of New Brunswick, 1987.

[9] Chu M. Study on super high efficiency operations of blast furnace based on multi-fluid model. D. Sendai: Tohoku University, 2004.

[10] Han Y H, Wang J S, Lan R Z. Kinetic analysis of iron oxide reduction in top gas recycling oxygen blast furnace. J. Iron-making and Steelmaking, 2012, 39(5): 313-317.

[11] Rasul M G, Tanty B S, Mohanty B. Modeling and analysis of blast furnace performance for efficient utilization of energy. Appl. Therm. Eng., 2007, 27(1): 78-88.

[12] Emre E M, Gürgen S. Energy balance analysis for Erdemir blast furnace number one. Appl. Therm. Eng., 2006, 26(11-12): 1139-1148.

[13] Ziebik A, Stanek W. Energy and exergy system analysis of thermal improvements of blast-furnace plants. Int. J. Energy Res., 2006, 30(2): 101-114.

[14] Ziebik A, Lampert K, Szega M. Energy analysis of a blast-furnace system operating with the Corex process and CO2 removal. Energy, 2008, 33(2): 199-205.

[15] Helle H, Helle M, Saxen H. Mathematical optimization of iron-making with biomass as auxiliary reductant in the blast furnace. ISIJ Int., 2009, 49(9): 1316-1324.

[16] Helle H, Helle M, Saxen H. Nonlinear optimization of steel production using traditional and novel blast furnace operation strategies. Chem. Eng. Sci., 2011, 66(24): 6470-6481.

[17] Yang T J, Gao B, Lu H S. Optimization aimed at the lowest coke consumption by model and analysis of the application. J. University. Sci. Techn. Beijing, 2001, 23(4): 305-307 (in Chinese).

[18] Zhang Q, Yao T, Cai J, Shen M. On the multi-objective optimal model of blast furnace iron-making process and its application. J. Northeastern(Nature Sci.), 2011, 32(2): 270-273 (in Chinese).

[19] Minoru A, Tasturo A, Michitaka S. Development of waste plastics injection process in blast furnace. ISIJ. Int., 2000, 40(3): 244-251.

[20] Dongsu K, Sunghye S, Seungman S. Waste plastics as supplemental fuel in the blast furnace process: improving combustion efficiencies. J. Hazard. Mater., 2002, 94(3): 213-222.

[21] Minoru A, Masahiko K, Hidekazu T. Evaluation of waste plastics particle for blast furnace injection. J. Japan Inst. Energy, 2012, 91(2): 127-133.

[22] Biswas A K. Principles of blast furnace iron-making: Theory and practice. Brisbane: Cootha, 1981.

[23] Kieatv B I, Yaroshenko Y G, Suchkov V D. Heat Exchange in Shaft Furnace. Oxford: Pergamon Press, 1967.

[24] Ding J, Gao B, Wang S, Zhang Q. Effect of silicon content in molten iron on carbon emission in blast furnace. J. Res. Iron Steel, 2011, 39(5): 1-3 (in Chinese).

[25] Na S R. Differentiation and Analyses of Iron-Making Calculations. Beijing: Metallurgical Industry Press, 2010 (in Chinese).

[26] Wang S L. Metallurgy of Iron and Steel (Iron-Making Part). Beijing: Metallurgical Industry Press, 2013 (in Chinese).

[27] Venkataraman P. Applied Optimization with MATLAB Programming (2nd ed.). Hoboken: John Wiley & Sons, Inc., 2009.

[28] Zhu Z B, Zhang K C. A new SQP method of feasible directions for nonlinear programming. Appl. Math. Comput., 2004(148): 121-134.

6

Model application for acid mine drainage treatment processes

Nantaporn Noosai, Vineeth Vijayan, Khokiat Kengskool

Department of Civil and Environmental Engineering, Florida International University, Miami, FL 33174, USA.

Abstract

This paper presents the utilization of the geochemical model, PHREEQC, to investigate the chemical treatment system for Acid Mine Drainage (AMD) prior to the discharge. The selected treatment system consists of treatment processes commonly used for AMD including settling pond, vertical flow pond (VFP) and caustic soda pond were considered in this study. The use of geochemical model for the treatment process analysis enhances the understanding of the changes in AMD's chemistry (precipitation, reduction of metals, etc.) in each process, thus, the chemical requirements (i.e., $CaCO_3$ and NaOH) for the system and the system's treatment efficiency can be determined. The selected treatment system showed that the final effluent meet the discharge standard. The utilization of geochemical model to investigate AMD treatment processes can assist in the process design.

Keywords: Acid mine drainage treatments; Acid mine drainage geochemical processes; PHREEQC model.

1. Introduction

Acid Mine drainage (AMD) is generally referred to an acidic metal-rich wastewater discharged from the mining industry. It has low pH and high concentrations of metals, which are the byproduct of the mining industries and/or chemically formed during the discharged process [1, 2]. The studies showed that the discharge of AMD causes environmental pollution in many countries having mining industries. Therefore, the AMD is required to be treated prior to the discharge by many countries. The treatment processes used to treat AMD are different from site to site depending on the water quality and its composition. However many studies reported that the combination of chemical treatment processes is the most effective technique used for AMD treatment [1-3]. That is because of their effectiveness in removing the metals out from the water and neutralization of the water pH [1, 2]. However as it was mentioned earlier, the treatment processes that work for one site may not work for another depending on AMD water quality at each site. Therefore, the investigation for treatment processes must be made for each site, thus, the suitable processes can be chosen for the site. Use of the geochemical, PHREEQC, model is a cost effective way for assessing the appropriate treatment processes for particular AMD water. PHREEQC (version 2) released by US Geological Survey (USGS) in 1999 is designed to perform a wide variety of aqueous geochemical calculations: speciation and saturation-index calculations, batch-reaction and one dimensional transport, etc [4, 5].The model can be used to estimate the efficiency and amount of chemical required for the treatment processes. This helps in supporting the decision making for selection of treatment processes. The objective of this study is to illustrate the use of PHREEQC model for AMD treatment processes assessment. The model was employed to estimate the amount of chemical required for the treatment and to determine the effectiveness of the selected treatment processes. The same method can be applied for any particular site where the selection of appropriate AMD treatment process is needed.

2. Scenario Study

The following scenario is hypothetical. It assumes that a reclaimed mining site has two discharges released from different mining process plants within the site. The water quality and flow rate data of the two hypothetical discharges are show in Table 1. The hypothetical discharged water quality data in Table 1 represent a typical AMD water quality, which has low pH, high sulfate and high concentrations of various heavy metals, in scenario, are iron, manganese, aluminum, cadmium and arsenic.

Table 1. Discharge characteristics [6]

Parameter	Discharge # 1	Discharge # 2
Design flow, liter per second	0.63	1.12
Average Flow (median), liter per second	0.45	0.86
Alkalinity, mg/L	5	4
pH	3.1	3.5
Ferric Iron; Fe^{3+}, mg/L	5.0	0.45
Ferrous Iron; Fe^{2+}, mg/L	46.8	32.4
Manganese, mg/L	14.2	18.2
Aluminum, mg/L	1.14	0.95
Cadmium, mg/L	1.10	1.00
Arsenic, mg/L	0.90	0.53
Uranium, mg/L	0.85	0.75
Sulfate, mg/L	580	950
Dissolved Oxygen, mg/L	5.30	4.6

3. Methodology

The geochemical model PHREEQC (Version 2) was used to assess and evaluate the effectiveness of the selected AMD treatment process. PHREEQC calculates geochemical reactions at equilibrium based on the available database using the activity and mass-action equation. Precipitation of newly formed solid phases could chemically control the fate of AMD contaminants in the neutralization reactions. This process may be predicted from supernatant solutions by a thermodynamic model and must be corroborated by characterization of final solid products. Equilibrium geochemical speciation/mass transfer model PHREEQC with the database of the speciation model MINTEQ was applied to determine aqueous speciation and saturation indices of solid phases [SI = log(IAP/KS), where SI is the saturation index, IAP is the ion activity product and KS is the solid solubility product]. Zero, negative or positive SI values indicate that the solutions are saturated, undersaturated and supersaturated respectively, with respect to a solid phase. For a state of subsaturation, dissolution of the solid phase is expected and supersaturation suggests precipitation.

The selected treatment system is the combination of different treatment processes put in order: Rock-lined ditch, settling pond, vertical flow pond (VFP) and caustic soda pond. This study assumed that the two AMD were produced from different plants within the mine with different flow rates and qualities (Table 1). The estimation of the chemical requirements for the selected treatment process to treat both discharges was conducted using the models. The final effluent is determined to meet the water quality discharge criterion.

3.1 Selected treatment processes

In order to save money, both discharges will be combined and treated with single treatment system. The schematic of the selected treatment system is shown in Figure 1.

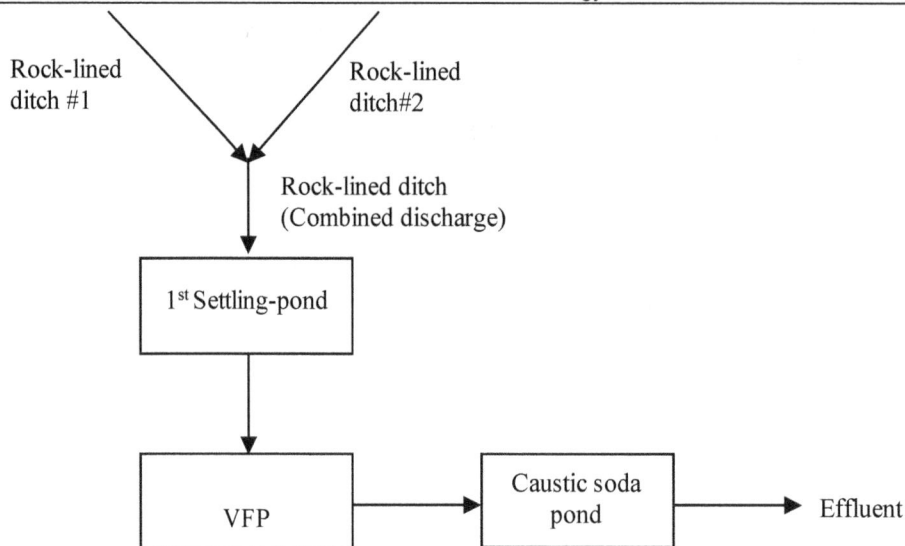

Figure 1. Schematic of selected AMD treatment process

3.2 AMD treatment processes description
Rock-line ditches

Two rock-lined ditches carry the discharges to the meeting point where the discharges are combined. The combined discharge then flows through another rock-lined ditched to the 1st settling pond. PHREEQC was used to calculate the precipitations and dissolutions that may occur after the waters are mixed. The results will be used for the settling pond design.

1st settling pond

This settling pond will hold the sludge volume that will be produced by the precipitation while maintaining a desired water retention time. The primary precipitation will be removed at this settling pond. The solution will then flow through the Vertical Flow Pond (VFP)

Vertical Flow Pond (VFP)

VFP or Vertical flow wetland, also known as Successive Alkalinity Producing System (SAPS), is designed to add alkalinity to net acidic discharges. The schematic of the VFP is shown in Figure 2.

The organic matter layer serves to remove dissolved oxygen (DO) from the water and promote the anaerobic environment with reducing conditions: that changes Fe^{3+} to Fe^{2+}, S^{6+} to S^{2-} and favors the precipitation of metal-sulfide [7, 8]. Reducing DO content in water will prevent the covering of limestone layer by the precipitated metals. The dissolution of limestone will then neutralize the acidity. The acidity of water is very important value for pond sizing design and sensitive to cost estimation. PHREEQC helped to determine the changes in water chemistry in reducing environment. That the calculation of the reductions (e.g., Fe^{3+} to Fe^{2+} and S^{6+} to S^{2-}) and the precipitation of metal-sulfide and other metals were made[7-9]. PHREEQC also used to estimate the amount of limestone needed to neutralize the acidity.

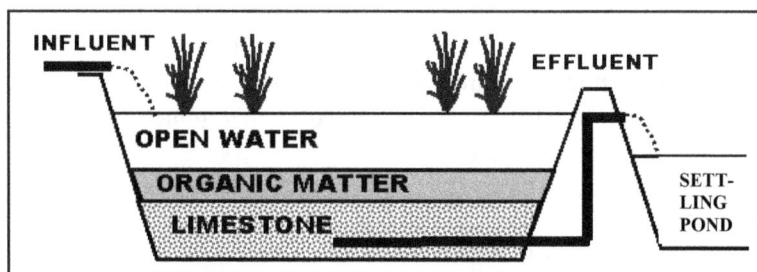

Figure 2. Schematic of Vertical Flow Pond (VFP) (modified after http://www.prp.cses.vt.edu)

<u>Caustic soda pond</u>
The purposes of caustic soda system are to be a backup system in case the VFP does not perform as expect and to remove the Mn since the VFP will not treat the Mn [8, 9, 10]. PHREEQC helped to calculate the caustic soda amount: the amount that will increase water pH to 9.5 where the precipitation of Mn occurs.

4. Results and discussions
The acidities of discharges #1 and #2 were calculated and shown in Table 2. The governing equations used in this model are shown in equations 1 to 3.

$$pH \text{ acidity (mg/L as CaCO}_3) = 50 \times 100 \times 10^{-pH} \tag{1}$$

$$Metal \text{ acidity (mg/L as CaCO}_3) = 50(\frac{2Fe^{2+}}{56} + \frac{3Fe^{3+}}{56} + \frac{3Al}{27} + \frac{2Mn}{55}) \tag{2}$$

$$Net \text{ acidity (mg/L as CaCO}_3) = 50(\frac{2Fe^{2+}}{56} + \frac{3Fe^{3+}}{56} + \frac{3Al}{27} + \frac{2Mn}{55} + 1000 x 10^{-pH}) \tag{3}$$

Table 2. The acidities of discharges

Parameters	Discharge #1	Discharge #2
pH acidity	39.72 mg/L as CaCO$_3$	15.81 mg/L as CaCO$_3$
Metal acidity	129.41 mg/L as CaCO$_3$	97.85 mg/L as CaCO$_3$
Net acidity	169.12 mg/L as CaCO$_3$	113.66 mg/L as CaCO$_3$

4.1 Combined discharge
PHREEQC was used to calculate the mixing of two discharges. Upon the mixing of these two discharges the pH behaved non-conservatively because of the release of $CO_2(g)$. The result of combined discharge is shown in Table 3.

Table 3. The combined discharge characteristics

Parameter	Combined discharge
Design flow, liter per second	1.75
pH	3.63
Acidity, mg/L as CaCO$_3$	148.84
Ferric Iron; Fe^{3+}, mg/L	35.37
Ferrous Iron, Fe^{2+}, mg/L	3.73
Manganese, mg/L	16.41
Aluminum, mg/L	1.00
Cadmium, mg/L	1.02
Arsenic, mg/L	0.65
Uranium, mg/L	0.77
Sulfate, mg/L	797.86

4.2 1st settling pond
The combined discharge entered the settling pond as an influent while the chemical changes upon the mixing slowly took place. The precipitation of iron hydroxide (Ferrihydrite) upon mixing leads to metal removal from the pond. Metals adsorbed on precipitated iron hydroxide and were removed from the water [1, 2, 7]. Table 4 shows the pond effluent, the precipitated minerals and, percentage removals.
Upon mixing, the acidity was decrease, that is because Fe^{3+} was precipitated out from the water, moreover, the precipitation of Fe(OH)$_3$ also released H$^+$ (equation 4) [1, 2, 7, 9], thus both reactions led to the decrease in pH (3.6 to 3.2).

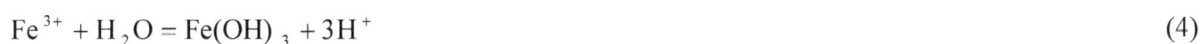

$$Fe^{3+} + H_2O = Fe(OH)_3 + 3H^+ \tag{4}$$

Table 4. The pond effluent, the precipitated minerals and, percentage removals

Parameter	Settling pond effluent	% Removal
Solution		
Design flow, liter per second	1.75	-
pH	3.24	-
Acidity	82.5 mg/L as $CaCO_3$	44.57%
Ferric Iron (Fe^{3+})	4.32 mg/L	87.80%
Ferrous Iron (Fe^{2+})	3.74 mg/L	-
Manganese	16.41 mg/L	-
Aluminum	1.00 mg/L	-
Cadmium	1.02 mg/L	-
Arsenic	0.29 mg/L	55.96%
Uranium	0.77 mg/L	
Sulfate	792.96 mg/L	0.61%
Precipitation		
Ferrihydrite, $Fe(OH)_3$	(SI = 0.9)	60.68 mg/L

4.3 VFP (Vertical Flow Pond)

The settling pond effluent then entered the VFP and the organic matter layer which has anaerobic condition (see Figure 2). The pe = -2 was assumed and fed to PHREEQC model in order to allow the occurrence of reducing condition, therefore, sulfate (S^{6+}) is reduced to sulfide (S^{2-}) and ferric (Fe^{3+}) to ferrous (Fe^{2+}) [9-11]. The effluent from organic matter layer then seeped through the limestone layer where the dissolution of limestone occurred and increased the pH of the discharge. This led to the precipitation of As-S, Cd-S and Fe-S and Al minerals [10-12]. The effluent from VFP treatment process is shown in Table 5.

Table 5. The results of VFP treatment process

Parameter	VFP influent	Organic matter layer effluent	Limestone layer effluent	% removal
Solution				
Design flow, liter per second	1.75	1.75	1.75	
pH	3.24	2.95	7.78	
Acidity mg/L as $CaCO_3$	82.5	91.53	0.95	> 98
Ferric Iron (Fe^{3+}),m mg/L	4.32	3.4×10^{-17}	~ 0.00	> 99
Ferrous Iron (Fe^{2+}), mg/L	3.74	0.008	3.6×10^{-6}	> 99
Manganese, mg/L	16.41	16.41	0.52	> 96
Aluminum, mg/L	1.00	1.00	1.8×10^{-4}	> 99
Cadmium, mg/L	1.02	3.98×10^{-4}	1.04×10^{-4}	> 99
Arsenic, mg/L	0.29	3.94×10^{-13}	3.9×10^{-13}	> 99
Uranium, mg/L	0.77	1.24×10^{-7}	1.24×10^{-7}	> 99
Sulfate (SO_4^{2-}), mg/L	792.96	~ 0.00	~ 0.00	
Sulfide (HS^-), mg/L	-	262.58	253.01	3.64%
Precipitation				
Greenockite, CdS, mg/L		1.31	-	
Orpiment, As_2S_3, mg/L		0.47	-	
Pyrite, FeS_2, mg/L		17.29	0.02	
Uraninite, UO_2, mg/L		0.88	-	
Diaspore, AlOOH, mg/L		-	2.26	
Greenockite, CdS, mg/L		-	0.00037	
MnS(green), mg/L		-	25.18	
Limestone needs, mg/L		-	461.1	

In this process, most of the SO_4 changed to HS^- and Fe^{3+} changed to Fe^{2+} in anaerobic condition resulting in removals of metal sulfide minerals. However, the rich HS^- in water decreased the water pH from 3.2 to 2.9. Water then flowed through the limestone layer. The dissolution of limestone increased the water pH to 7.7. The model calculated the amount of limestone needs by allowing limestone to dissolve in water until its saturation index (SI) reached 0, where the water is saturated with $CaCO_3$. The amount of limestone required was 461.1 mg/L. The increase in pH led to the precipitation of Al, Mn, Cd and Fe minerals thus these precipitated minerals were then removed out from the water [11-13]. VFP treatment increased the water pH and removed most of the metals from the water. However, the amount of Mn in VFP effluent was still greater than the discharge standard (Mn < 0.2 mg/L). Therefore, the further treatment is required.

4.4 Caustic Soda Pond

Recall that the purpose of caustic soda pond is to increase pH to 9.5 (based on the titration to 8.3) to remove Mn. The pond is an open air pond (pO_2 = 0.21 atm) therefore, the water in this treatment process has an aerobic condition. The effluent and the metal removals by this process are shown in Table 6.

Table 6. Treatment results of caustic soda pond

Parameter	influent	Effluent	% removal	Discharge Standard
Solution				
Design flow, liter per second	1.75	1.75		
pH	7.78	8.34		6.5 – 8.5
Acidity, mg/L $CaCO_3$	0.95	0.0002	> 99	
Ferric Iron (Fe^{3+}), mg/L	~ 0.00	5.6×10^{-9}	-	< 1
Ferrous Iron (Fe^{2+}), mg/L	3.6×10^{-6}	~ 0.00	-	
Manganese, mg/L	0.52	5.36×10^{-12}	> 96	<0.2
Aluminum, mg/L	1.8×10^{-4}	1.8×10^{-4}	-	-
Cadmium, mg/L	1.04×10^{-4}	1.04×10^{-4}	-	<0.01
Arsenic, mg/L	3.9×10^{-13}	3.9×10^{-13}	-	<0.05
Uranium, mg/L	1.24×10^{-7}	1.24×10^{-7}	-	<0.1
Sulfate, SO_4^{2-}, mg/L	~ 0.00	736.03	-	<2500
Sulfide, HS^-, mg/L	253.01	~ 0.00	-	-
Precipitation				
Calcite, $CaCO_3$, mg/L	-	21.3		
Hematite, Fe_2O_3, mg/L	-	5.3×10^{-6}		
Pyrolusite, MnO_2, mg/L	-	0.82		
NaOH needs, mg/L		10		

Since the water is aerated (pO_2 = 0.21 atm), Fe^{2+} was oxidized to Fe^{3+} and HS^- as S^{2-} to SO_4^{2-} as S^{6+} [13-15]. The 10 mg/L of NaOH was needed to rise the pH to 9.5. At water pH 9.5, some minerals; $CaCO_3$, Fe_2O_3 and MnO_2, were precipitated out from the water and this led to the decrease in water pH that precipitation of CaCO3 released $CO_2(g)$ thus the pH decreased from 9.5 to 8.34 [11, 15, 16]. This treatment removed 96.8% of Mn out from the water. Thus, the final effluent met the discharge standard.

5. Conclusion

Using the geochemical models help to support the AMD treatment system design. The study points out that iron can be removed via the oxidation process in the settling pond. Most of metals were removed in the VFP. Although most of Mn was removed via VFP but in order to meet the discharge standard requirement the caustic soda pond was required. With employing the PHREEQC model, the optimum amount of chemical requirements for the treatment processes; to neutralize the pH of water and to remove the metals, could be calculated. The similar analysis method with the help of the PHREEQC model can be used to support the decision making for the most suitable treatment processes and system for particular AMD water quality, thus, the final effluent can meet the discharge standard requirement.

References

[1]   Johnson, D. B. and Hallberg, K. B. Acid mine drainage remediation options: a review. Science of the Total Environment. 2005, 338, 3-14.

[2]   Akcila, A. and Koldas, S. Acid mine drainage (AMD): causes, treatment and case studies. Journal of Cleaner Production. 2006, 14, 1139-1145.

[3]   McCauley, C. A., O'Sullivan, A. D., Milke, M.W., Weber, P. A., and Trumm, D. A. Sulfate and metal removal in bioreactors treating acid mine drainage dominated with iron and aluminum. Water Research. 2009, 43, 961-970.

[4]   Parkhurst, D.L., Appelo C.A.J. User's Guide to PHREEQC (Version 2) A Computer Program for Speciation, Batch-Reaction, One-Dimensional Transport, and Inverse Geochemical Calculations, USGS Water-Resources Investigations, Denver, Colorado, 1999.

[5]   Macíasa, F., Caraballoa, M.A., Nietoa, J.M., Röttingb, T.S., Ayora, C. Natural pretreatment and passive remediation of highly polluted acid mine drainage. Journal of Environmental Management. 2012, 104, 93-100.

[6]   Bain, J.G., Mayera, K.U., Blowesa, D.W., Frinda, E.O., Molsona, J.W.H., Kahntb, R., Jenkb U. Modelling the closure-related geochemical evolution of groundwater at a former uranium mine. Journal of Contaminant Hydrology. 2001, 52, 109-135.

[7]   Sheoran, A.S. and Sheoran, V. Heavy metal removal mechanism of acid mine drainage in wetlands: A critical review. Minerals Engineering. 2006, 19, 105-116.

[8]   Hallberg, K.B. New perspectives in acid mine drainage microbiology. Hydrometallurgy. 2010, 104(3-4), 448-453.

[9]   Battaglia-Brunet, F., Dictor, M.C., Garrido, F., Crouzet, C., Morin, D., Dekeyser, K. A simple biogeochemical process removing arsenic from a mine drainage water. Geomicrobiol. J. 2002, 23, 201–211.

[10]   Younger, P.L., Jayaweera, A., Elliot, A., Wood, R., Amos, P., Daugherty A.J. Passive treatment of acidic mine waters in subsurface-flow systems: exploring RAPS and permeable reactive barriers. Land Contam Reclam. 2003, 11, 127–135.

[11]   Hashim, M.A., Mukhopadhyay, S., Sahu, J.N., Sengupta, B. Remediation technologies for heavy metal contaminated groundwater. J. Environ. Manag. 2011, 27, 2355–2388.

[12]   Caraballo, M.A., Rötting, T.S., Silva V. Implementation of an MgO-based metal removal step in the passive treatment system of Shilbottle, UK: column experiments. J. Hazard. Mater. 2010, 181, 923–930.

[13]   Mayes, W.M., Batty, L.C., Younger, P.L., Jarvis, A.P., Koiv, M., Vohla C. Wetland treatment at extremes of pH: a review. Sci. Total. Environ. 2009, 407 (13), 3944–3957.

[14]   Nyquist, J., Greger, M. A field study of constructed wetlands for preventing and treating acid mine drainage. Ecol. Eng. 2009, 35, 630–642.

[15]   Kalin, M. Passive mine water treatment: the correct approach? Ecol. Eng. 2004, 22, 299–304.

[16]   Chenga, H., Hub, Y., Luoc, J., Xua, B., Zhao, J. Geochemical processes controlling fate and transport of arsenic in acid mine drainage (AMD) and natural systems. Journal of Hazardous Materials. 2009, 165, 13-26.

7

Optimization of long-term performance of municipal solid waste management system: A bi-objective mathematical model

Hao Yu[1], Wei Deng Solvang[1], Shiyun Li[1,2]

[1] Department of Industrial Engineering, Narvik University College, Postboks 385 Lodve gate 2, 8505 Narvik, Norway.
[2] College of Mechanical Engineering, Zhejiang University of Technology, No. 18 Caowang Road, 310016 Hangzhou, P.R.China.

Abstract

Management of municipal solid waste has becoming an extremely important topic for any urban authorities in recent years due to the rapidly increasing solid waste quantity and potential environmental pollution. In this paper, a bi-objective dynamic linear programming model is developed for decision making and supporting in the long-term operation of municipal solid waste management system. The proposed mathematical model simultaneously accounts both economic efficiency and environmental pollution of municipal solid waste management system over several time periods, and the optimal tradeoff over the entire studied time horizon is the focus of this model. The application of the proposed model is also presented in this paper, and the computational result and analysis illustrate a deep insight of this model.

Keywords: Waste management; Municipal solid waste; Multi-criteria analysis; Dynamic programming; Environmental pollution.

1. Introduction

Solid waste management has becoming a challenging task for any municipal authorities due to rapidly increasing waste amount, increasing concern for environmental pollution, more complex waste composition, as well as limited capacity for waste treatment and disposal [1]. In order to operate municipal solid waste management system in a cost efficient and sustainable manner, the decision-makers should look at the "overall picture" from long-term perspectives. On one hand, the system operating cost should be minimized so that the increasing amount of solid waste can be efficiently and effectively treated and disposed, and this is especially important for developing countries where the fast increase of solid waste due to the rapid urbanization and industrialization has become a burden for both municipalities' infrastructure and the community [2]. On the other hand, the concern of environmental pollution and risk (e.g. contamination of surface water and ground water from landfill, air pollution from incineration, etc.) from the public have been significantly increased in recent years, furthermore, the emission of greenhouse gases from the treatment and disposal of increase quantity of municipal solid waste is also accused as one of the primary contributors to global warming and climate change [3, 4]. However, the cost objective and environmental pollution/risk objective are conflict with one another, the optimal scenario for one objective usually lead to a bad solution for the other [5]. Therefore, the optimal balance between economic efficiency and environmental pollution is of significance in determining the long-term performance of municipal solid waste management system.

Previously, a large number of studies focused on the optimization of municipal solid waste management system [6]. Son [2] proposes a computational model for vehicle routing problem of waste collection, and the model is resolved through combining chaotic particle swarm optimization with global information system. The waste collection problem is also focused by Ghiani et al. [7] who develop a two-stage location model. The first step is to determine the number and locations of waste collection bins in a residential area, and the second step is to decide the service zone of each waste collection bin and optimal route of waste collection vehicles. Eiselt and Marianov [8] report a bi-objective optimization model for determining the most appropriate location of waste treatment and disposal facilities, and the tradeoff between economic efficiency and environmental issue is the focus of this location model. Badran and El-Haggar [9] propose a mixed integer programming model for determining the optimal configuration of a multi-echelon municipal waste management system through minimizing the overall cost, and a real-world case at Port Said, Egypt, is also presented in the study. Zhang and Huang [10] develop a single objective model in order to mitigate greenhouse gas emissions associated with municipal solid waste management system, and fuzzy possibilistic integer programming is employed for dealing with uncertain parameters. Alcada-Almeida et al. [11] investigate a multi criteria approach for locating incineration plant in Portugal. The tradeoff among overall system cost, total impact, maximum average impact and impact to individuals is optimized in this study, and the overall system cost is comprised of annualized investment and processing cost. A multi-objective approach for determining the optimal configuration of waste management system is developed by Galante et al. [12]. In order to optimize the tradeoff of total cost and environmental impact, a combination of mathematical tools including fuzzy multi-objective programming, weighed sum as well as goal programming is applied in this study. Dai et al. [13] formulate a mixed integer linear programming model with interval parameters for the optimization of municipal solid waste management system, and a support-vector-regression approach is developed as well. Mavrotas et al. [14] propose a bi-objective integrated optimization model for simultaneously minimizing the overall system cost and greenhouse gas emissions related to the transportation and treatment of municipal solid waste. A generic cost-minimization formula for the network design and planning of municipal solid waste management system is investigated by Eiselt and Marianov [15], and the location selection of landfill and transfer station is especially emphasized in this study.

Generally, the location problem related to municipal solid waste management system has played a predominant role in previous studies, and different mathematical tools such as linear programming, nonlinear programming, goal programming, mixed integer programming, multi-objective programming, etc., have been extensively applied for formulating and resolving the location problems of municipal solid waste management system. However, the scope of previous studies is limited to the network design, expansion and development of municipal solid waste management system, and the optimal and most sustainable operation planning of existing waste management systems is rarely mentioned. In this paper, different from previous literature, the location problem of waste treatment and disposal facilities is not taken into consideration, but the optimal operation planning of municipal solid waste management system over a set of continuous time periods is focused, and a bi-objective dynamic optimization model is developed to determine the optimal operation plan of the municipal solid waste management system within the studied time horizon. Moreover, the solution method and numerical experimentation of this model are also presented latter in this paper, and the computational result and analysis illustrate a deep insight of this model.

2. The model

Based upon the reverse waste supply chain network developed by Zhang et al. [16], municipal solid waste management system is constituted by three levels of facilities, namely local waste collection center, regional distribution center as well as treatment and disposal facility, and Figure 1 illustrates a simplified framework of municipal solid waste management system. Local waste collection can be considered as the initial step of municipal solid waste management system, and the locally collected waste will then be sent to regional distribution center at which separation and pre-treatment of solid waste are performed in order to provide appropriate "input resources" to the subsequent waste treatment and disposal plants. Finally, different types of municipal solid waste will be treated or properly disposed through corresponding treatment methods i.e. recycling, incineration, composting, mechanical biological treatment, landfill, etc.

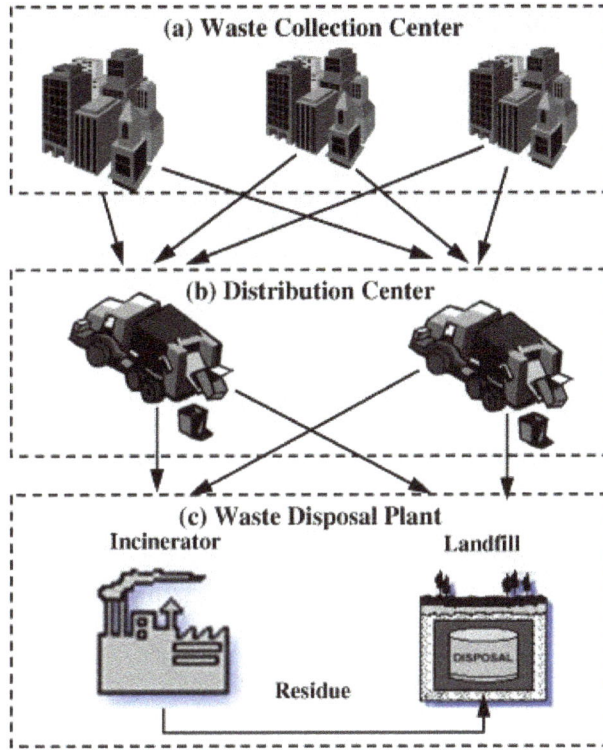

Figure 1. Municipal solid waste management system [16]

2.1 Objective function

The overall cost of municipal solid waste management system within the studied time horizon is expressed in Eq. (1). The first four parts in this equation represent the annualized investment and flexible operating cost of waste collection, distribution, treatment and disposal, respectively. The other three parts formulate the inter-facility transportation cost from waste collection center to distribution center, from distribution center to treatment plant, and from distribution center to landfill. The flexible facility operating cost and inter-facility transportation cost are linearly associated with the quantity of solid waste.

$$
\begin{aligned}
\text{Min } cost = & \sum_{1}^{s}\sum_{1}^{c}(AI_{c(s)} + WCC_{c(s)}QT_{c(s)}) + \sum_{1}^{s}\sum_{1}^{dt}(AI_{dt(s)} + WDtC_{dt(s)}QT_{dt(s)}) \\
& + \sum_{1}^{s}\sum_{1}^{t}(AI_{t(s)} + WTC_{t(s)}QT_{t(s)}) + \sum_{1}^{s}\sum_{1}^{d}(AI_{d(s)} + WDC_{d(s)}QT_{d(s)}) \\
& + \sum_{1}^{s}\sum_{1}^{c}\sum_{1}^{dt} WTpC_{c/dt(s)}QTp_{c/dt(s)} + \sum_{1}^{s}\sum_{1}^{dt}\sum_{1}^{t} WTpC_{dt/t(s)}QTp_{dt/t(s)} \\
& + \sum_{1}^{s}\sum_{1}^{dt}\sum_{1}^{d} WTpC_{dt/d(s)}QTp_{dt/d(s)}
\end{aligned}
\tag{1}
$$

The environmental pollution of municipal solid waste management system is formulated in Eq. (2). The environmental pollution indicator illustrates the pollution level and potential risk of each plant. The environmental pollution related to waste distribution, treatment and disposal linearly increases with the increase of solid waste quantity, while it linearly decreases with the increase of the distance between population center and waste management facility. It is noteworthy that the distance between existing plants and communities is fixed and not changes with time, so the periodic adjustment is not applied for this parameter, however, the environmental pollution indicator may be changed within the studied period due to technological upgrade or other developments. Besides, the population of each affected area is introduced to pollution-minimization objective as an important adjustment factor in order to minimize the environmental pollution to the most populated communities.

$$\text{Min } pollution = \sum_1^s \sum_1^{af} POL_{af(s)} \left(\sum_1^{dt} \frac{EP_{dt(s)}QT_{dt(s)}}{DS_{dt/af}} + \sum_1^t \frac{EP_{t(s)}QT_{t(s)}}{DS_{t/af}} + \sum_1^d \frac{EP_{d(s)}QT_{d(s)}}{DS_{dt/af}} \right) \qquad (2)$$

It is prerequisite that all the waste collected at each defined time period is totally treated or disposed, so the cost and environmental pollution related to waste storage at each period is not taken into consideration.

2.2 Composite objective function

The model is formulated through multi-period linear programming for simultaneously minimizing the overall system cost and environmental pollution of municipal solid waste management system. In order to combine cost-minimization and pollution-minimization objective, the challenge brought by different measure of units of those two objective functions must be first resolved. In this paper, a weighted sum utility method developed from Nema and Gupta [17] is introduced in Eq. (3), and similar method for combining multi-objective functions with different units is also provided by Hu et al. [18] and Yu et al. [19]. The optimal solution of cost-minimization and pollution-minimization can be first found out through solving the single objective linear function, and the unit of $\frac{Cost\ objective}{Min\ cost}$ and $\frac{Pollution\ objective}{Min\ pollution}$ can then be eliminated. In Eq. (3), ∂_C and ∂_p indicate the importance of relevant objective function, and they follow the relation $\partial_p = 1 - \partial_C$.

$$\text{Min } objective = \partial_C \frac{Cost\ objective}{Min\ cost} + \partial_p \frac{Pollution\ objective}{Min\ pollution} \qquad (3)$$

2.3 Constraints

The waste amount collected at each community by local collection center cannot be more than the maximum collecting and storage capacity in each period (Eq. (4)). For waste collection center, the entire input waste amount are totally processed, and it also equals to the summation of waste transported to all distribution centers in each period (Eq. (5)). Those two constraints are conflict with each other when the waste amount generated in one community exceed the capacity of local waste collection center, and expansion of limited waste collection capacity must be planned under such condition so that the result solved by this model is meaningful.

$$QT_{c(s)} \le MAX_{c(s)}, \text{For } 1, \dots, c, 1, \dots, s \qquad (4)$$

$$\sum_1^{dt} QTp_{c/dt(s)} = QT_{c(s)} = SW_{c(s)}, \text{For } 1, \dots, c, 1, \dots, s \qquad (5)$$

For each waste distribution center in each period, the maximum capacity and minimum quantity constraints must be fulfilled (Eqs. (6) and (7)). For waste distribution center, treatment plant as well as disposal facility, the minimum waste processing amount is required so as to maintain the economic efficiency for opening and operating the waste management facilities. If the utilization of waste management facility is very low, the annualized investment will constitute a significant share in the overall system operating cost, and the spare capacity will become a big economic burden for the waste management companies. Besides, the summation of input waste from local collection centers equal to the summation of waste transported to the treatment plants and disposal facilities at each regional distribution center in each period (Eq. (8)).

$$QT_{dt(s)} \le MAX_{dt(s)}, \text{For } 1, \dots, dt, 1, \dots, s \qquad (6)$$

$$QT_{dt(s)} \geq MIN_{dt(s)}, \text{For } 1, \dots, dt, 1, \dots, s \tag{7}$$

$$\sum_{1}^{c} QTp_{c/dt(s)} = QT_{dt(s)} = (\sum_{1}^{t} QTp_{dt/t(s)} + \sum_{1}^{d} QTp_{dt/d(s)}), \text{For } 1, \dots, dt, 1, \dots, s \tag{8}$$

Similarly, the maximum processing capacity and minimum required waste amount at treatment plant and disposal facility in each period are restricted by Eqs. (9), (10), (12) and (13), respectively. Eqs. (11) and (14) regulate the input waste amount equals to the waste quantity processed at treatment plant and disposal facility in each period. In addition, the numerical values of all the parameters and decision variables in this bi-objective multi-period optimization model for municipal solid waste management system are positive.

$$QT_{t(s)} \leq MAX_{t(s)}, \text{For } 1, \dots, t, 1, \dots, s \tag{9}$$

$$QT_{t(s)} \geq MIN_{t(s)}, \text{For } 1, \dots, t, 1, \dots, s \tag{10}$$

$$\sum_{1}^{dt} QTp_{dt/t(s)} = QT_{dt(s)}, \text{For } 1, \dots, t, 1, \dots, s \tag{11}$$

$$QT_{d(s)} \leq MAX_{d(s)}, \text{For } 1, \dots, d, 1, \dots, s \tag{12}$$

$$QT_{d(s)} \geq MIN_{d(s)}, \text{For } 1, \dots, dt, 1, \dots, s \tag{13}$$

$$\sum_{1}^{dt} QTp_{dt/t(s)} = QT_{d(s)}, \text{For } 1, \dots, dt, 1, \dots, s \tag{14}$$

3. Application of the model

In this section, the proposed model is applied to determine the optimal waste allocation plan of a municipal solid waste management system in a continuous five time periods. The studied area includes three communities, and the municipal solid waste management system is constituted by three local collection centers, two regional distribution centers, two incineration plants and one landfill. The parameters of local waste collection centers are presented in Table 1. It is noteworthy that all the numerical values of the parameters in this illustrative example are unitless.

Table 1. Parameters of local waste collection center

Parameter	Community	Period				
		s=1	s=2	s=3	s=4	s=5
$AL_{c(s)}$	c=1	3500000	3750000	3900000	4050000	4200000
	c=2	5000000	5300000	5550000	5800000	6300000
	c=3	3200000	3300000	3400000	3500000	3600000
$SW_{c(s)}$	c=1	85500	92000	94500	99200	102500
	c=2	106000	113500	121000	132000	135800
	c=3	68000	68500	69200	70150	72000
$WCC_{c(s)}$	c=1	35	38	41	45	51
	c=2	32	34	37	40	43
	c=3	35	37	40	42	45
$MAX_{c(s)}$	c=1	105000	105000	105000	105000	105000
	c=2	120000	120000	120000	120000	120000
	c=3	85000	85000	85000	85000	85000
$POL_{af(s)}$	af=1	32133	33110	33575	34123	35501
	af=2	45101	45893	46355	46908	47366
	af=3	26105	27122	27833	28206	28633

In this example, all the three communities are influenced by the municipal solid waste management system, so the set of communities (c) equals to the set of affected areas (af). The parameters of regional waste distribution centers, incineration plants as well as landfill are illustrated in Tables 2, 3 and 4, respectively. For those three levels of facilities, the environmental pollution indicator is also given so that the environmental pollution of the municipal solid waste management system can be calculated. The population of each affected community introduced in Table 1 adjusts the overall negative environmental impact and risk to relevant communities, and this will push the environmental pollution objective tightening towards the minimum impact on most populated areas.

Table 2. Parameters of regional waste distribution center

Parameter	Distribution	Period				
		s=1	s=2	s=3	s=4	s=5
$AL_{dt(s)}$	dt=1	5500000	5650000	5800000	6000000	6150000
	dt=2	4500000	4600000	4700000	4800000	4900000
$WDtC_{dt(s)}$	dt=1	25	27	28	30	31
	dt=2	27	29	30	32	33
$MAX_{dt(s)}$	dt=1	155000	155000	185000	185000	185000
	dt=2	135000	135000	135000	135000	135000
$MIN_{dt(s)}$	dt=1	70000	70000	70000	70000	70000
	dt=2	65000	65000	65000	65000	65000
$EP_{dt(s)}$	dt=1	1.5	1.5	1.5	1.65	1.65
	dt=2	1.3	1.3	1.3	1.3	1.3

Table 3. Parameters of waste treatment plant

Parameter	Treatment	Period				
		s=1	s=2	s=3	s=4	s=5
$AL_{t(s)}$	t=1	10250000	10350000	10500000	10750000	10900000
	t=2	8500000	8800000	8900000	9050000	9200000
$WTC_{t(s)}$	t=1	18	20	20	21	21
	t=2	19	19	22	22	22
$MAX_{t(s)}$	t=1	110000	110000	110000	110000	110000
	t=2	90000	90000	90000	90000	90000
$MIN_{t(s)}$	t=1	70000	70000	70000	70000	70000
	t=2	60000	60000	60000	60000	60000
$EP_{t(s)}$	t=1	2.6	2.6	2.7	2.7	2.7
	t=2	2.2	2.3	2.3	2.3	2.4

Table 4. Parameters of waste disposal facility

Parameter	Treatment	Period				
		s=1	s=2	s=3	s=4	s=5
$AL_{d(s)}$	d=1	4500000	4550000	4600000	4650000	4700000
$WDC_{d(s)}$	d=1	13	14	15	16	17
$MAX_{d(s)}$	d=1	250000	245000	230000	220000	210000
$MIN_{d(s)}$	d=1	50000	50000	50000	50000	50000
$EP_{d(s)}$	d=1	4.5	4.9	5.3	5.7	6.2

Table 5 presents the distance between local waste collection centers to other downstream facilities within municipal solid waste management system. Table 6 gives the unit inter-facility transportation cost of solid waste. The waste locally collected will be first sent to regional distribution center for separation and pre-treatment, and the direct transportation of waste between local collection center to treatment plant or landfill is therefore impossible, and this type of unit transportation cost of municipal solid waste is not listed in this table.

Table 5. Distance between different facilities

Community	Distribution		Treatment		Disposal
	dt=1	dt=2	t=1	t=2	d=1
c=1	8	10	16	32	45
c=2	12	10	20	29	34
c=3	18	6	18	19	30

Table 6. Parameters of inter-facility transportation of municipal solid waste

Facility		Distribution		Treatment		Disposal	Period				
		dt=1	dt=2	t=1	t=2	d=1	s=1	s=2	s=3	s=4	s=5
Community	c=1	√					14	15	15	17	18
	c=1		√				11	12	13	14	14
	c=2	√					17	18	19	22	22
	c=2		√				12	13	15	16	17
	c=3	√					23	25	27	28	28
	c=3		√				10	14	15	17	18
Distribution	dt=1			√			8	9	10	11	11
	dt=1				√		10	10	11	12	14
	dt=1					√	15	16	17	18	19
	dt=2			√			13	14	15	16	17
	dt=2				√		8	9	9	10	11
	dt=2					√	13	13	13	14	14

The mathematical model is programmed in Lingo package and run at a personal laptop. Due to the small size of the question, the optimal solution of cost objective, environmental pollution objective as well as the composite objective can be calculated within 1 second. The cost optimization and environmental pollution optimization are first solved individually, and waste allocation of both individual objective functions in the studied period is presented in Tables 7 and 8. The optimal individual cost over the studied time horizon is 401421800, and it is 26602910000 for the optimal individual environmental pollution.

Table 7. Optimal waste allocation for cost-minimization objective

Transportation of waste	Period				
	s=1	s=2	s=3	s=4	s=5
$QTp_{c=1/dt=1(s)}$	85500	92000	94500	99200	102500
$QTp_{c=1/dt=2(s)}$					
$QTp_{c=2/dt=1(s)}$	39000	47000	55200	67150	72800
$QTp_{c=2/dt=2(s)}$	67000	66500	65800	64850	63000
$QTp_{c=3/dt=1(s)}$					
$QTp_{c=3/dt=2(s)}$	68000	68500	69200	70150	72000
$QTp_{dt=1/t=1(s)}$	110000	74000	84700	101350	110000
$QTp_{dt=1/t=2(s)}$		65000	65000	65000	65300
$QTp_{dt=2/t=1(s)}$					
$QTp_{dt=2/t=2(s)}$	65000				
$QTp_{dt=1/d=1(s)}$	14500				
$QTp_{dt=2/d=2(s)}$	70000	135000	135000	135000	135000

Table 8. Optimal waste allocation for pollution-minimization objective

Transportation of waste	Period				
	s=1	s=2	s=3	s=4	s=5
$QTp_{c=1/dt=1(s)}$					
$QTp_{c=1/dt=2(s)}$	85500	92000	94500	99200	102500
$QTp_{c=2/dt=1(s)}$	87000	86500	115800	96200	103300
$QTp_{c=2/dt=2(s)}$	19000	27000	5200	35800	32500
$QTp_{c=3/dt=1(s)}$	68000	68500	69200	70150	72000
$QTp_{c=3/dt=2(s)}$					
$QTp_{dt=1/t=1(s)}$	65000	65000	70000	110000	85300
$QTp_{dt=1/t=2(s)}$	90000	90000	90000		90000
$QTp_{dt=2/t=1(s)}$	5000	5000			24700
$QTp_{dt=2/t=2(s)}$				90000	
$QTp_{dt=1/d=1(s)}$			25000	56350	
$QTp_{dt=2/d=2(s)}$	99500	114000	99700	45000	110300

A significant difference of periodic waste allocation can be observed in those two different scenarios. For the local waste collection center at community $c=3$, all the collected solid waste is sent to distribution center $dt=2$ in individual cost optimization scenario due to the predominant advantage of the low unit transportation cost between those two facilities, however, the short distance between them also lead to a much higher value of $\frac{EP_{dt(s)}QT_{dt(s)}}{DS_{dt/af}}$ in the environmental pollution objective, and because of this reason, all the collected waste at community $c=3$ are allocated to distribution center $dt=1$ in the individual environmental pollution optimization scenario even through the environmental pollution indicator of $dt=1$ is slightly greater than that in $dt=2$.

In individual cost optimization scenario, most waste at distribution center $dt=1$ is distributed to the incineration plants due to the much lower unit transportation cost, however, because of the lower unit processing cost of landfill, it becomes the primary destination of the waste at distribution center $dt=2$ where the unit transportation cost to incineration plants and landfill are similar. In individual environmental pollution optimization scenario, the waste treated at incineration plant $t=1$ is minimized due to the large value of $\frac{EP_{t(s)}QT_{t(s)}}{DS_{t/af}}$ resulting from the small distance between incineration plant $t=1$ and affected communities. Besides, the allocation of waste to landfill is less in the individual environmental pollution optimization scenario due to the large value of environmental pollution indicator of landfill.

The optimal value of individual cost and individual environmental pollution can then be brought into the composite objective function Eq. (3), and the optimal value of composite objective can be calculated with given ∂_C and ∂_p. Those two adjustment parameters determine the relative importance of system cost and environmental pollution of the municipal solid waste system, which significantly influence the decision-making of long term allocation of solid waste to different facilities. In this paper, ten different scenarios with incremental value of ∂_C are defined, and it equals to 0.1, 0.2, 0.3, 0.4, 0.5, 0.6, 0.7, 0.8 and 0.9, respectively. Figure 2 illustrates the comparison of the optimal value of the composite objective functions in those ten defined scenarios.

As shown in the figure, the value of the composite objective function increases with the increase of the value of parameter ∂_C. Besides, the optimal value of Eq. (3) equals to 1 when ∂_C equals to 0 or 1, and that represents the individual cost optimization and individual environmental pollution optimization. The long-term performance of municipal solid waste management system becomes much better when the optimal value of the composite objective function approaches to 1, so for this illustrative case, the system performance becomes much better when the environmental pollution objective plays more important role in the decision-making of the long-term waste allocation plan.

The focus on environmental pollution of municipal solid waste management system may lead to extremely high cost, and the optimal balance of cost objective and environmental pollution is therefore emphasized. Herein, a compromising scenario with ∂_C equals to 0.5 is detailed in Table 9. As shown in the table, there is a significant difference of waste allocation over the five periods from that in individual cost objective and individual environmental pollution objective, and a more even allocation of waste to different facilities in the studied time horizon can be observed in this scenario. The balance of those two objective functions is optimized for the given numerical value of ∂_C. Therefore, the proposed model provides an effective solution for the long-term operational planning of the municipal solid waste management system.

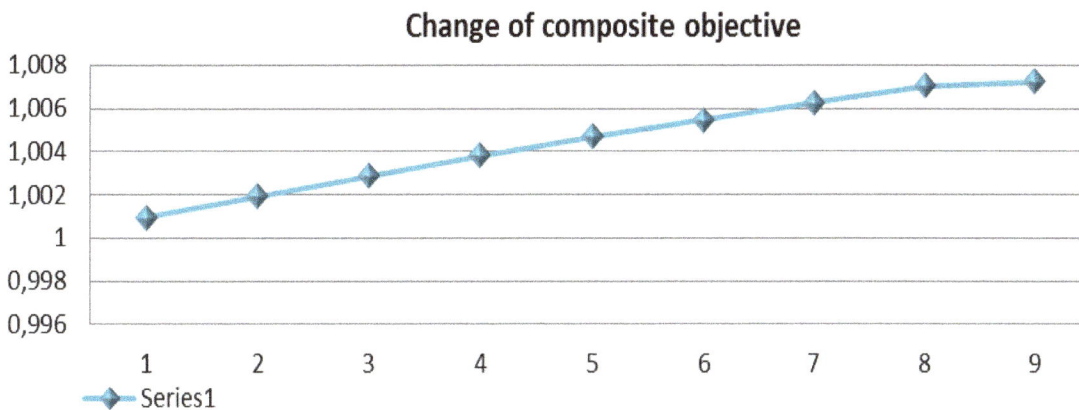

Figure 2. Comparison of the optimal value of the composite objective functions in the defined ten scenarios

Table 9. Optimal waste allocation when ∂_C equals to 0.5

Transportation of waste	Period				
	s=1	s=2	s=3	s=4	s=5
$QTp_{c=1/dt=1(s)}$					
$QTp_{c=1/dt=2(s)}$	85500	92000	94500	99200	102500
$QTp_{c=2/dt=1(s)}$	56500	70500	80500	96200	103300
$QTp_{c=2/dt=2(s)}$	49500	43000	40500	35800	32500
$QTp_{c=3/dt=1(s)}$	68000	68500	69200	70150	72000
$QTp_{c=3/dt=2(s)}$					
$QTp_{dt=1/t=1(s)}$	70000	70000	70000	110000	11000
$QTp_{dt=1/t=2(s)}$		69000	79700	56350	65300
$QTp_{dt=2/t=1(s)}$					
$QTp_{dt=2/t=2(s)}$	90000	21000	10300	33650	24700
$QTp_{dt=1/d=1(s)}$	54500				
$QTp_{dt=2/d=2(s)}$	45000	114000	124700	101350	110300

4. Conclusion

This paper has presented a bi-objective dynamic optimization model for long-term planning of municipal solid waste management system. Previously, most literature focuses on the methods and models for the network design and location problems of waste treatment facilities (e.g. incinerator, landfill, etc.) and transfer station, but this study aims to develop navel methods and computation model for determining the optimal long-term operation plan of municipal solid waste management system. The model developed in this study is a bi-objective linear programming model which simultaneously optimizes the system operating cost and environmental pollution of municipal solid waste management system, and an illustration is also presented for a deep insight of the model application.

Future improvement can be focused on two aspects. First, the consideration of the entire reverse supply chain of waste management should be taken into account. With the promotion of sustainable development, many types of municipal solid waste has been considered as the "raw material" of the reverse supply chain, and more alternatives for waste treatment, recycling, reuse and remanufacturing have dramatically increased the complication and complexity of the reverse network of municipal solid waste management system. Therefore, the development of decision support tools for the entire reverse supply chain of waste management is initially suggested. Second, some parameters are impossible to be predicted precisely for the given time periods, and methods for effectively dealing with the uncertain parameters are therefore important for the decision support model and suggested for further improvement.

Nomenclature
Subscripts

s	Number of defined time periods;
c	Number of local waste collection centers;
dt	Number of regional waste distribution centers;
t	Number of waste treatment plants;
d	Number of disposal facilities;
af	Number of affected communities;

Parameters (The meaning of the parameters subjects to the subscripts)

AI	Annualized investment;
WCC	Unit collection and processing cost at local waste collection center;
$WDtC$	Unit processing cost at regional waste distribution center;
WTC	Unit processing cost at waste treatment plant;
WDC	Unit processing cost at waste disposal facility;
$WTpC$	Unit waste transportation cost;
QT	Waste amount processed;
QTp	Waste amount transported;
POL	Population of affected community;
EP	Environmental pollution indicator;
DS	Distance between waste management facility and affected community;
MAX	Maximum capacity;
MIN	Minimum required waste quantity;
SW	Waste generation at each community;

Acknowledgements
This research was supported by National Natural Science Foundation of China (Grand No. 71201144).

References
[1]   Srivastava, A.K., Nema, A.K. Fuzzy parametric programming model for multi-objective integrated solid waste management under uncertainty. Expert Systems with Application 2012, 39, 4657-4678.

[2]   Son, L.H. Optimizing municipal solid waste collection using chaotic particle swarm optimization in GIS based environments: A case study at Danang. Expert Systems with Applications 2014, 41, 8062-8074.

[3]   Batool, S.A., Chuadhry, M.N. The impact of municipal solid waste treatment methods on greenhouse gas emissions in Lahore, Pakistan. Waste Management 2009, 29, 63-69.

[4]   He, L., Huang, G.H., Lu, H. Greenhouse gas emissions control in integrated municipal solid waste management trough mixed integer bilevel decision-making. Journal of Hazardous Materials 2011, 193, 112-119.

[5]   Yu. H., Solvang, W.D., Yuan, S. A multi-objective decision support system for simulation and optimization of municipal solid waste management system. Proceeding of 3rd IEEE International Conference on Cognitive Info communications. Kosice, Slovakia, 2012, pp: 193-199.

[6]   Khan, S., Faisal, M.N. An analytical network process model for municipal solid waste disposal options. Waste Management 2008, 28, 1500-1508.

[7]   Ghiani, G., Manni, A., Manni, E., Toraldo, M. The impact of an efficient collection sites location on the zoning phase in municipal solid waste management. Waste Management 2014, 34, 1949-1956.

[8]   Eiselt, E.A., Marianov, V. A bi-objective model for the location of landfills for municipal solid waste. European Journal of Operational Research 2014, 235, 187-194.

[9] Badran, M.F., El-Haggar, S.M. Optimization of municipal solid waste management in Port Said – Egypt. Waste Management 2006, 26, 534-545.

[10] Zhang, X., Huang, G. Municipal solid waste management planning considering greenhouse gas emission trading under fuzzy environment. Journal of Environmental Management 2014, 135, 11-18.

[11] Alcada-Almeida, L., Coutinho-Rodriues, J., Current, J. A multi objective modeling approach to locating incinerators. Socio-Economic Planning Sciences 2009, 43, 111-120.

[12] Galante, G., Aiello, G., Enea, M., Panascia, E. A multi-objective approach to solid waste management. Waste Management 2010, 30, 1720-1728.

[13] Dai, C., Li, Y.P., Huang, G.H. A two-stage support-vector-regression optimization model for municipal solid waste management – A case study of Beijing, China. Journal of Environmental Management 2011, 92, 3023-3037.

[14] Mavrotas, G., Skoulaxinou, S., Gakis, N., Katsouros, V., Georgopoulou, E. A multi-objective programming model for assessment the GHG emissions in MSW management. Waste Management 2013, 33, 1934-1949.

[15] Eiselt, H.A., Marianov, V. Location modeling for municipal solid waste facilities. Computers & Operations Research 2014. doi:10.1016/j.cor.2014.05.003

[16] Zhang, Y., Huang, G.H., He, L. A multi-echelon supply chain model for municipal solid waste management system. Waste Management 2014, 34, 553-561.

[17] Nema, A.K., Gupta, S.K. Optimization of regional hazardous waste management systems: an improved formulation. Waste Management 1999, 19, 441-451.

[18] Hu, T.L., Sheu, J.B., Huang, K.H. A reverse logistics cost minimization model for the treatment of hazardous wastes. Transportation Research Part E 2002, 38, 457-473.

[19] Yu, H., Solvang, W.D., Chen, C. A green supply chain network design model for enhancing competitiveness and sustainability of companies in high north arctic regions. International Journal of Energy and Environment 2014, 5(4), 403-418.

8

Feasibility investigation and combustion enhancement of a new burner functioning with pulverized solid olive waste

Bounaouara H.[1,2], Sautet J.C.[1], Ben Ticha H.[2], Mhimid A.[2]

[1] CORIA UMR 6614 CNRS, Université et INSA de ROUEN, Avenue de l'Université, BP 12, 76801 Saint Etienne du Rouvray, Cedex, France.
[2] LESTE, Ecole Nationale d'Ingénieurs de Monastir, 5019 Monastir, Tunisie.

Abstract
This article describes an experimental study on solid olive residue (olive cake) combustion in form of pulverized jet. This is a contribution to the valorization of olive residue as a source of renewable energy available in the majority of mediterranean countries. A sample of olive cake from Tunisian origin is prepared for the experiment; this sample is crushed, dried and sifted in order to obtain the desired particles form. A new burner made up of a coaxial cylindrical tube is especially designed and fabricated. In order to start the combustion of olive cake and maintain the main flame, two types of pilot flame were used: a central premixed flame of methane/oxygen and an annular diffusion flame of methane. This paper shows the conditions for an efficient olive cake burner operation in free air. The effects of particle size and pilot flame position have been discussed. The olive cake combustion is possible only with particles at a size below 200 μm. Moreover, the combustion maintained by the annular pilot flame ensures better burning conditions than the central pilot flame. Finally, the inserted preheating system has improved the olive cake combustion.

Keywords: Pulverized jet; Olive cake combustion; Solid fuels; Biomass.

1. Introduction and state of the art

The upward technological and demographic developments imply an increase in demand of energy in the world. In addition, the air pollution due to the intense use of fossil energies threatens the environment and human habitat. The dangerous emissions of these energy carriers create an increasing environmental awareness. Because of the mentioned issues, studies and researches on new and renewable energies as energy solutions for the future have been encouraged. The present study carries an interest on the biomass valorization as a renewable energy, more particularly the solid olive residue. The conscious use of the olive cake in the energy production allows solving two problems: clean energy production and an effective treatment of waste coming from the oil mills. Olive cake can be burned alone or can contribute in combustion with other biomass and coal [1]. Olive residue is regarded as a good alternative to other fossil energies in the Mediterranean countries. In particular, in Tunisia, there are more than 67 million olive trees [2] and the amount of solid olive residue is about 350-450 thousands of tonnes per years [3]. Thus, Tunisia can benefit from the very significant renewable energy source. Olive cake is obtained after oil extraction from olive fruits. The types of the olive residue depend on the treatments that they underwent. Thus, the olive cake coming from traditional oil mills (by press) contains between 8 and 12 weight percent (wt%) of oil, otherwise the exhausted olive cake, which underwent a double extraction by pressing and by extraction with hexane, contains a quantity less than 5wt% oil. The raw olive cake stored in open sky has a water content varying from 15 to 50 wt%. This solid olive residue contains between 40 and 60 wt% of carbon and its lower heating value ranging from 12,500-26,000 kJ/kg; these values are comparable with the heating values of the other biomasses like wood ($\approx$17,000 kJ/kg) and some types of coal (23,000-26,000 kJ/kg [4-6]). The range of sulfur content in the olive cake is 0.05-0.1 wt% [4, 6].

1.1 Processes of olive cake valorization

The energy valorization of the olive cake can be carried out according to three principal processes:

- The first is the pyrolysis which is a thermochemical decomposition of solid matter in an inert atmosphere (low in oxygen). The principal gases generated are CO, CO_2, H_2, CH_4, C_2H_4 and C_2H_6 [3]. In general, pyrolysis produces gas (volatiles) and liquid products (tars) and leaves a solid char residue with high carbon content. The greatest commercial interest of the solid products is due to its low sulphur and phosphorus contents [7].

- The gasification technology: This process is used for fuel gas production by treatment of solid fuel. Gasification is achieved by oxidation of organic material at high temperatures without combustion and thereafter gaseous species like CO, H2 and CO2 are produced. Vera et al. [8] have obtained a fuel gas which has a low calorific value around 4350 kJ/kg. Moreover, gasification of the char can improve its porous structure to produce activated carbon, which is widely used as adsorbent [6].

- The third process is the direct combustion of the olive cake. In this process, a combustion reaction occurs and a flame is generated. In general, the combustion of solid can be subdivided in five steps: heating up, devolatilization, combustion of volatiles, combustion of char leaving ash and inert heating of ash. Technologies of solid fuel combustion are: fixed bed fluidized bed and pulverized jet.

1.2 Previous studies on olive cake combustion

Several studies have an interest concerning the energy production starting from the direct combustion of olive cake (alone or mixed with other fuels). Abu-Qudais [9] has studied the olive cake combustion of 0.53 mm average size in a fluidized bed reactor. The temperature of the fluidized bed varied between 775 and 935°C for mass ratios air/olive-cake ranging between 2.30 and 4.30. In addition, this author has noted that the quantity of unburned oil cakes can be decreased and the fuel output has been improved from 85 to 95% for optimal air velocities.

Alwidyan et al. [10] have studied the direct combustion of olive cake in pulverized form in a vertical tube furnace. The maximum flame temperature reached 980°C and the combustion efficiency was about 82%. Moreover, Alkhamis and Kablan [11] studied the effect of the olive cake grain size on its calorific value. The highest value of the calorific heat was 37.300 kJ/kg and it corresponds to size particles between 125 and 250 μm. Thus, olive cake can be considered as a good fuel and a potential source of energy.

Atimtay and Topal [6] have studied the co-combustion of olive cake (diameter d=2.3mm) with lignite coal (diameter d=0.46mm) in a circulating fluidized bed. The results have shown that as the percentage of olive cake in the mixture increases, the combustion mainly occurs in the upper regions of the main column. The maximum temperature reaches 900 °C. The minimum carbon monoxide emissions are observed at an excess air ratio about 1.5. The mass percentage of olive cake in the mixture fuel is suggested to be below 50 wt% in order to be within the European Union limits for emissions. Their results suggest that olive cake is a good fuel that can be mixed with lignite coal for cleaner energy production. Later, Atimtay and Varol [12] studied the co-combustion of coal and olive cake in a bubbling fluidized bed with secondary air injection. They determined that as the amount of olive cake in the fuel mixture increases, sulfur dioxide emissions decrease because of the very low sulphur content of olive cake. The conditions of optimum operating with respect to NOx and sulfur dioxide emissions were found to be 35% for primary excess air and 30L/min for secondary air flow rate when a fuel mixture composed of 75 wt% olive cake and 25 wt% coal is burned in the bed. The system reached its highest combustion efficiency of 99.8 % when 25 wt% olive cake and 75 wt% coal mixture is combusted with a primary excess air of 70% and a secondary air mass flow of 40L/min.

Abu-Qudais and Okasha [5] have realized an important experimental study of the direct combustion of a diesel and olive cake slurry in the form of a jet-blast atomizer in a vertical cylindrical water-cooled combustor. These authors have approved that olive cakes can be a significant source of energy. The combustion efficiency was improved as the percentage of olive cake in the fuel mixture was increased to

7wt%. Stable flames were observed for the combustion of fuel mixture composed of up to 20 wt% of olive cake.

In addition, olive cake is used as a fuel in several tunisian applications: boilers in oil mills, driers in factories of oil extraction from olive cake, furnaces in brickyards. Masghouni and Elhassayri [13] have studied the substitution of No. 2 heavy fuel (a heavy fuel oil containing 4% of sulphur) by olive cake in a static furnace of a tunisian brick factory. The lower heating value of No. 2 heavy fuel and olive cake were 43.500 kJ/kg and 16.500 kJ/kg, respectively. A financial comparison of the costs of No. 2 heavy fuel and olive cake shows a reduction of 63.8% in the cost of energy. The combustion of olive cake was less polluting than the combustion of No. 2 heavy fuel;

solid particles, black carbon, carbon monoxide and sulphur oxides in the flue gas are negligible [13].

In order to set up favorable conditions of olive cake combustion it's necessary to have reliable knowledge concerning olive cake properties and major processes such as ignition, drying, devolatilization, and volatile and char reactions.

Chouchene et al. [14] have studied the thermal degradation behavior of olive solid waste by thermogravimetric experiments. The phenomenon of olive cake pyrolysis under oxidative conditions (10% O_2 and 90% N_2) has proceeded with three major stages: the first weight loss is due to drying of the sample. The second stage corresponds to volatiles release. A char is formed following this step. The last stage corresponds to the oxidation of the formed char. At the end, ashes and some residual char were remained. For the sample having a size below 0.5mm, the second step named volatilization is carried out for the temperatures range 180-290°C. Thereafter, the char oxidation step has taken place between 290°C and 470°C. Chouchene et al [14] noted that the amounts of remaining ashes at the end of pyrolysis with particle sizes ranging between 1 mm and 1.5 mm is close to that obtained during agriculture residues combustion. Based on both study of Chouchene et al [14] and Gani and Naruse [15] the behavior of pyrolysis and the combustion of biomass have a large similarity. Moreover, kinetic parameters during devolatilization step and char oxidation step for different particle size and under inert and oxidative atmosphere are presented in reference [14].

1.3 Interest of pulverization technique

At present, the majority of coal power plants use the technique of pulverized coal combustion. In pulverization method solid fuel is pulverized into fine particles of several tens of micrometers in diameter. For the great installations combustion (>100MW) of coal and of co-combustion (mixture of coal and biomass), the more practical technology is the pulverization [16]. Pulverization technique is privileged because it improves the economic and environmental performance of these plants by reducing combustible consumption and increasing productivity and minimizing pollutants emissions. In fact, the combustion becomes faster thanks to the increase in the contact surface between the particles and the oxidant. In addition, the problem of the incomplete combustion of solid fuel that causes poor gas emission can be solved by using very small particles [17]. In this context, Alwidyan et al. [10] concluded that under certain local conditions, olive cake on its pulverized form can be considered as the most practical renewable energy among solid fuels.

In this paper, an experimental study of feasibility of a burner functioning with the pulverized olive cake in free air is presented. The experimental conditions to enhance combustion of pulverized olive cake in new burner at open air have been studied. In order to achieve this aim, the olive cake is preliminary prepared, a special burner is fabricated and a new pilot plant is designed and associated. Thus, a turbulent and reactive two-phase jet of air loaded with olive cake particles is formed. Particle size of olive cake samples is below 200 μm. In order to start the combustion and ensure the persistence of the main flame in ambient air, two types of pilot flame are used: a central premixed flame of methane/oxygen and an annular diffusion flame of methane. A preheater is inserted in the inlet air line in order to study its effect on the main flame.

2. Experimental study
2.1 Presentation of experimental device

The experiment principle consists of producing a jet of air loaded with olive cake particles having sizes below 200 μm the two-phase jet discharged in free air. The pulverized jet flame is ignited by a pilot flame which can be placed at the port center or around the main port. Figure 1 illustrates the schematic diagram of the pilot plant used for pulverized olive cake combustion.

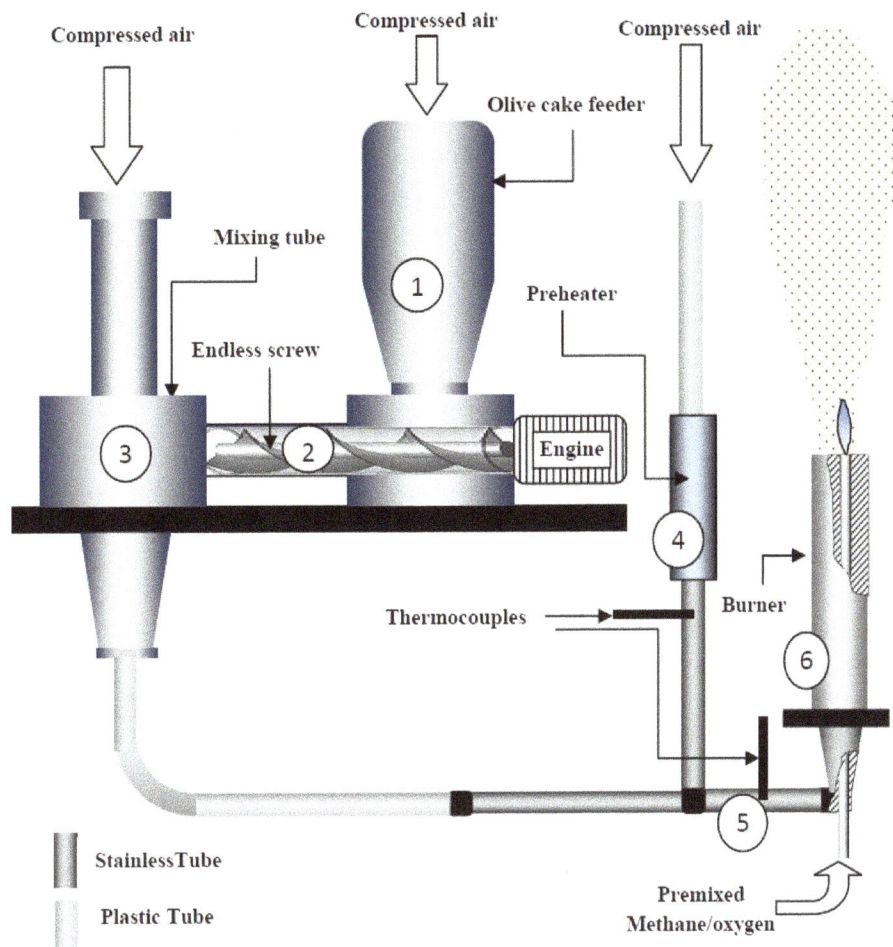

Figure 1. Schematic diagram of the pilot plant used for pulverized olive cake combustion

The olive cake particles are initially stored in the storage tank (1). Then, they are moved by an endless screw (2) towards mixing tube (3). Compressed air is injected into the particles feeder (1) with an aim to improve the movement of particles. Another flow rate of compressed air is injected into the tube (3) in order to entrain the particles towards the burner (6). If the preheating is necessary, a hot air coming from the preheater (4) should be mixed with the cold mixture of air and olive cake. In the both case, the mixture of air and olive cake goes in the burner (5). The main injector tube (5) has an inner diameter of 30 mm and a height of 210 mm. The mixture of air and olive cake passes through tubes with different geometry before being introduced in the straight main injector tube.

Each gas line is consisted by the following essential elements assembled in series: a valve, a pressure regulator, a manometer and a colsonic.

2.2 Fuel preparation

In order to carry out the experiment, the primary particles of olive cake have to be converted on small particles having sizes lower than 200 µm. Firstly, several alternate operations of crashing and garbling are carried out. Crash operation is done by a grinding mill with steel plates. The desired size is obtained by a screen with granulometry having meshes of 200µm in accordance with AFNOR French norm.

Researches of Kurose et al. [18] show the negative effects of moisture on the coal combustion effectiveness. Based on the above mentioned result, we carried out the drying of the olive cake particles until a water content of 1.85 %. Drying is carried out by a conventional furnace.

A sample of olive cake particles thus sifted and dried underwent a microanalysis (see Table 1).

The results of ultimate analysis allow calculating the volume of air needed to burn 1 kg of olive cake. The stoichiometric air/fuel ratio is about 4.45 Nm^3 of air per kg of olive cake. The High and the Lower Heating of olive cake are HHV=18800 kJ/kg and LHV=17510 kJ/kg, respectively.

Table 1. Properties of olive cake (Sample coming from the Tunisian Sahel region, 2011)

Parameter		Value
Ultimate analysis	Carbon	47.15 wt%
	Hydrogen	6.03 wt%
	Oxygen*	41.3 wt%
	Nitrogen	1.34 wt%
	Sulfur	< 1 wt%
Ash		2.36 wt%
Moisture**		1.85 wt%
Oil content**		2.25 wt%
Lower Heating Value		17510 kJ/kg

* Value determined by difference
**Analysis carried out just after drying stage

2.3 First configuration with a central pilot flame

This configuration represents the selected parameters running the first test. The air which is supplied into the main injector tube represents the mass flow rate calculated at the stoichiometry.

The pulverized jet thus created is highly diluted with a volume fraction of olive cake is about 0.032%. The mass loading of the discrete phase is equal to 0.15.

In this configuration, combustion reaction is induced by a central pilot flame; it is a small premixed flame of methane and oxygen located in the center of the main injector tube. Experimental parameters are listed in Table 2.

Table 2. Experimental conditions

Flame type	Parameter	value
Premixed pilot flame	O_2 flow rate (m^3/s)	10.5×10^{-6}
Diffusion/ Premixed pilot flame	CH_4 flow rate* (m^3/s)	4.5×10^{-6}
	Thermal input of CH_4* (W)	145
Pulverized jet	Air flow rate (m^3/s)	2.01×10^{-3}
	Velocity of pulverized jet (m/s)	2.84
	Reynolds number, Re (-)	5838
	Pulverized olive cake feed rate (kg/s)	3.67×10^{-4}
	Thermal input of olive cake (W)	6426
	Mass loading of olive cake (-)	0.15
	Volume fraction of olive cake (%)	0.032

*same value for the two types of pilot flame

The inner and outer diameters of the pilot flame pipe are 2 and 4 mm, respectively. In order to vary the velocity of pilot flame gases, nozzles of diameters ranging between 0.2 to 1.5 mm as shown in Figure 2. A nozzle diameter of 1 mm has used for this configuration and that can increase the pilot flame velocity until 19.10 m/s.

2.4 Second configuration with annular pilot flame

A new pilot flame was designed and it is installed outside of the main burner. This flame is fed by methane which is supplied to the annular slit having a width of 0.5mm.

The methane jet is inclined towards center at an angle of 60 degrees to horizontal in order to allow best contact with pulverized particles (see Figure 3).

2.5 Third configuration: Burner with a preheating system

This part of the present study aims to improve the combustion by using a preheating system. Indeed, preheat olive cake particles before being supplied to the burner allows them a drying stage and the combustion becomes more quickly. Pulverized particles can also begin the devolatilization stage inside tubes.

The inserted preheater is equipped with regulator and it aimed at providing hot air for the air and olive cake mixture. In this configuration, the line of combustion air was divided into two fractions; one of these crosses the preheater and the other fraction drives the olive cake particles. The two fractions meet after preheater and before being supplied in burner tube.

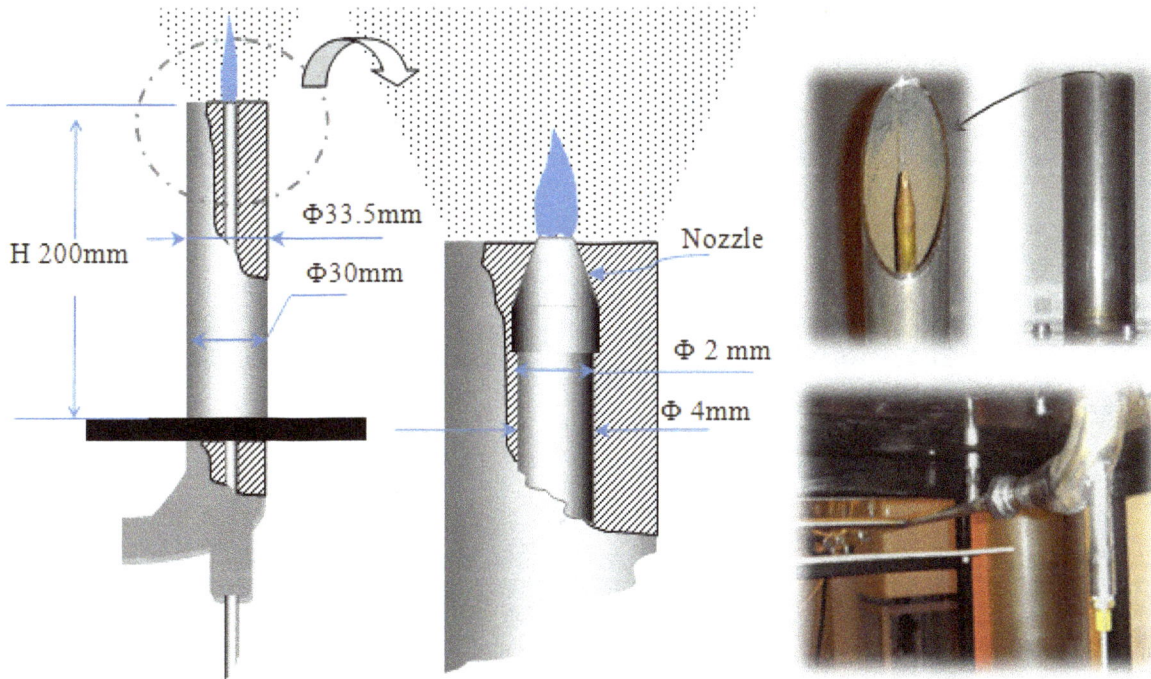

Figure 2. Dimensional details of the main burner and the central premixed pilot flame

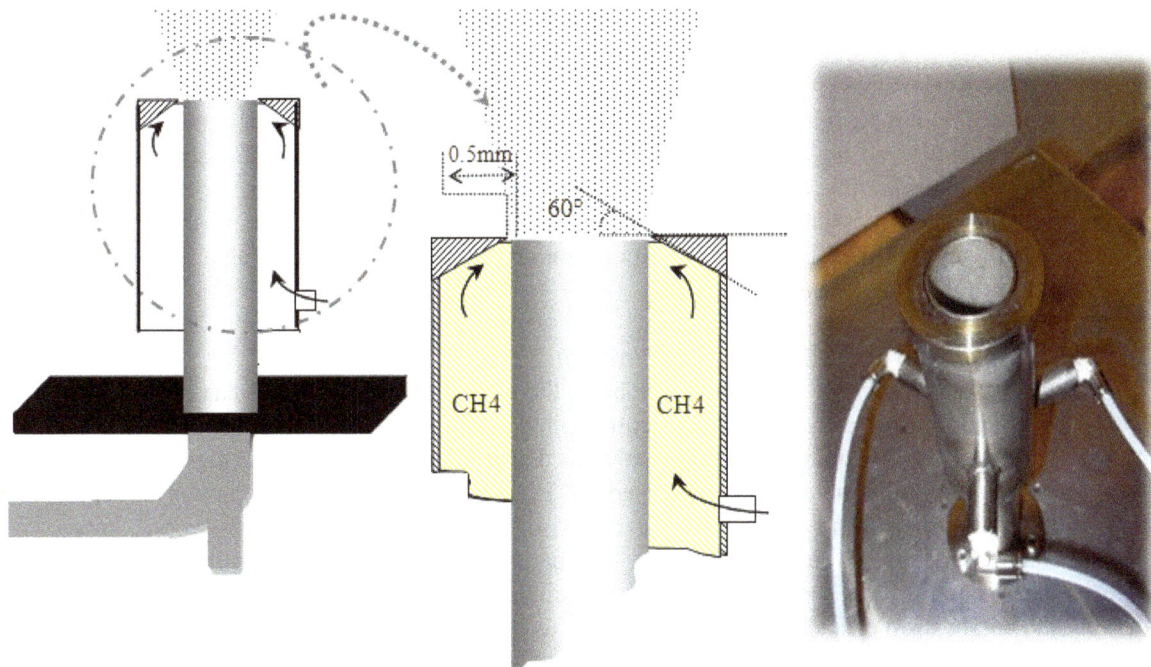

Figure 3. Schematic of the annular diffusion pilot flame

3. Results and discussion

3.1 Test of feasibility (First configuration)

With a central pilot flame, the flames generated by the combustion of the pulverized olive cake jet with particles having sizes below 200 μm are quite stable. Figure 4 shows a real pictures series of the flame, recorded in an interval of time equal to 1 second. These pictures confirm that the obtained flame is not an impulsive but it is a persistent flame. However, the pilot flame is vital to ensure the persistence of the olive cake flame. The reading order of the photos is from the left to the right and from the top to the down.

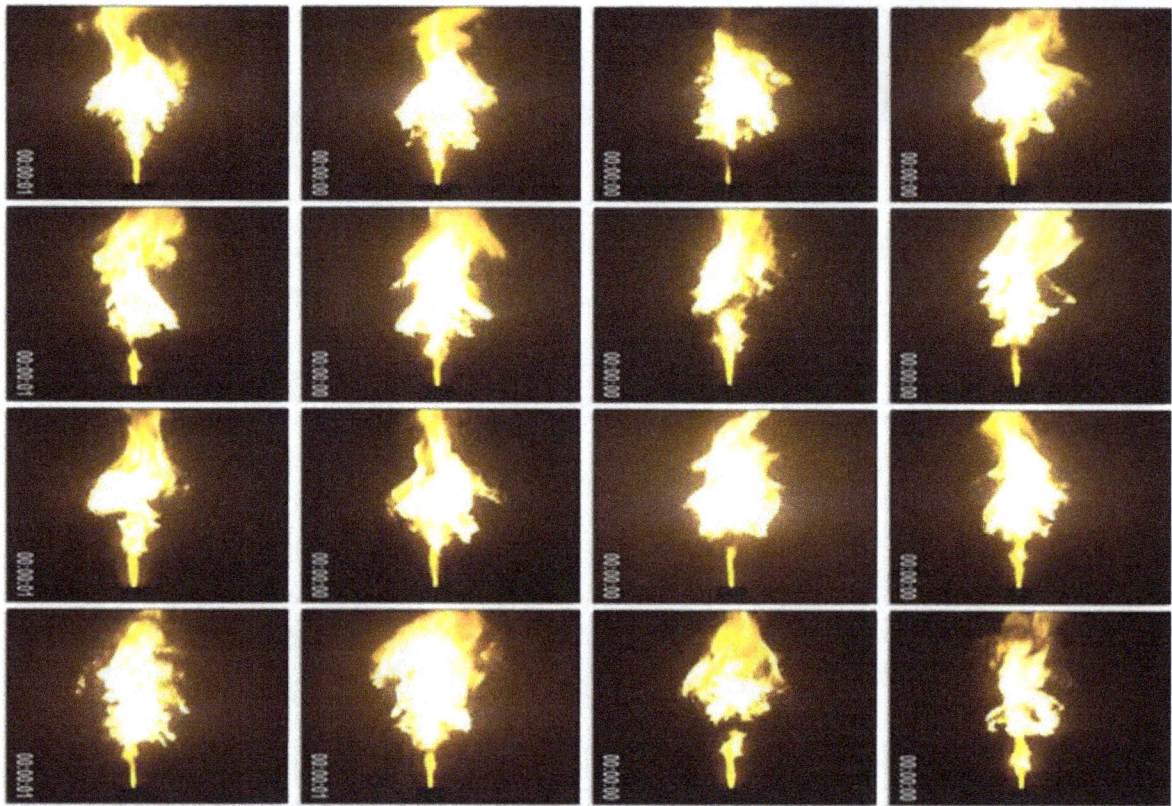

Figure 4. Photographs series of olive cake flame

In order to investigate the combustion enhancement of the pulverized jet flame, not only the feasibility of the designed burner, position of the pilot flame nozzle have been tested. In fact, combustion efficiency can be improved by sliding the tube of pilot flame inside the main injector. Thus, the combustion begins inside the burner. The quantity of unburned olive cake particles is greatly reduced and therefore more energy is recovered in the main jet flame. It's important to note that the most considerable heat losses are due to the bad combustion near the burner periphery.

3.2 Effect of the annular pilot flame (Second configuration)
In order to enhance combustion near the burner periphery, the new annular pilot flame has been installed. The second configuration aims at replacing the central pilot flame by another annular. The new used configuration is a methane diffusion flame. Tests showed that this pilot flame with a thermal power of 145 W was able to ignite and maintain the pulverized jet. Figure 5 represents a comparison between of the jet flame obtained for the second configuration and the first one.
The new main flame is wider and more powerful as compared to that one obtained with a central pilot flame. This is explained by the increase of the burned particles coming from burner periphery. Consequently, the combustion efficiency was increased by using an annular pilot flame.

3.3 Effect of jet velocity and the combustible feed rate
During tests of the second configuration, the effect of air velocity on the flame structure has been studied. Figure 6 shows direct photographs of olive cake flame for various input conditions; jet velocity (V_{jet}) and combustible feed rate (m_{OC}).
According to Figure 6 (a and b), we have noticed that as the mixture velocity is decreased, the pulverized jet is wider and the flame becomes brighter and more attached to the burner. Figure 6c represents a flame for a lower combustible feed rate and a high jet velocity: it's a short flame and less attached to the burner.

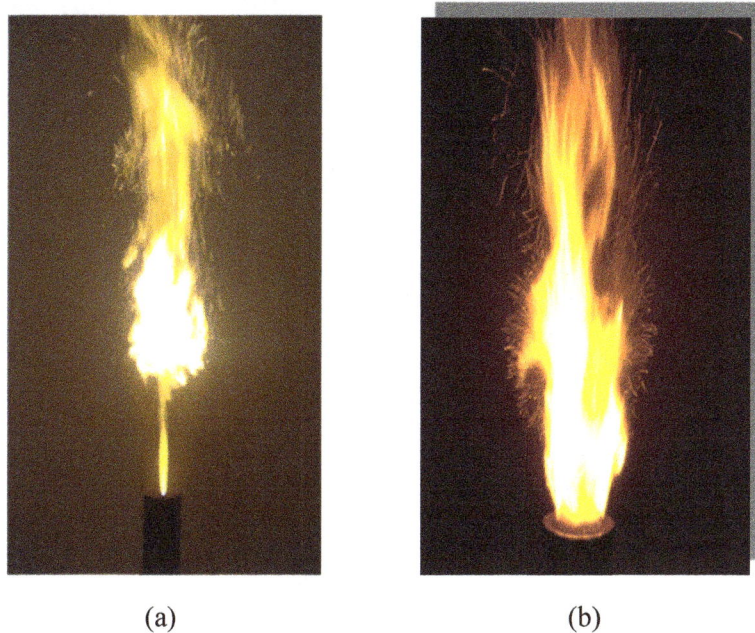

(a)   (b)

Figure 5. Olive cake flame controlled by a central pilot flame (a), Olive cake flame controlled by an annular pilot flame (b)

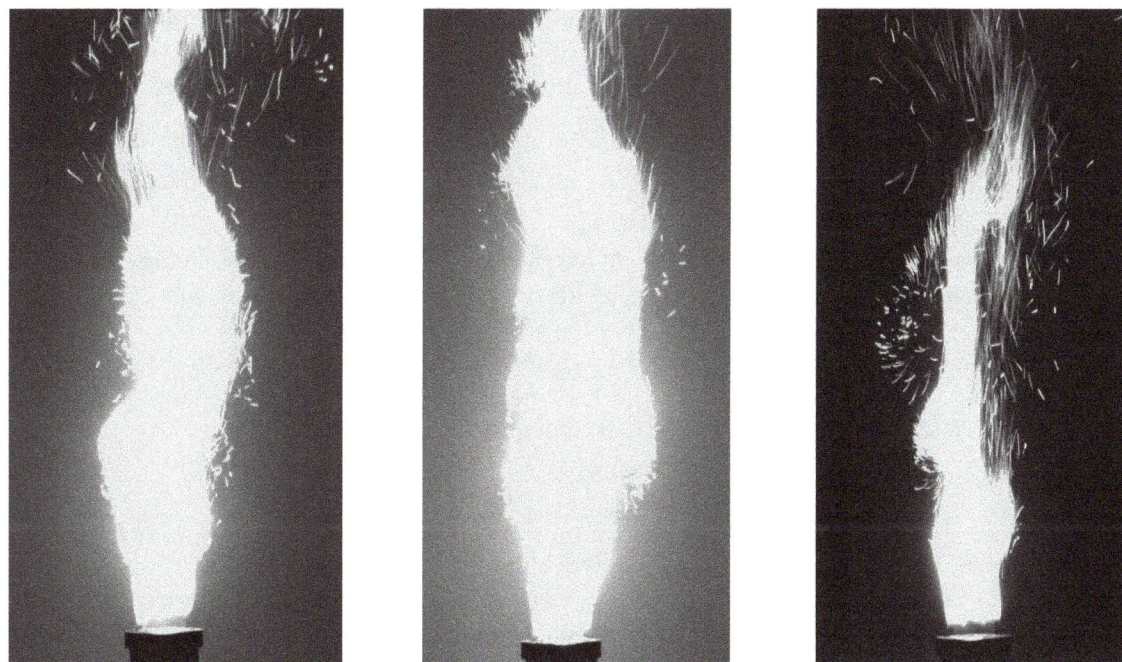

(a) V_{jet}= 3.31m/s (Re = 6770),   (b) V_{jet}= 1.91m/s (Re= 3898),   (c) V_{jet}= 3.31m/s (Re = 6770),
   m_{OC}=22.96g/min m_{OC} =22.96g/min m_{OC} =13.79g/min

Figure 6. Direct photographs of olive cake flame for various input conditions

One can conclude that the combustion is enhanced by lowering the jet velocity. This is can be explained by the increase in the particles residence time and the importance of the transverse thermal diffusion compared to the axial thermal convection.

3.4 Effect of particle size

Particles size is a significant parameter in the study of two-phase jets and pulverized solids combustion. Indeed, new experiments are carried out with particles sizes above 200 microns.

The generated flame as shown in Figure 7 is very short and limited to the zone near the pilot flame.

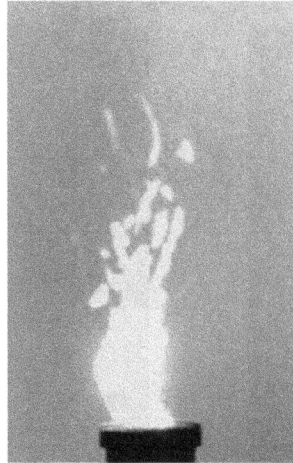

Figure 7. Combustion of olive cake particles having sizes ranging between 200μm and 350 μm

The particles having a size above 200 μm are difficult to be burned in this burner under the present experimental conditions. Thus, by comparison between combustion of particles below 200 μm and particles above 200 μm, it can be concluded that small particles can be burned better than the big particles. This is due to the effect of the thermal inertia and dynamic of the big particles.

4. Conclusions
A burner of pulverized olive cake particles was successfully implemented. A stable and controlled flame is obtained. The pilot plant used for experiments is mainly constituted by the burner and, a controlled drive system of solid particles. In addition, a system of preheating is set up and two types of pilot flame have been made and tested.

This study has led to the following conclusions:
- Combustion is more effective by using an annular pilot flame than a central flame. The preheating system can enhance the combustion of pulverized jet.
- Sizes of olive cake particles must be below 200 μm to obtain an effective combustion in the free air.
- The adequate drying of particles is one of the most favorable conditions for the combustion enhancement.

We will use the present experimental device to carry out many investigations on the olive cake flame: dynamic measurements, thermal behavior and control of pollutant emissions. In the future, this pilot plant will be considered as a preferential device to validate our numerical models.

References
[1]   Kalembkiewicz J. and Chmielarz U., Ashes from co-combustion of coal and biomass: New industrial wastes, Resources-Conservation and Recycling, 2012, 69, pp 109-121.
[2]   Direction Générale des Etudes et du Développement Agricole (DGEDA), Patrimoine Oleicole Tunisien, Mai-9-2012 from website http://www.onh.com.tn, Ministry of Agriculture, Republic of Tunisia.
[3]   Chouchene, A., Etude expérimentale et théorique de procédés de valorisation de sous-produits oléicoles par voies thermique et physico-chimique, Ph. D thesis, Ecole Nationale d'Ingénieurs de Monastir (Tunisia) and Université de Haute-alsace (France), Tunisia-France, 2010.
[4]   Topal, H., Atimtay A.T., Dormaz, A. Olive cake combustion in a circulating fluidized bed, Fuel, 2003, 82, pp 1049-1056.
[5]   Abu-Qudais, M., Okasha, G., Diesel Fuel and Olive-Cake Slurry: Atomization and Combustion Performance, Applied Energy, 1996, 54, pp 315-326.
[6]   Atimtay A.T., Topal H., Co-combustion of olive cake with lignite coal in a circulating fluidized bed. Fuel, 2004, 83, 859-867.
[7]   I. Aljundi, N. Jarrah., A study of characteristics of activated carbon produced from Jordanian olive cake. J. Anal. Appl. Pyrolysis, 2008, 81; 33-36.
[8]   Vera D.., De Mena B., Jurado F., Schories G., Study of a downdraft gasifier and gas engine fueled with olive oil industry wastes, Applied Thermal Engineering, 2013, 51, pp119-129.

[9]   Abu-qudais, M., Fluidized-bed combustion for energy production from olive cake, Energy, 1996, 21, pp 173-181.

[10] Al widyan, M. I., Tashtoush, G., Hamasha, A. M., Combustion and emissions of pulverized olive cake in tube furnace, Energy Conversion and Management, 2006, 47, pp 1588-1596.

[11] Alkhamis, T. M., Kablan M.M., Olive cake as an energy source and catalyst for oil shale production of energy and its impact on the environment, Energy Conversion & Management, 1999, 40, pp1863-1870.

[12] Atimtay, A.T., Varol, M., Investigation of co-combustion of coal and olive cake in a bubbling fluidized bed with secondary air injection, Fuel, 2009, 88, pp1000-1008.

[13] Masghouni, M., Elhassayri, M., Energy applications of olive-oil industry by-products: -I. The exhaust foot cake, Biomass and Bioenergy, 2000, 18, pp 257-262.

[14] Chouchene, A., Jeguirim, M., Khiari, B., Zagrouba, F., Trouve, G., Thermal degradation of olive solid waste: Influence of particle size and oxygen concentration, Resources-Conservation and Recycling, 2010, 54, pp 271-277.

[15] Gani A, Naruse I. Effect of cellulose and lignin content on pyrolysis and combustion characteristics for several types of biomass. Renewable Energy, 2007, 32:649-61.

[16] T. FLOREA., Simulation numérique de la combustion du bois dans une chaudière automatique de 400 kW, Ph. D Thesis, Université de Valenciennes et du Hainaut Cambrésis, France, 2010.

[17] Y.C. Guo, C.K. Chan, K.S. Lau, Numerical studies of pulverized coal combustion in a tubular coal combustor with slanted oxygen jet, Fuel, 2003, 82; 893-90.

[18] Kurose, R., Tsuji, H., Makino, H., Effect of moisture in coal on pulverized coal combustion characteristic, Fuel, 2001, 80, pp 1457-1465.

9

Thermal insulation capacity of roofing materials under changing climate conditions of Sub Saharan regions of Africa

Julien G. Adounkpe[1], Clement Ahouannou[2], O. Lie Rufin Akiyo[3], Augustin Brice Sinsin[1]

[1] Laboratory of Applied Ecology, Faculty of Agronomic Sciences, University of Abomey Calavi, 03 BP 3908 Cotonou Republique du Benin.
[2] Département de Génie Mécanique et Energétique Ecole Polytechnique d'Abomey Calavi, Université d'Abomey Calavi, 03 BP 1175 Cotonou Republique du Benin.
[3] Department of Geography of the University of Parakou, Republic of Benin BP 123 Université de Parakou Republic of Benin.

Abstract

Climate change is affecting human indoor thermal comfort. Human habitat roof's thermal insulation capacity may play key role in reducing the discomfort resulting from climate change. In the present study, six roof materials are analyzed for their thermal insulation capacity: aluminum-iron (Al-Fe) sheet, Al-Fe sheet with outer face white painted, Al-Fe sheet with various straw thick, white tile, red tile and gray tile. Solar radiations, ambient temperature, wind speed, roof inner and indoor temperatures were daily measured during April and June. Measured roof inner wall temperatures for each type of material agreed with the model set forth. The indoor temperature showed, under the same atmospheric conditions, Al-Fe sheet at a maximum of 51.4°C ; Al-Fe sheet with outer face white paint at 40.3°C; Al-Fe sheet with 3cm thick of straw at 41.2°C; and Al-Fe with 6cm thick of straw at 36.8°C, making the latter the better roof at day time. For the inner wall temperatures of the roof without ceilings, Al-Fe sheet has a maximum at 73°C; Al-Fe sheet with outer wall white paint at 48.1°C; Al-Fe sheet with 3cm straw thick at 45.2°C; and Al-Fe with 6cm straw thick at 37.9°C, red tile at 51.3°C; white tile at 41.6°C and grey tile at 51.6°C. This study enlightens the change that can be made on the traditional roof to improve indoor thermal comfort in changing climate conditions.

Keywords: Climate change; Indoor thermal comfort; Roof materials; Solar radiation; Thermal insulation.

1. Introduction

The thermal behavior of a building depends on the construction materials used to build it. Recently, a study related to construction materials and thermal comfort in hot zones of Cameroon dealt with local materials such as woods, clay bloc, compared to cement bloc [1]. Obviously, the study was about the building's wall materials. Wall materials play key role in a building comfort. However, a building roof, the one receiving the most direct solar radiation, contributes a lot to the heat in the building. According to a study, the heat contribution from roof to the overall heat of a room is about 70% depending on the type of the material [2]. Various roof materials are in use, depending on climatic conditions, people's desire, and finance. The primary concern in selecting roof material is its durability [3]. Most of homeowners select the roof material based on how resistant to weather conditions it may be. However, other homeowners look for roof material to match their homes where aesthetic is the focus. Other non negligible points are the availability of installation technique, the maintenance effort and the ease to repair.

A human habitat roof plays keys role in both protecting against rain, wind, sun and others [4] and providing certain comfort [5]. In Sub-Sahara Africa, traditional and modern roofs are in use depending on the areas (cities or villages), and the financial situation.

Traditional roofs made of local materials, even though proven to be thermally comfortable to some extends suffer of a relatively short life which forces the users to reconstruct basically almost after every rainy season. [4, 5].

To efficiently address the issues of house roof frequent wear, modern roofs are developed even in remote areas of Sub-Saharan African Countries [4]. The modern roofs may consist of metal sheets, reinforced concrete, for the economically sufficient people. The less fortunate will use metal sheets made of galvanized iron, aluminum, and zinc. But in the relatively wet zones, under intensive rainfalls, high relative humidity, and coastal zones of Sub-Sahara Africa, even those roofs suffer frequent rusting and need to be renewed very often. Some alternatives are found such as painting the galvanized iron with anti-corrosive substances, or using various types of tile, etc.

However, as the issues of frequent roof wear seem to be resolved, everything being equal elsewhere, users face another crucial issue: the thermal comfort in modern roof houses at low cost under changing climate [6].

Various studies related to the electric consumption and the perceptions of the indoor comfort by inhabitants in low income cities such as Mexicali, in Mexico, other 11 countries and 36 sustainable and institutional buildings were undertaken [7]. Links were established with electric consumption and the perceptions of the indoor comfort and building design and building material [8].

However, few studies deal with the specific thermal contribution of the roof to the house.

Recently, a model has been developed to measure, taking into account the climate parameters, the heat flux transfer to the room through its roof [4].

The present study is a comparative analysis of the thermal insulation capacity of various types of roof materials used in Sub-Sahara African regions in order to propose the thermally efficient roof material that would contribute in low heat transfer to room, thus contributing to climate change adaptation and reducing climate change impact on people of those regions.

2. Method and material

2.1 Study site characteristics

The method and the material employed in the present study have been extensively described elsewhere [4]. Succinctly, the study was conducted near the University of Ougadougou (Burkina Faso), a city located in the Soudano-Sahelian zone. The climate of the zone is characterized by a unimodal precipitation regime, with a rainfall of about 600 mm per year. The rainy season stretches from mid- May to mid-October, with an average temperature of 30°C. June, July, August, and September totalize more than 80 % of the seasonal rainfall. The cold season runs from December to January, with a minimum temperature of 19°C. The maximum temperature during the hot season, which runs from March to May, can reach 45°C. Relative air humidity varies from 20% in March during dry season to 80% in August during rainy season with a mean of 49% [9].

The geographic characteristics of the study site are: 12°22′46.19″N, 1°29′58.77″W.

2.2 Material

Experimental boxes made of wood on top of which the roofing material is placed to serve as roof were used. The boxes were put at a height of 1m above the ground at points where no shade influence from building, trees or other affects the measurements.

The galvanized iron sheet produced by Qingdao Hengcheng Steel Co., Ltd. (Thickness: 0.5mm, length: 1m, width: 1m), was purchased at Ouagadougou's market place. Straw was harvested on the campus field and dried for thirty days prior to the experimentation. The white paint coat was uniformly applied on the galvanized iron sheet; red, gray, and white tile samples were obtained from a roof promoter in Ouagadougou.

Solar radiations were measured using a CMP- pyranometer placed close to the box at the same height from the ground.

Temperature gauges were positioned for the ambient temperature, the room temperature, the external and internal roof wall temperatures, etc. The ambient temperature is measured at a height of 5m above the ground. The gauges were electrically connected to the temperature sensor that gives a simultaneous read-out of all temperatures with decimal absolute error.

A rotating cup anemometer was employed to measure the wind speed.

3. Results and discussion

The atmospheric parameters such as solar irradiation, wind speed, and ambient temperature were recorded on clear sky days of April and June (Figure 1). At night time, between 7:00PM and 7:00AM while solar irradiation is zero, roof temperatures are not similar to the ambient ones to reflect the nullity of solar energy (results not shown). This is basically due not only to the thermal inertia that characterizes some roof materials, but also the fact that there was no openings made in the experimental boxes to serve as window through which heat could be evacuated to the ambient. Thus, it is solely reported in the present study data for which the solar radiation is not null so that solar influence on the roof temperature can be seen for proper analysis.

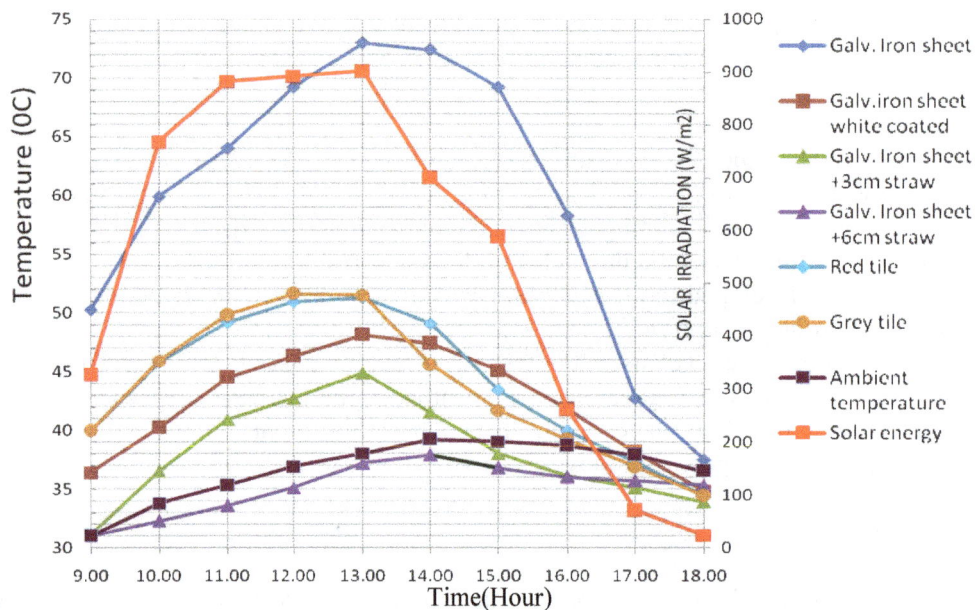

Figure 1. Hourly solar radiation was measured and represented in red on the second vertical axis; roof inner wall temperatures for each roof material were simultaneously recorded. Al-Fe sheet roof appears to be the thermally unfavorable roof material followed by red and grey tiles, and then comes the Al-Fe white painted. Increasing straw thick on the Al-Fe considerably reduces heat transfer to the room

Burkina Faso, the country where the study was conducted, Mali, Niger, the northern part of Benin, Togo and Nigeria in West Africa are located in the bands between $15°$ and $35°$ north and south around the earth where the greatest amount of solar energy is received [10, 11]. The direct solar energy recorded in the present study matches the one observed in various cities of Nigeria [12] and elsewhere [13-15].

Table 1 reports on the daytime temperatures of the inner wall of various roof materials. All the roof materials have their maximum temperature towards 1:00 PM, which corresponds to the time the solar radiation is approximately at its maximum (Table 2). The galvanized aluminum-iron sheet (Al-Fe) reached the highest temperature of about $73°C$, way above the ambient temperature of $39°C$. Among all the roof materials, only Al-Fe+6cm provides temperature inferior to the ambient one all day long. Second to the Al-Fe + 6cm, in terms of low temperature roof material, is the Al-Fe + 3cm straw thick. Therefore, it can be concluded that the thicker the straw, the lower the temperature. In fact, house roofs in villages are made solely with straw which thick can reach 60cm. The experiences show that the room temperatures in such houses are below the atmospheric one. However, the thick of the straw is not only for low room temperature but also to render the roof waterproof. Roof waterproofing is the first intent of great straw thick. The present study has the merit of showing that the roof thick, when straw is used, contributes to lowering room temperature.

Table 1. Ambient temperature, Galvanized aluminum-iron (Al-Fe) sheet, Al-Fe sheet +3cm straw, Al-Fe sheet+6cm straw, Red tile, and Grey tile roof inner wall temperatures in °C are recorded on an hourly basis

Time	Ambient Temp.	Al-Fe	Al-Fe white painted	Al-Fe +3cm 3cm st straw	Al-Fe +6cm straw	Red Tile	Grey Tile
09:00	31	50,2	36,4	31,1	31	39,9	39,9
10:00	33,8	59,9	40,2	36,5	32,3	45,8	45,9
11:00	35,3	64	44,5	40,9	33,6	49,2	49,8
12:00	36,9	69,2	46,3	42,7	35,1	50,9	51,6
13:00	38	73	48,1	44,9	37,2	51,3	51,5
14:00	39,2	72,4	47,4	41,5	37,9	49,1	45,6
15:00	39	69,2	45,1	38	36,8	43,4	41,7
16:00	38,7	58,3	41,8	36,1	36	39,9	39,2
17:00	37,9	42,7	38,1	35,1	35,7	37,3	36,9
18:00	36,5	37,5	34,8	33,9	35,3	34,4	34,4

Table 2. Roof material thermal and physical properties

Roof material	Absorp. coef ε_{pa}	Int. emis. coef ε_i	Ext. emis. coef ε_e	Therm. conduct. $\lambda(w/m^2)$	Roof thick e (mm)
Al-Fe	0.8	0.3	0.3	45	0.5
Al-Fe white coat	0.3	0.3	0.9	45	0.6
Al-Fe + straw	0.7	0.3	0.9	0.045	30-60
Red tile	0.68	0.9	0.9	0.5	8
Grey tile	0.58	0.85	0.85	0,5	8
White tile	0.3	0.9	0.9	0.5	8

The third roof material that shows low roof temperature among the roof materials employed is the Al-Fe white painted. It is common to people living in coastal zone or wet atmospheric conditions to use anti corrosive materials to paint their house roof. The goal is to avoid a fast rusting of the sheet. In our experiment, we have had a close look at the impact of a white coating over the galvanized sheet on the roof temperature. Amazingly, the white painted roof shows temperature distribution in a Gaussian shape comparable to the shape previously obtained [4]. The temperature behavior observed is a gain of about 25°C just by painting white the Al-Fe. The reduced comfort cost with roof material had been discussed elsewhere. In the Baltimore, a significant summer cooling cost reduction along with health benefits by liquid coating of roof (roof painting) is reported [16]. White or light-colored roofs can cut-off air-conditioning costs by up to 20 percent and even lower indoor temperatures inside buildings without air conditioning. In Washington, areas where cool roofs were installed, heat-related deaths declined by 6 percent to 7 percent [16].

In the present study, at sun set, between 5:00PM and 6:00 PM, where the solar radiation is almost zero, it can be noticed that all of the roof materials, except for the Al-Fe, cools down more rapidly than the atmosphere. However, between 4:00PM and 6:00 PM, Al-Fe + 6 cm exhibit higher temperatures than Al-Fe+3cm. This means that the former roof cools down more slowly than the later. Again, the thermal inertia due to the thick of the roof material is at the origin of this observation. Heat evacuation appears to be easier with smaller straw thick.

Red tile and grey tile exhibit the same temperature behaviors between 9:00AM and 1:00PM. However, from 1:00PM to 6:00PM, grey tile roof shows temperatures lower than those from the red tile. This can be explained by the absorption of wave lengths that is color dependent. In the present study, grey tile absorbs solar radiation less than the red tile. From the thermal insulation stand point, the galvanized Al-Fe sheet appears to be the roof material that transfers more heat to a room. A maximum temperature of 73°C was attained, thermally making this type of roof material the most undesirable one. The high absorption coefficient of the Al-Fe material compared to the other roof materials explains this observation. However, when slight modifications such as white painting or using various straws thick are made on the Al-Fe material, this greatly improves the thermal behavior of the Al-Fe.

It is expected that the room temperatures under various roof materials be lower than their respective roof temperatures. Curves from Figure 2 give the Al-Fe room temperature to vary from 40° C by 9:00 AM to a maximum of 57°C in the early afternoon. Rooms under grey and red tiles have the same temperature behaviors, their temperatures varying approximately from 37°C in the morning to a maximum of 52°C in the early afternoon; thus yielding, compared to the Al-Fe roof, a 5°C reduction at room temperature. However, comparing tile roof with three others as for their thermal behavior, it was found that tile roof transfers less energy to the room [9]. Al-Fe white coated shows relatively low temperatures compared to the former three roof materials. The room temperatures under white painted Al-Fe vary from 34° C in the morning to a maximum of 40°C in the afternoon, showing a better temperature comfort compared to Al-Fe and the tiles. Under the same atmospheric conditions, Al-Fe +3 or 6cm straw thick shows the lowest room temperatures. The room temperatures under such a roof material vary from approximately 30°C in the morning to a maximum of 37°C in the afternoon while the ambient temperature is at a maximum of 39°C. Again, as expected, these types of modifications on Al-Fe are very beneficial for the thermal comfort in hot and dry zones of Sub-Sahara African regions. However, as the sun starts to set, Al-Fe + 3cm gives lower temperatures than even Al-Fe + 6cm.

Some important factors have also been looked at during the course of the present study. To improve the indoor comfort, it is recommended to plant trees at a giving distance from the openings of a building such as at the doors, windows, etc. We have compared the thermal behavior of Al-Fe /Al-Fe white painted roof material under the influence of the shade from a tree.

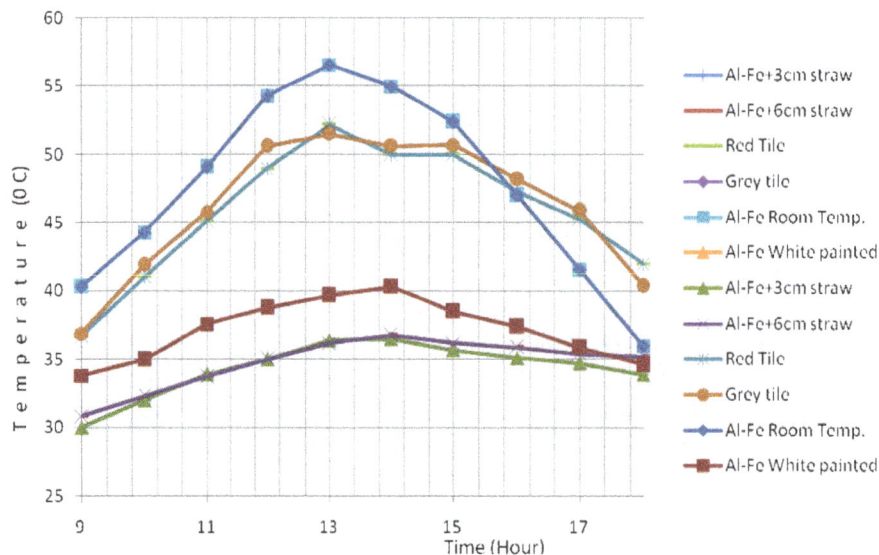

Figure 2. The temperatures of the rooms under various roof materials

The graphs of Figure 3 show that, up to 11:00 AM, the influence of the tree was not noticeable. However after 11:00 AM, a temperature gradient is observed with both the Al-Fe and Al-Fe white painted roof materials. The Al-Fe and the white painted Al-Fe show a temperature difference varying from 1 to 6o C due to the shading from a tree. This is the proof that tree plays key role in term of standing hot weather period. The presence of tree suitably located in a house lowers significantly the room temperatures. It has been demonstrated that protection of the buildings from the sun, primarily by shading, but also by the appropriate treatment of the building cover, that is, the use of reflective colors and surfaces, reduces the temperature of the room [10].

Another observation is the impact of water on the roof temperature. Four of the roof materials have been compared for the impact water has on their temperature behavior. All of them have been watered in the morning the way plants and flowers are in a house.

Al-Fe shows no change in temperature patterns, even though watered the same way as the others. Thus this roof material keeps its position of the most thermally unfavorable roof material. But it is worth to acknowledge that after watering the roof around 6:30AM, all of them have the same temperature trends up to 9:00 AM when each one yields its appropriate curve. Here is a possible classification as far as thermal insulation:

- From 9:00AM to around 11:00 AM, grey tile roof offer the lowest temperature followed by the red tile, and lastly the Al-Fe white painted.
- From 11:00AM to 1:00 PM, the Al-Fe white painted becomes the second best roof material after the grey tile
- From 1:00PM to 5:00PM, Al-Fe white painted becomes the best roof material followed by the grey and red tile respectively.

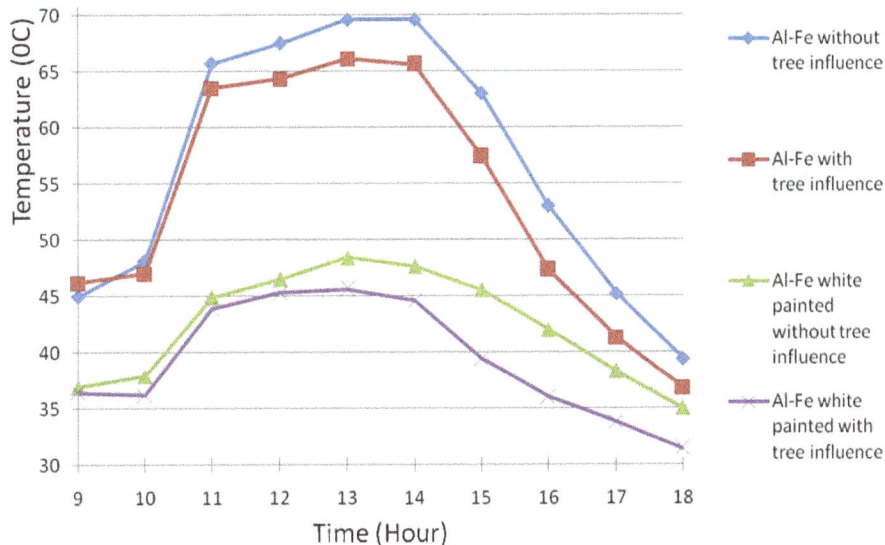

Figure 3. Al-Fe and Al-Fe white painted inner wall temperatures were measured with/out the influence of shading coming from a tree. A gain of 1 to 5°C was observed with the influence from the tree

The thermal behavior change observed so far is due to water retention capacity of grey and red tile versus the Al-Fe white painted which has none. In fact, the retained water will use the solar incoming energy to vaporize, thus lowering the temperature of such roofs. As it can be seen from the graphs of Figure 4, as soon as the retained water is vaporized, the roof materials recover their normal behavior. From this, it can be said that the vaporization of the retained water lasts from 9:00AM to 1:00PM with temperature gains with the grey tile of 28°C, 9°C, and 3°C respectively on Af-Fe, red tile, and Al-Fe white painted. A good application of this observation, basically in hot zones where grey tile is used is to water the roof twice a day (on sunny days), one by 6:30 AM and the second by 1:00PM. This will keep a gradient of about 20°C on average below the ambient temperature. The cost of using water to cool down the room must be compared to any other room air refreshment cost.

4. Conclusion

The present study has the merit of comparing various roof materials thermal behaviors under the same climatic conditions. While Al-Fe + x cm straw thick yields lower room temperatures, grey and red tile are good thermal material to consider in buildings, not only because of their thermal behavior, but also because of their aesthetic, their mechanical resistance and their water retention capacity. Painting white the Al-Fe largely improve the room temperature, giving the proof that under the tropics, slight modification on the most common used roof material may bring thermal comfort. However, grey tile, if adopted, with the watering possibility will turn out to be the best thermal roof in Sub-Saharan Africa regions.

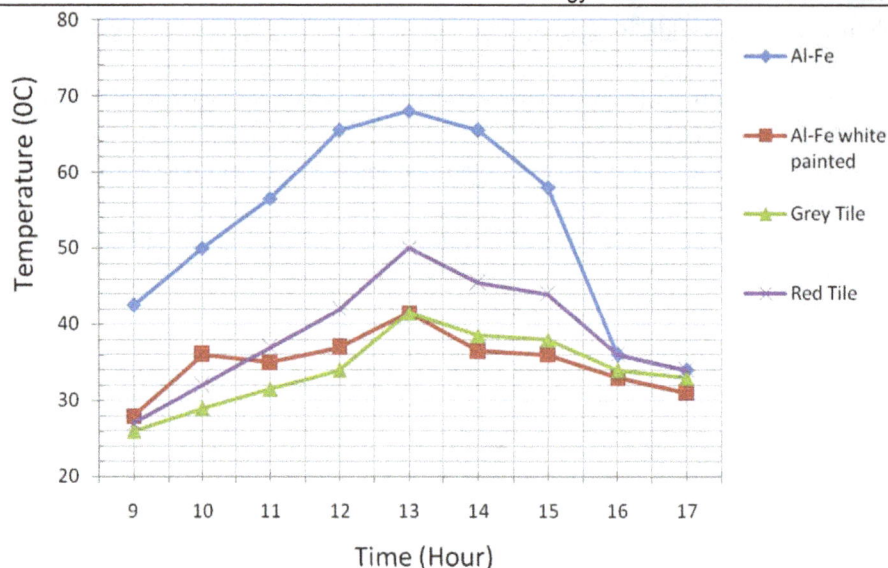

Figure 4. Effect of watering roof materials on roof inner wall temperature

References

[1] A. Kemajou et L. Mba, Matériaux de construction et confort thermique en zone chaude Application au cas des régions climatiques camerounaises, Revue des Energies Renouvelables Vol. 14 N°2 (2011) 239–248 (Article).

[2] K. C. K. Vijaykumar, P. S. S. Srinivasan, and S. Dhandapani, "A performance of hollow clay tile (HCT) laid reinforced cement concrete (RCC) roof for tropical summer climates," Energy and Buildings, vol. 39, no. 8, pp. 886–892, 2007. (Article).

[3] Ignasio NGOMA and Mauro SASSU: Sustainable African housing through traditional techniques and materials: a proposal for a light seismic roof, 13th World Conference on Earthquake Engineering Vancouver, B.C., Canada August 1-6, 2004, Paper No. 170.

[4] Julien G. Adounkpe, A. Emmanuel Lawin, Clément Ahouannou, Rufin Offin Lié Akiyo, and Brice A. Sinsin: Modeling Solar Energy Transfer through Roof Material inAfrica Sub-Saharan Regions ISRN Renewable Energy Volume 2013, Article ID 480137, 8 pages http://dx.doi.org/10.1155/2013/480137 (Article).

[5] GRET Toitures en Zones Tropicales Arides, vol. 1, SOFIAC, Paris, France, 1984. (Book).

[6] X. Bai, "Integrating global environmental concerns into urban management: the scale and readiness arguments," Journal of Industrial Ecology, vol. 11, no. 2, pp. 15–29, 2007.(Article).

[7] G. Baird and C. Field, "Thermal comfort conditions in sustainable buildings-results of a worldwide survey of users perceptions," Renewable Energy, vol. 49, pp. 44-47, 2013.(Article).

[8] R. A. Romero, G. Boj´orquez, M. Corral, and R. Gallegos, "Energy and the occupant's thermal perception of low-income dwellings in hot-dry climate: Mexicali, M´exico," Renewable Energy, vol. 49, pp. 267–270, 2013 (Article).

[9] L. L. Bayala, Monographie de la Ville de Ouagadougou, Ministere de l'Economie, 2009 (Book).

[10] J.A. Duffie: Solar Engineering of Thermal Processes, John Willey &Sons, New York, 1980 (Book).

[11] F. Kreith and J. F. Kreider: Principles of Solar Engineering, Hemisphere Publishing, New York, 1978, p 17 (Book).

[12] Okundamiya M.S., and Nzeako A. N. IRSN Renewable Energy Estimation of Diffuse Solar Radiation for Selected Cities in Nigeria Vol 2011, p 6-12 doi: 10.5402/2011/439410 (Article).

[13] Shahidul Islam Khan & Asif Islam Smart Grid and Renewable Energy Performance Analysis of Solar Water Heater, 2011, 2, 396-398 (Article).

[14] Hernandez-Gomez V.H., Contreras Espinosa J.J., Morillon-Galvez D., Fernandez-ZayasJ.L., Gonzalez-OrtizG.: Ingeniera Investigacion y Tecnologia Analytical Model to Describe the Thermal Behavior of a Heat Discharge System in Roofs, 2012 Vol III, Num 1, 33-42 (Article).

[15] Fabrice Motte, Gilles Notton, Christian Cristofari, and Jean-Louis Canaletti Renewable Energy A building integrated solar collector: Performances characterization and first stage of numerical calculation 2013 49 pp 1-5 (Article).

[16] Timothy B. Wheeler Baltimore Sun: White or light coatings reduce energy costs, last longer, October 15, 2013, retrieved from http://articles.baltimoresun.com/2013-10-15/features/bs-gr-cool-roofs-report-20131014_1_roofs-air-conditioning-baltimore-sun/2.

10

Exergy analysis of CO_2 heat pump systems

A. Papadaki, A. Stegou-Sagia

School of Mechanical Engineering, Department of Thermal Engineering, National Technical University of Athens, 9 Iroon Polytechniou Str., Zografou 15780, Athens, Greece.

Abstract

Carbon dioxide (CO_2, R744), a natural refrigerant of beneficial properties found everywhere in our ambiance, can provide answer to the environmental problems caused by other refrigerants' use. The intention of this work is to outline the variation of exergy efficiency factor, COP and exergy flow related to the use of CO_2 in two stage and single stage heat pumps. The relevant mathematical models to the thermodynamic cycles were developed and an attempt was made for our efficiency and exergy losses results to be displayed. Moreover, fundamental process and system design issues of the applicable CO_2 heat pumps cycles were inaugurated, along with their properties and characteristics, comparing CO_2 use to that of R22 and its substitutes R407C and R410A applied in relevant conditions. Since exergy analysis is important theoretical basis for optimizing the systems operation and minimizing the losses, the results of this paper will advance the systems' design and performance.

Keywords: Exergy; Carbon dioxide; Heat pumps; Exergy analysis; Single stage cycle; Two stage cycle.

1. Introduction

Carbon dioxide is one of the most feasible answers to the contribution of the fluorocarbon refrigerants to global warming and ozone depletion, being a natural refrigerant with zero ODP (Ozone depletion potential), negligible GWP (Global warming potential), and very low cost. Global warming effect is considered to be the most prominent problem of the world climate. Refrigerants that are utilized in the heat and cooling systems have quite higher GWP than CO_2. Even refrigerants that were considered ozone layer friendly, such as HFC-134a, have GWP of many times greater than CO_2's (in HFC-134a is 1300 times) [1, 2]. In addition carbon dioxide (CO_2) is not toxic, flammable or corrosive. It is inexpensive and readily available. After the Montreal Protocol the interest for CO_2 cycles was so great that a large number of research developments have been commenced for the production of carbon dioxide's refrigeration system components.

1.1 CO2 properties

Carbon dioxide furthermore has two exceptional properties, its most remarkable one being its low critical temperature T_{crit}, of 31.1°C, compared to conventional refrigerants and working close or even above the critical pressure P_{crit} of 73.8 bar in vapour compression systems functioning in normal ambient temperatures [3, 4].

In a subcritical heat pump cycle, such low critical temperature is considered an inconvenience as heat cannot be delivered at temperatures greater than the critical temperature limiting consequently the operating temperature range. Additionally, heating capacity and the performance of the system are relegated at temperatures inferior but close to T_{crit}, since the enthalpy of vaporization then is reduced [5], making the operation of a conventional heat pump avoidable at a heat rejection temperature near T_{crit}. Carbon dioxide's low critical temperature provides the opportunity to operate in a transcritical manner. In a transcritical heat pump, heat rejection (gas cooler) is operated above the critical pressure, heat delivery temperatures are no longer limited by T_{crit} and the evaporator is operated below that and for this reason the cycle is identified as transcritical.

The other unique property of CO_2 is the high working pressure required to use under typical heat pump conditions. Heat pump systems, both sub and transcritical, using CO_2, work at greater pressures than with the majority of other refrigerants. The operational pressures of subcritical CO_2 heat pumps reach as high as 60–70 bar, whereas for the transcritical pressures vary from 80 to 110 bar or even more. Although high pressure defies compressors' capability and components' robustness, it presents some benefits as well, providing to CO_2 a relatively high vapor density and an equally high volumetric heating capacity. This attribution offers the option for CO_2 to have a smaller working volume cycled in order to attain the same heating demand which permit the use of smaller components and more compact systems [3].

Nevertheless, the most important disadvantage of CO_2 cycle is that owing to huge expansion loss compared to conventional refrigerants' cycle it presents lower COP making the modifications of the cycle crucial [6]. Lorentzen [4] described more than a few customized cycles comprising of two-stage internal 'subcooling' and expansion options. By modifying the basic single-stage transcritical cycle a lot can be achieved. Some adaptations that are promising are dividing of flows, expansion via work generation instead of throttling, staging compression and expansion and the use of internal heat exchange. Trying to obtain higher efficiency values, we will employ the modification of the two-stage compression of the CO_2 with intercooling. Then we will compare these results to the equivalents of the single stage CO_2 and conventional vapour compression cycles. In order to model the total systems, and thereby investigate the possible operating conditions with replacement refrigerant mixtures, a computer code was created.

1.2 First and second law analysis

Studying the inefficiencies of existing systems our work focuses on the understanding of heat pumps cycles, their efficiencies and potentials for improvement, based on First and Second Laws of Thermodynamics. COP is used to evaluate performance of air-conditioning or heat pump from the viewpoint of the First Law of Thermodynamics. Exergy, being presented in an amount of works [7-13] corresponds quantitatively to the useful part of energy, the maximum possible amount of work a system, a flow of matter or energy can produce as it comes to equilibrium with an appointed reference environment. Exergy analysis combines the conservation of mass and energy principles with the second law of thermodynamics for the design of more efficient and environmental friendly systems. While efficiencies using energy are ambiguous for not being measures of "an approach to an ideal", exergy efficiencies are considered as such, measuring, in a way, the potential of the system for improvement [12].

2. Modelling of operation

2.1 Conventional heat pump's model

Figure 1 shows the heat pump's vapour / transcritical CO_2 compression cycle flow chart. The working fluid moves from the evaporator, which is connected to the low-temperature heat source into the compressor as a superheated vapour. Following, the compressed vapour, flows into the condenser which is connected to the high-temperature heat sink and respectively to the gas cooler for the CO_2. Here it condenses and afterwards, as a liquid, it undergoes expansion in the throttling valve. The throttled two-phase mixture, which is liquid for the most part, moves into the evaporator from which ensues the vapour that is then superheated and directed to the compressor to complete the flow cycle.

2.2 Transcritical two-stage CO_2 cycle with intercooling

Figure 2 shows the two stage CO_2 transcritical heat pump cycle with intercooling used. Here the saturated working fluid of state 2 moves from the evaporator into the low pressure (LP) compressor where it's compressed to state 3 before it enters the intercooler. There takes place the cooling, by external fluid, of the vapour which increases the mass of CO_2 vapour entering the high pressure (HP) compressor. Ambient air is taken as the external fluid. The saturated vapour from the intercooler at state 4 is compressed to state 5 and afterwards the super-critical vapour is cooled in the gas cooler to state 6.

CO_2 vapour is further cooled in the internal heat exchanger to state 7. CO_2 then expands in the expansion device to state 8 and evaporates to state 1 producing cooling effect. The internal exchanger in the system exists for system thermal efficiency improvement [14].

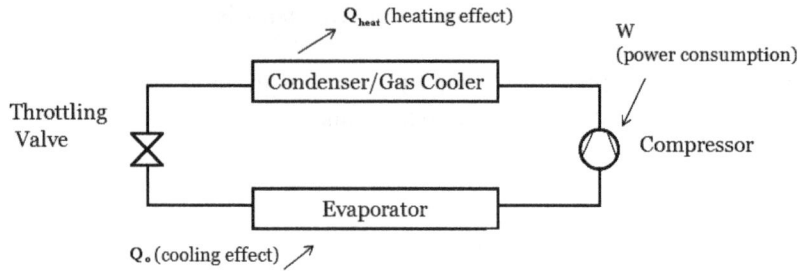

Figure 1. Schematic diagram of the single stage heat pump cycle

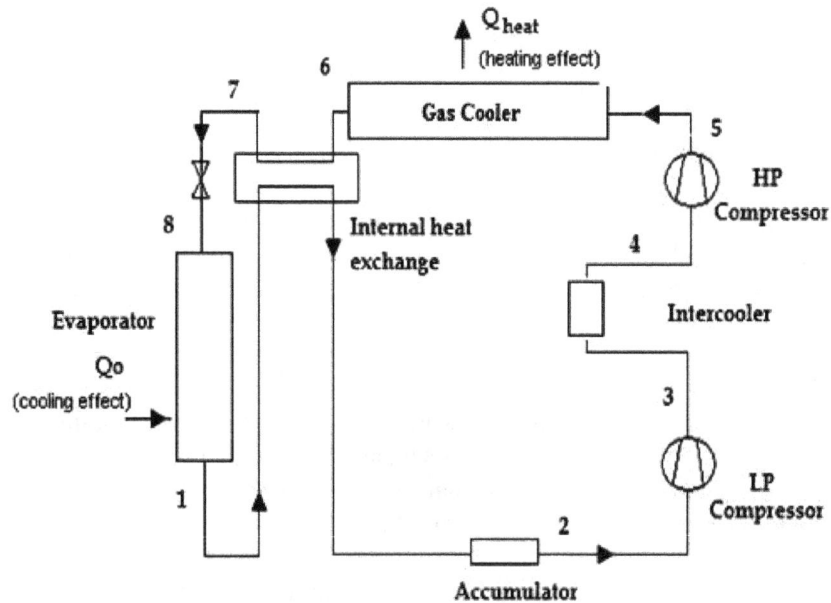

Figure 2. Schematic diagram of the two stage heat pump cycle with intercooling

2.3 Thermodynamic analysis
2.3.1 Single stage cycle
Based on the known equations for the exergy and energy analysis [15-17] of a heat pump cycle, as the one shown in Figure 1, we have:
The exergy efficiency factor ζ is

$$\zeta = COP\left(\frac{T_w - T_a}{T_w}\right) \tag{1}$$

with the coefficient of performance (COP) of the system being

$$COP = \frac{q}{\Delta e_{abs} + \sum \Delta e_{loss}} =$$

$$\frac{(h_2 - h_4)}{(h_2 - h_4)\left(1 - \dfrac{T_a}{T_w}\right) + \sum \Delta e_{loss}} \tag{2}$$

Exergy losses, for each component of the system are:
• Compression losses:

$$\Delta e_{comp} = T_a(s_2 - s_1) \tag{3}$$

• Additional losses due to the compressor motor:

$$\Delta e_{mot} = w \frac{1 - \eta_{mot}}{\eta_{mot}} \frac{T_a}{T_w} \tag{4}$$

where η_{mot} is the compressor motor efficiency factor, w the specific compression power demand (h_2-h_1) and the heat from the heat pump motor absorbed by the heated substance [16].
• Condensation / gas cooler losses:

$$\Delta e_{cond} = (h_2 - h_4)\frac{T_a}{T_w} - T_a(s_2 - s_4) \tag{5}$$

• Evaporation losses:

$$\Delta e_{evap} = T_a(s_1 - s_5) - (h_1 - h_5) \tag{6}$$

• Throttling (isenthalpic process) losses:

$$\Delta e_{thr} = T_a(s_5 - s_4) \tag{7}$$

Therefore, summing up we obtain the total exergy loss:

$$\sum \Delta e_{loss} =$$
$$\Delta e_{comp} + \Delta e_{mot} + \Delta e_{cond} + \Delta e_{evap} + \Delta e_{thr} =$$
$$(h_5 - h_1) + \tag{8}$$
$$\left[(h_2 - h_4) + (h_2 - h_1)\frac{1 - \eta_{mot}}{\eta_{mot}} \right] \frac{T_a}{T_w}$$

The exergy efficiency factor is consequently given by the equation (1):

$$\zeta = \frac{(h_2 - h_4)(1 - \frac{T_a}{T_w})}{(h_2 - h_1)\left(1 + \frac{1 - \eta_{mot}}{\eta_{mot}} \frac{T_a}{T_w} \right)} \tag{9}$$

The refrigerants compared to R744 are R407C and R410A and R22.
A variety of sources were used [18-24] to ensure the consistent application of property. The differences observed were minimal. It is taken into consideration in all relevant calculations the fact that R407C and R410A are non-azeotropic, since they show a different behaviour from pure substances [25].
Firstly, due to different evaporator and condenser inlet/outlet temperatures, we have to select condenser inlet temperature in opposition to the warm space temperature taking care of the condenser inlet and outlet temperatures to be sufficient so as to reject heat and finally liquid enthalpy at the expansion device and related property data being in position to achieve the suitable evaporator inlet temperature. The fluid behaves normally in all other points. Undeterred by the fact that this method of evaluation occupied in

this study is not fully representative of a dynamic operation of a heat pump system, yet it sets up the foundations for understanding its thermodynamic performance.

2.3.2 Two stage cycle

The two-stage CO_2 transcritical heat pump cycle with intercooling is modelled modularly incorporating each individual process of the cycle. The state points in Figure 2 are defined as the conditions of the refrigerant characterized by its temperature, mass flow rate and quality.

Exergy losses, for each component of the system are:

• Compression losses:

$$\Delta e_{comp} = \Delta e_{comp1} + \Delta e_{comp2} = T_a(s_3 + s_5 - s_2 - s_4)$$

(10)

• Additional losses due to the compressors motors:

$$\Delta e_{mot} = w \frac{1 - \eta_{mot}}{\eta_{mot}} \frac{T_a}{T_w}$$

(11)

where η_{mot} is the compressors motor efficiency factor, w the specific compression power demand (h_5-h_4+h_3-h_2) and the heat from the heat pump motor absorbed by the heated substance.

• Intercooler losses

$$\Delta e_{ic} = (h_3 - h_4)\frac{T_a}{T_w} - T_a(s_3 - s_4)$$

(12)

• Gas cooler losses:

$$\Delta e_{gc} = (h_5 - h_6)\frac{T_a}{T_w} - T_a(s_5 - s_6)$$

(13)

• Evaporation losses:

$$\Delta e_{evap} = T_a(s_1 - s_8) - (h_1 - h_8)$$

(14)

• Expander valve (isenthalpic process) losses:

$$\Delta e_{ex} = T_a(s_8 - s_7)$$

(15)

• Internal heat exchanger:

$$\Delta e_{ihe} = T_a[(s_7 - s_6) - (s_1 - s_2)]$$

(16)

Therefore, summing up we obtain the total exergy loss:

$$\sum \Delta e_{loss} =$$
$$\Delta e_{comp} + \Delta e_{mot} + \Delta e_{gc} + \Delta e_{evap} + \Delta e_{ihe} =$$
$$(h_8 - h_1) +$$
$$\left[\begin{array}{c} (h_3 - h_4 + h_5 - h_6) + \\ (h_5 - h_4 + h_3 - h_2)\dfrac{1 - \eta_{mot}}{\eta_{mot}} \end{array} \right] \frac{T_a}{T_w}$$

(17)

The exergy efficiency factor is consequently given by the equation (1):

$$\zeta = \frac{\left(h_3 - h_4 + h_5 - h_6\right)\left(1 - \dfrac{T_a}{T_w}\right)}{\left(h_5 - h_4 + h_3 - h_2\right)\left(1 + \dfrac{1 - \eta_{mot}}{\eta_{mot}} \dfrac{T_a}{T_w}\right)} \tag{18}$$

2.4 Assumptions

It is renowned that CO_2 refrigeration and air conditioning systems shows cooling COP more sensitive to ambient temperature variation than conventional systems, being therefore superior at sensible and low ambient temperature, and to some extent poorer at very high temperature. Consequently, it would be deceptive to base a comparison of CO_2 with the other refrigerants on design point conditions, which typically are at an extreme ambient temperature while the use of average seasonal conditions is wiser [26].

Our exergy efficiency, COP and exergy losses diagrams of the mixtures under consideration are schematized in comparison with the CO_2 and are plotted based on calculations, having taken into consideration the following assumptions:

The environmental temperature (T_a) is equal to 273 K [8, 27], while the temperature of the warm place (T_w) is considered 308 K [28]. The temperatures T_{con} and T_e are taken: T_{con} at the inlet of the condenser at the vapour saturation curve for R22, R407C and R410A and at the inlet of the gas cooler for R744, while T_e at the exit of the evaporator in the superheat region. Pressure drops in evaporator are for R22 [29] and R407C [30, 31] 135 kPa, for R410A [32] 85 kPa and for R744 [33] 100kPa, while during condensation the pressure drop varies for R22 (Judge et al, 2001) from 46 to 52 kPa, for R407C [30, 31] from 40 to 46 kPa and for R410A [31, 32] from 32 to 35 kPa, lessening with increasing condensation temperature; whereas correspondingly for R744 [28, 34] the already small (1 to 3 kPa as shown in [35]) pressure drop during cooling process of supercritical CO_2 decreases as inlet pressure of gas cooler increases having a temperature glide of approximately 61K [28]. In addition the isentropic compressor efficiency factor is chosen as 0,75 and the compressor motor efficiency factor as 0,85 in a endeavor to maintain a logical price for the evaluation.

The evaporator temperature (T_e) is taken as 263K for all condensing temperatures whilst condensation temperature (T_{con}) for the mixtures and the outlet temperature of the CO_2 gas cooler is ranging from 313 to 328 K. Accordingly the temperature ratio $\tau = (T_{con}/T_e)$ or $\tau = (T_{gc}/T_e)$ varies within the range of 1.19 to 1.25.

The featured two-stage CO_2 transcritical cycle configuration is solely a theoretical one to present the basis for performance comparison with other refrigerants. It is simulated and its performance is evaluated on the basis of maximum combined COP to obtain the optimum gas cooler and in-between pressures. These values are obtained for various operating conditions along with simultaneous variation of the compressors discharge pressure and intermediate pressure having a step size of 0.5 bar for each. The performance is evaluated on various evaporator temperatures T_e from 223 K to 243 K) and gas cooler outlet temperatures T_{gc} (308 K to 333 K) [36].

3. Results and discussion

The results attained in this analysis are comparison of refrigerants for exergy efficiency, COP and exergy losses (Figures 3 to 5). Properties of R22 are illustrated in plots by bold continuous lines, while R407C by thin discontinuous lines, R410A by thin continuous lines, R744 (single stage) by dotted lines and R744 (two stage) bold dotted lines.

Figure 3 shows the exergy efficiency factor as a function of temperature ratio τ. Exergy efficiency decreases when the temperature ratio τ increases. The curves' hollows are facing upwards.

The single stage heat pump working with R744 has the least favourable exergy behaviour with an exergy efficiency of 13% at the temperature ratio of $\tau = 1.25$ and 28% at the temperature ratio of $\tau = 1.19$. While the R744 of the two-stage transcritical heat pump features far better exergy performance compared to the latter, with an exergy efficiency of 31% at a temperature ratio of $\tau = 1.25$ and 38% at a temperature ratio of $\tau = 1.19$, demonstrating less variation on its performance with the change of temperature ratio.

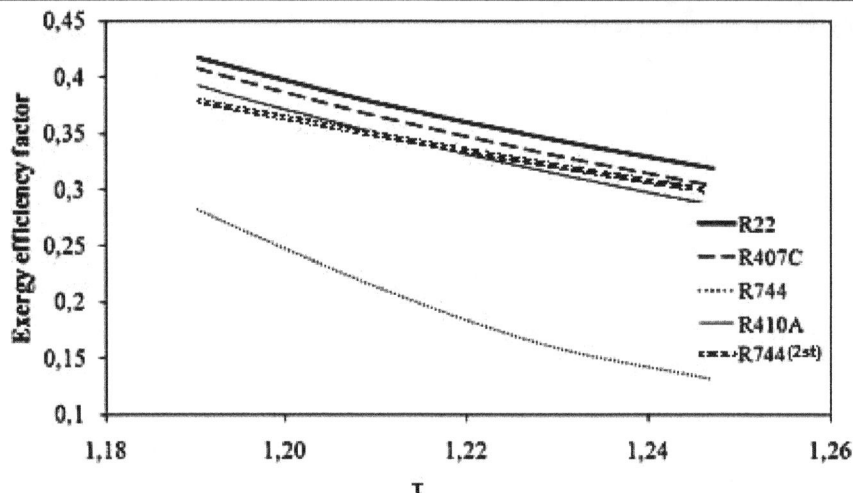

Figure 3. Variation of exergy efficiency factor for various temperature ratios τ

R22, on the other hand, presents the best exergy behaviour of all with an exergy efficiency of 42% at a temperature ratio of τ=1.19 and 33% at τ=1.25, followed by R407C (ζ = 41% at τ=1.19 and ζ = 32% at τ=1.25) and R410A (ζ= 40% at τ=1.19 and ζ = 29.5% at τ=1.25).

Figure 4 shows the disparity of COP of the heat pump system for each working refrigerant related to the temperature ratio τ, decreasing while the latter lifting as exergy efficiency factor does. COP ranges from 1.15 at temperature ratio of τ=1.19 (for R744) to 3.77 (for R22) at τ=1.25. There is a pointed increase in COP for the two-stage R744 system compared to the single stage one. Here, the single stage working R744 has likewise the worst behaviour, with COP to vary between 2.48 (at τ=1.19) and 1.15 (at τ=1.25), whilst the R744 of the two-stage transcritical heat pump features once more better comportment, with COP fluctuating amid 3.40 (at τ=1.19) and 2.70 (at τ=1.25).

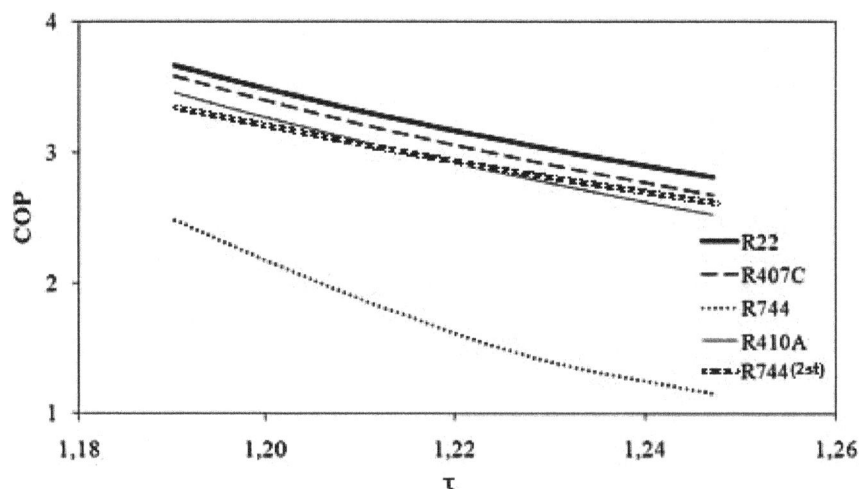

Figure 4. Variation of COP for various temperature ratios τ

The optimum performance is displayed yet again by R22, with COP of 3.77 at a temperature ratio of τ=1.19 and 2.88 at τ=1.25, followed by R407C with COP of 3.70 at a temperature ratio of τ=1.19 and 2.74 at τ=1.25 and R410A with COP of 3.57 at a temperature ratio of τ=1.19 and 2.58 at τ=1.25. R22 may seem more attractive to use from the efficiency aspect, however we have to bear in mind that it constitutes a harmful effect on the ozone layer with the result of extreme UV levels conducing to further environmental damage and several deadlines have been arranged depending on the country for complete R22 replacement in accordance to the terms established by the Montreal Protocol meetings.

The prices for COP and exergy efficiency factor are in agreement with those of Robinson and Groll [37] at the equivalent conditions' region.

Figure 5 presents the percentage of the major exergy losses for the two CO_2 systems. These are of the gas cooler and of the compressor and we can conclude that for the two stage CO_2 transcritical cycle the losses lessen dramatically. For the single stage heat pump working with R744 the compressor accounts for approximately 49% of the total cycle irreversibility and the gas cooler for the 25%, while respectively the percentage of exergy losses in the two-stage transcritical heat pump is 32% for the compressor and 20% for the gas cooler.

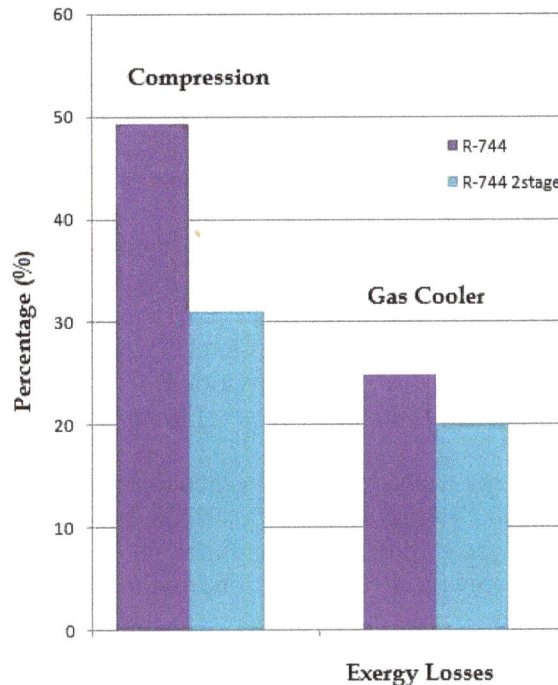

Figure 5. Exergy losses of the systems' components

As pointed out by Dincer and Rosen [38] "Exergy efficiency weights energy flows by accounting for each in terms of availability. It stresses that both losses and internal irreversibilities need to be dealt with to improve performance" and by Moran and Shapiro [39] "Exergy analysis is particularly suited for furthering the goal of more efficient energy use, since it enables the locations, types, and true magnitudes of waste and lost to be determined". Following the above described study the behaviour of the system can be improved, minimising individual exergy loss of each component and maximising efficiencies. Compressor efficiency is a major factor in enhancing the performance of the system, the smaller the compressor, the more prominent the compression losses. Generally speaking throttling losses can be reduced minimising the temperature difference before and after the throttling valve, as well as by decreasing the temperature differences in evaporator and condenser. This would also produce lower compression losses.

4. Conclusions

In this report we have made an effort to elucidate the diversity of the alternatively used refrigerant mixtures R407C and R410A replacing R22, and R744 replacing all of them in the field of exergy efficiency, COP and exergy losses depending on temperature ratio τ, for constant warm place temperature. The best exergy behaviour of all is presented by R22, with an exergy efficiency of 42% at a temperature ratio of $\tau=1.19$. R744 may seem to fall short in comparison to the rest refrigerants for some conditions, nevertheless it is the most environment friendly of all and based on that and on its beneficial potentialities its use signifies a "new" ecological era for the field. As stated before, one of the downsides associated with transcritical cycles is that the system operates at a very high discharge pressure. There is a sharp reduction in optimum discharge pressure by adopting staging in compression. Inter-stage pressure is one of the most critical parameters for optimizing COP values. Moreover, by using highly efficient system components, the transcritical two-stage CO2 systems can be used more effectively. Two-stage transcritical heat pump working with R744 features far better exergy performance compared to the single stage cycle with a pointed increase in COP for the two-stage R744 system. Furthermore for the two stage R744 transcritical cycle the losses lessen dramatically.

The evolution of exergy efficiency factor and COP are illustrated and collated in diagrams so as to clarify the differences of alternative refrigerants more accurately.

References

[1] Bellstedt, M., Elefsen, F., Jensen, S.S. Application of CO2 (R744) refrigerant in industrial cold storage plant. EcoLibrium. 2002, 1(5), 25-30.

[2] Laipradit, P., Tiansuwan, J., Kiatsiriroat, Aye, L. Theoretical performance analysis of heat pump water heaters using carbon dioxide as refrigerant. International Journal of Energy Research. 2008, 32(4), 356-366.

[3] Cavallini, A., Cecchinato, L., Corradi, M., Fornasieri, E., Zilio, C. Two-stage transcritical carbon dioxide cycle optimisation: A theoretical and experimental analysis. International Journal of Refrigeration. 2005, 28(6), 1274-1283.

[4] Lorentzen, G. Revival of carbon dioxide as a refrigerant. International Journal of Refrigeration. 1994, 17(5), 292-301.

[5] Kim, M.H., Pettersen, J., Bullard, C.W. Fundamental process and system design issues in CO2 vapor compression systems. Progress Energy Combustion Science. 2004, 30(2), 119-174.

[6] Sarkar, J., Bhattacharyya, S., Ramgopal, M. Transcritical CO2 heat pump systems: Exergy analysis including heat transfer and fluid flow effects. Energy Conversion Management. 2005, 46(13-14), 2053-2067.

[7] Bejan, A. Fundamentals of exergy analysis, entropy generation minimization, and the generation of flow architecture. Int. Journal of Energy Research. 2002, 26, 545-565.

[8] Dincer, I. The role of exergy in energy policy making. Energy Policy. 2002, 30, 137-149.

[9] Frangopoulos, C.A. Exergy, Energy System Analysis, and Optimization, vol. 1. EOLSS Publishers Co Ltd. Ramsey/Is, 2009.

[10] Kotas, T.J. The Exergy Method of Thermal Plant Analysis, Reprint edition, Krieger. Malabar, FL, 1995.

[11] Moran, M.J., Sciubba, E. Exergy analysis: principles and practice. Journal of Engineering for Gas Turbines and Power. 1994, 116, 285–290.

[12] Rosen, M.A. Clarifying thermodynamic efficiencies and losses via exergy. Exergy, an International Journal. 2002, 2, 3-5.

[13] Szargut, J. Component efficiencies of a vapour-compression heat pump. Exergy, an International Journal. 2002, 2, 99-104.

[14] Chen, Y., Gu, J. The optimum high pressure for CO2 transcritical refrigeration systems with internal heat exchangers. International Journal of Refrigeration. 2005, 28, 1238–1249.

[15] Baehr, H.D. Thermodynamik. Siebente Auflage, Springer-Verlag. Berlin/Heidelberg, 1989.

[16] I.I.R. Compression Cycles For Environmentally Acceptable Refrigeration, Air Conditioning and Heat Pump Systems, 1992.

[17] Stegou-Sagia, A., Papadaki, A. Exergy modelling of vapour compression heat pumps using refrigerant mixtures, International Journal of Exergy. 2006, 3(3), 304-321.

[18] ASHRAE. Fundamentals handbook. American Society of Heating Refrigerating and Air-conditioning Engineers. New York, 2001.

[19] Desideri, U., Sorbi, N., Arcioni, L., Leonardi, D. Feasibility study and numerical simulation of a ground source heat pump plant, applied to a residential building. Applied Thermal Engineering. 2011, 31(16), 3500-3511.

[20] Lemmon, E.W., McLinden, M.O., Huber, M.L. NIST Reference Fluids Thermodynamic Properties - REFPROP, Ver. 7.0. NIST Standard Reference Database 23. NIST, Gaithersburg, ML, U.S.A., 2002.

[21] Span, R., Wagner, W. A new equation of state for carbon dioxide covering the fluid region from the triple point temperature to 1100 K at pressures up to 800 MPa. Journal of Physical and Chemical Reference Data. 1996, 25(6), 1509-1596.

[22] Stegou-Sagia, A. Thermodynamic property formulations and heat transfer aspects for replacement refrigerants R123 and R134a. Int. Journal of Energy Research. 1997, 21, 871-884.

[23] Stegou-Sagia, A., Damanakis, M. Thermophysical property formulations for R32/R134a mixtures. Int. Journal of Applied Thermodynamics. 1999, 2(3), 139-143.

[24] Stegou-Sagia, A., Damanakis, M. Binary and ternary blends of R134a as alternative refrigerants to R-22. Int. Journal of Energy Conversion and Management. 2000, 41, 1345-1359.

[25] Smith, J.M., Van Ness, H.C. Introduction to chemical engineering thermodynamics, third edition. Chemical Engineering Series. McGraw-Hill, New York, 1975.

[26] Neksa, P. CO2 as a refrigerant for systems in transcritical operation: principles and technology. EcoLibrium. 2004, 3(9), 26-31.

[27] Lorentzen, G. Heat pumps - where are improvements possible? An exercise in exergy. Rev. Int. Froid. 1986, 9, 105-107.

[28] Brown, J.S., Kim, Y., Domanski, P.A. Evaluation of carbon dioxide as R-22 substitute for residential air-conditioning. ASHRAE Trans, HI-02-13-3. 2002, 108(2), 3–13.

[29] Judge, J., Hwang, Y., Radermacher, R. A Transient and Steady State Study of Pure and Mixed Refrigerants in a Residential Heat Pump. Prepared for the U.S. Environmental Protection Agency Office of Research and Development. EPA Washington, D.C., 2001.

[30] Choi, J.Y., Kedzierski, M.A., Domanski, P.A. A Generalized Pressure Drop Correlation for Evaporation and Condensation of Alternative Refrigerants in Smooth and Micro-fin Tubes. NISTIR 6333. Building and Fire Research Laboratory National Institute of Standards and Technology. Gaithersburg, MD, 1999.

[31] Spatz, M.W, Motta, Yana S.F. An evaluation of options for replacing HCFC-22 in medium temperature refrigeration systems. International Journal of Refrigeration. 2004, 27, 475-483.

[32] Hsieh, Y.Y., Lin, T.F. Evaporation heat transfer and pressure drop of refrigerant R-410A flow in a vertical plate heat exchanger. ASME Trans. Journal of Heat Transfer. 2003, 125(October), 852-857.

[33] Yin, J.M., Bullard, C.W., Hrnjak, P.S. R744 gas cooler model development and validation. International Journal of Refrigeration. 2001, 24, 692-701.

[34] Petrov, N.E., Popov, V.N. Heat transfer and resistance of carbon being cooled in the supercritical region. Therm Eng. 1985, 32(3),131–134.

[35] Yoon, S. H., Kim, J.H., Hwang, Y.W., Kim, M.S., Min, K., Kim, Y. Heat transfer and pressure drop characteristics during the in-tube cooling process of carbon dioxide in the supercritical region. International Journal of Refrigeration. 2003, 26, 857–864.

[36] Stoecker, W.F. Industrial Refrigeration Handbook. International Edition. McGraw-Hill, New York, 1998.

[37] Robinson, D.M., Groll, E.M. Efficiencies of transcritical CO2 cycles with and without an expansion turbine. International Journal of Refrigeration. 1998, 21(7), 577-589.

[38] Dincer, I., Rosen, M.A. Thermodynamic aspects of renewable and sustainable development. Renewable and Sustainable Energy Reviews. 2005, 9, 169-189.

[39] Moran, M.J., Shapiro, H.N. Fundamentals of Engineering Thermodynamics, 4th ed. Wiley, New York, 2000.

11

Monitoring of air pollution spread on the car-free day in the city of Veszprém

Georgina Nagy, Anna Merényi, Endre Domokos, Ákos Rédey, Tatiana Yuzhakova

Institute of Environmental Engineering, Faculty of Engineering, University of Pannonia, 10 Egyetem St., Veszprém, Hungary H-8200.

Abstract

One of the major factors which adversely affect environmental quality in many cities all over the world is air pollution, with profound negative effects on human health [1]. Apart from the health risks through the inhalation of gases and particles, urban air pollution is the source of other problems such as accelerated corrosion and deterioration of materials, damage to historical monuments and buildings and damage to vegetation in and around the city [2]. In this study, we aimed to investigate the effect of the vehicle related emissions which are a significant source of air pollutants. The research was conducted during the European Mobolity Week (EMW) on the Car-Free Day (CFD). For the characterization of the air quality the generally accepted indicators – O_3, CO, SO_2, NO/NO_2/NO_x, PM_{10}, Benzene (B), Toluol (T), Etil-benzene (E), m-, p-Xilol (MP), o-Xilol (O) concentrations – were used, which well characterizes the changes in air pollution. The average concentrations measured on the car free day for O_3 was 64,5 $\mu g/m3$, for NO_2 was 6,76 $\mu g/m3$, for CO was 127,12 $\mu g/m3$, for SO_2 was 5,19 $\mu g/m3$, for PM_{10} was 10,88 $\mu g/m3$, for Benzene was 0,38 $\mu g/m3$, for Toluol was 0,58 $\mu g/m3$, for Etil-benzene was 0,22 $\mu g/m3$, for MP-Xilol was 1,64 $\mu g/m3$ and for O-Xilol was 2,93 $\mu g/m3$. The results clearly shows that the daily fluctuation of the air pollutants depending on the traffic.

Keywords: Air quality; Air pollution monitoring; Car-free day; Traffic.

1. Introduction

Nowadays the most serious atmospheric problems are the pollution of the air, the destruction of the tropical rain forests, the weather changes, the acid rain, the green house gasses, the warming up of the Earth and the breaking of the ozone layer. There has been more and more damage to the environment, which proceeds faster and faster, since the increase of the world's population and industry. The growing number of cars [3] on roads contributes much to the air pollution. Factories produce large quantities of carbon dioxide [4], sulphur and nitrogen which get into the air too. Different chemicals have damaged the ozone layer and as well as caused the green-house effect, which have led to an increase on temperature levels [5]. To summarize this changed weather has a harmful effect on human health, on flora and on fauna, and that is the reason why we need to build a monitoring system, measure the air pollution particles and deliver strict environmental regulations.

The purpose of this study was to follow the change in the urban air quality. We aimed to investigate the effect of the vehicle related emissions which are a significant source of air pollutants. The monitoring campaign of air pollutants was conducted during the European Mobolity Week (EMW) on the Car-Free

Day (CFD). For the characterization of the air quality the generally accepted indicators – O3, CO, SO2, NO/NO2/NOx, PM10, Benzene (B), Toluol (T), Etil-benzene (E), m-, p-Xilol (MP), o-Xilol (O) concentrations – were used.

2. Experimental

Our experimental design focused on the detection and assessment of the transfer of pollutants of the locality of Veszprém, in the Central Transdanubien Region (Figure 1). Veszprém is not a large urban area in Hungary. There are no factories with significant air pollutant emission. Therefore pollution originating from public transportation and domestic heating are the most representative pollution sources. The growing number of motor vehicles and the related air pollution cause an increasing problem from year to year [6].

Figure 1. M1 is the place where the mobile measurement laboratory was stand (source: Google Earth)

To the determination of the air pollutants different analyzers and measurement methods were used in the course of the research. The concentration of carbon-monoxide measured by a non-dispersive infrared analyzer. The reference measurement method of carbon-monoxide can be found in the MSZ MSZ ISO 4224:2003 standard: "Ambient air - Determination of carbon monoxide - Non-dispersive infrared spectrometric method [7, 8]".

The concentration of nitrogen-oxides measured by chemiluminescence analyzer. The reference measurement method of nitrogen-oxide and nitrogen-dioxide can be found in the MSZ ISO 7996 standard: "Ambient air. The determination of nitrogen-oxides' mass concentration. The method of chemiluminescence [9]".

The concentration of sulphur-dioxide measured by UV fluorescence analyzer, which measurement method can be found in the MSZ 21456-37 standard: "Examination of the air pollutant. The determination of sulphur-dioxide by UV fluorescence method" [10].

To the determination of ozone ultraviolet photometric analyzer was used. The reference measurement method of ozone which was used called MSZ 21456-26:1994 – Determination of ozone with ultraviolet photometric method [11].

To the measurement of the particulate matters dust filter systems were used, which working on gravimetric way (e.g.: Beta-ray particulate monitor). The MSZ ISO 10473:2003 - determination of the mass of particulate metters ona filter, Beta-ray absorption method were used as reference measurement method [12].

The concentration of benzene-toluene-xylene (BTX) measured by infrared analyzer. The reference measurement method of benzene can be found in the MSZ EN 14662-3:2005 standard: "Ambient air quality - Standard method for measurement of benzene concentrations - Part 3: Automated pumped sampling with in situ gas chromatography" [13].

3. Results and discussion

The monitoring of air pollutants was conducted on the Car-Free Day (CFD) on 20 of September and on a Control Day (CD) on 27 of September 1 week later. In Figure 2 the average hourly temperature and humidity values are visible . The temperature values are variable on the CFD between 7.8°C to 18°C and 9.5°C to 13.2°C on the CD. The humidity values are between 45.3 % to 87.6 % on the CFD and 61 % to 82.1 % on the CD.

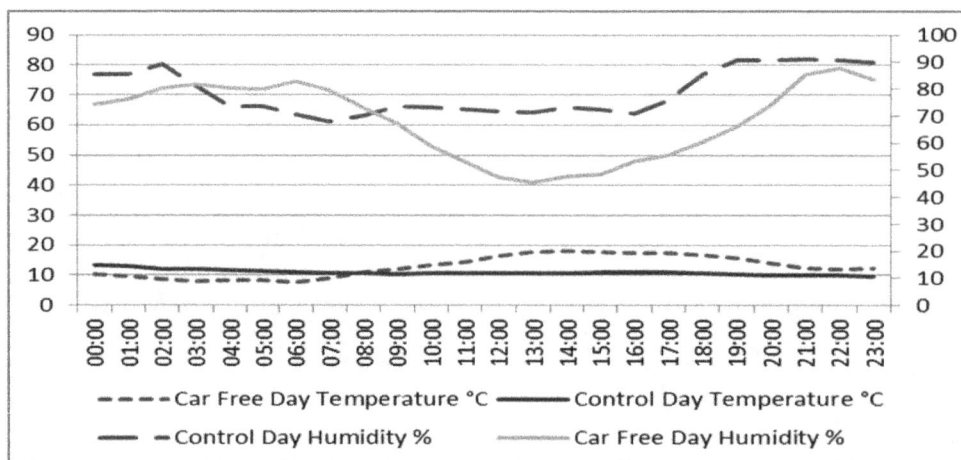

Figure 2. Average hourly temperature and humidity values on the car free day and the control day

Figures 3, 4 and 5 present the daily average concentration values of the two measurement periods. Figure 3 shows the daily average concentration values of No, No2, Nox, O3, So2, PM and Figure 4 shows the daily average concentration values of CO and Figure 5 shows the Benzene (B), Toluol (T), Etil-benzene (E), m-, p-Xilol (MP), o-Xilol (O) concentrations.

It should be noted that in case of NO_2, SO_2 the measured value on the car free day was higher than on the control day. The measured value of NO_2 was 6.76 $\mu g/m^3$ on the car free day and 6.22 $\mu g/m^3$ on the control day. On the car free day the measured value of SO_2 was 5.19 $\mu g/m^3$ and on the control day it was 3.20 $\mu g/m^3$.

The possible reason of the higher values of the car free day was that next to the Warta Vince street on the Victor Hugo Street there is an elementary school (visible at Figure 6.), and in the morning (between 7 am and 8 am) and in the afternoon (between 3 pm to 5 pm) when the parents were taking the children to school than from school at the intersection of Warta Vince Street and Hóvirág Street there were a traffic jam and the exhaust gases stemming from the machines are added to the typical air pollution level.

The measured value of the particulate matters was also higher on the car free day compare with the control day. On the car free day it was 10.88 $\mu g/m^3$ and on the control day it was 10.82 $\mu g/m^3$. On the car free day in the forenoon (between 10 and 12 o'clock) it was organized a program for elementary and high school students, called "Clean Air-Run for it" and because of that the dust concentration were increased and reached a higher level.

As it is visible on Figure 4, the measured daily average concentration of carbon monoxide was much higher on the control day (200.3 $\mu g/m^3$), than on the car free day (127.12 $\mu g/m^3$). According to the source (http://www.idokep.hu/hirek/talajmenti-fagy-is-lehetett) the reason of the higher concentration was that in the morning the temperature was under excelsior and because of that the exhaust gases stemming from the family houses stokers increased the pollution level.

It should be noted that on the Figure 5 the case of m-, p-Xilol (MP), o-Xilol (O) the measured value on the car free day was higher than on the control day. The measured value of m-, p-Xilol (MP) was 1.64 $\mu g/m^3$ on the car free day and 1.36 $\mu g/m^3$ on the control day. On the car free day the measured value of o-Xilol (O) was 2.93 $\mu g/m^3$ and on the control day it was 1.58 $\mu g/m^3$.

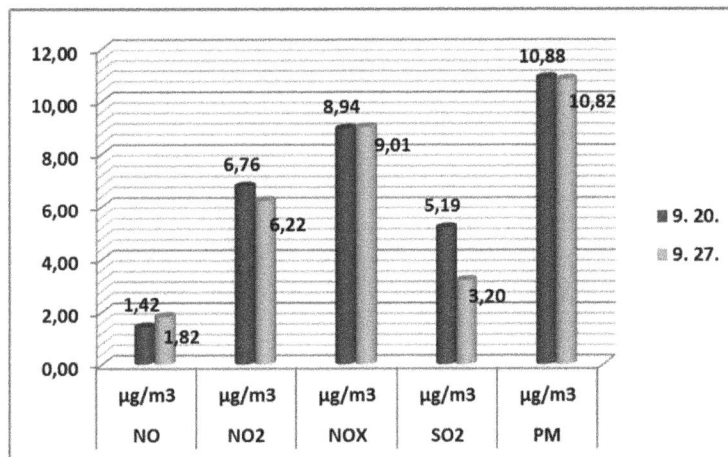

Figure 3. Daily average measured concentration of No, No$_2$, No$_x$, O$_3$, So$_2$, PM

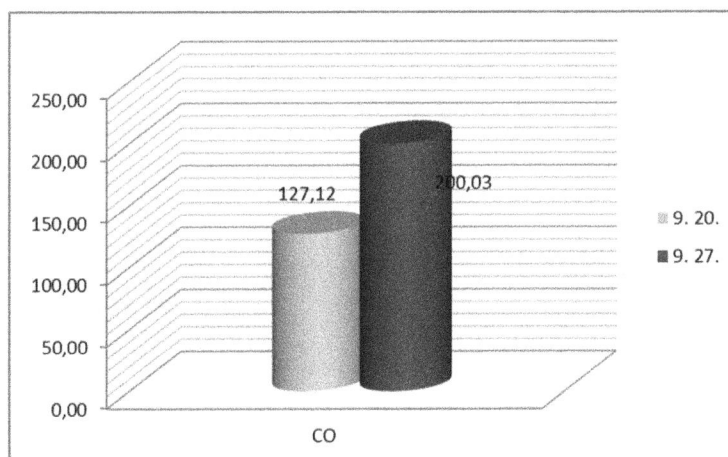

Figure 4. Daily average measured concentration of CO

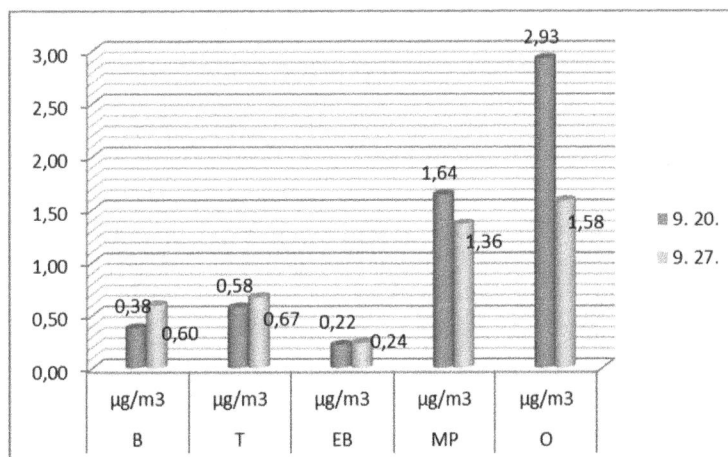

Figure 5. Daily average measured concentration of Benzene (B), Toluol (T), Etil-benzene (E), m-, p-Xilol (MP), o-Xilol (O)

Figure 6. M1 is the place where the mobile measurement laboratory was stand, the orange dashed lines signed the closed area, the green arrow shows the place of the elementary school, the white lines signed the way of the traffic and the red no entrance symbol shows where the traffic jam was (source: Google Earth)

4. Conclusion

Air quality control is one of those fields where many steps were taken by the European Union recently. Aim of the Committee is to establish a comprehensive strategy through which the air quality might be preserved for a long time [14]. These policies not only deal with technology and infrastructure, they also underline the importance of awareness-raising, citizen's engagement and people-focused planning processes [15]. European Mobility Week is an annual campaign on sustainable urban mobility, which runs from 16 to 22 September every year since 2000, organised with the support of the Directorates-General for the Environment and Transport of the European Commission. The aim of the campaign, is to encourage European local authorities to introduce and promote sustainable transport measures and to invite their citizens to try out alternatives to car use [15].

Within the confines of the Car Free Day the University of Pannonia carried out a series of air pollutant measurement with the generally accepted indicators – O3, CO, SO2, NO/NO2/NOx, PM10, Benzene (B), Toluol (T), Etil-benzene (E), m-, p-Xilol (MP), o-Xilol (O). According to the measurements it has been found that the daily average concentrations of the air pollutants are higher at those measuring times where the direct impact of traffic on the air pollutant concentrations is significant.

References

[1]   Ch. Vlachokostas, Ch. Achillas, N. Moussiopoulos, G. Banias (2011): Multicriteria methodological approach to manage urban air pollution, Atmospheric Environment, Volume 45, Issue 25, Pages 4160-4169.

[2]   Ch. Vlachokostas, S. Nastis, Ch. Achillas, K. Kalogeropoulos, I. Karmiris, N. Moussiopoulos, E. Chourdakis, G. Banias, N. Limperi (2010): Economic damages of ozone air pollution to crops using combined air quality and GIS modelling, Atmospheric Environment, Vol.44, pp.3352–3361.

[3]   Ralf Kurtenbach, Jörg Kleffmann, Anita Niedojadlo,Peter Wiesen, 2011. Primary NO2 emissions and their impact on air quality in traffic environments in Germany. Environmental Sciences Europe, Volume 23.

[4]   Roelof D Schuiling, Poppe L de Boer, 2013. Six commercially viable ways to remove CO2 from the atmosphere and/or reduce CO2 emissions. Environmental Sciences Europe, Volume 25.

[5]   Wilhelm Kuttler, 2011. Climate change in urban areas, Part 1, Effects. Environmental Sciences Europe, Volume 23.

[6]   Csom V., Kovács J., Szentmarjay T. (2009), Urban air quality, in view of the characteristics of suspended particulate matter, published in Chemical Engineering Days, 21-23th of April, 2009.

[7]   MSZ ISO 4224:2003 standard: "Ambient air - Determination of carbon monoxide - Non-dispersive infrared spectrometric method".

[8]   MSZ ISO 7996 standard: "Ambient air. The determination of nitrogen-oxides' mass concentration. The method of chemiluminescence.

[9]   MSZ 21456-37 standard: "Examination of the air pollutant. The determination of sulphur-dioxide by UV fluorescence method".

[10]   MSZ 21456-26:1994 – "Determination of ozone with ultraviolet photometric method".

[11]   MSZ ISO 10473:2003: "Determination of the mass of particulate metters ona filter, Beta-ray absorption method".

[12]   MSZ EN 14662-3:2005 standard: "Ambient air quality - Standard method for measurement of benzene concentrations - Part 3: Automated pumped sampling with in situ gas chromatography".

[13]   Csom V., Kovács J., Szentmarjay T. Doomokos E. (2010), Study of traffic-related urban PM pollution at different locations, Hungarian Journal of Industrial Chemistry, Volume 38, pp.15-19.

[14]   European Union offcial webpage, available at: http://europa.eu/citizens-2013/en/news/european-mobility-week-citizens-and-local-authorities-join-forces-create-new-urban-mobility-cul.

[15]   European mobility week official webpage, available at: http://www.mobilityweek.eu/join-us/about/.

12

Najaf, new Saffron homeland

Hashim R. Abdol Hamid

Environment Dept., International Energy and Environment Foundation, Najaf, P.O.Box 39, Iraq.

Abstract
Since its foundation, the International Energy and Environment Foundation (IEEF) interested in improving the situation of the environment in Iraq, where it has several projects and activities in this regard. Among these pilot projects, comes agriculture and indigenization of saffron plant project in Najaf, Iraq. Saffron is a spice derived from the flower of Crocus Sativus, commonly known as the saffron crocus. Crocus is a genus in the family Iridaceae. This project is implemented in two stages: The first stage: Pioneering Agriculture: where is planting limited area and under inspection for the purpose of follow-up agricultural conditions and took the basic data for the stages of plant growth. The second stage: Production Agriculture: where is planting large areas and productive quantities. This project is an important achievement, which is the first of its kind in Iraq in general and in the holy city of Najaf, in particular, where large tracts of land will be invested to cultivate this important and useful plant, and in addition to its commercial benefits, it will contribute stabilize the topsoil and preservation of erosion and improve the climate in the adjacent areas.

Keywords: Saffron, Najaf city, Iraq, Agriculture, IEEF.

1. Introduction
Crocus sativus, commonly known as saffron crocus, is a species of flowering plant of the Crocus genus in the Iridaceae family. It is best known for the spice saffron, which is produced from parts of the plant's flowers. The cormous autumn-flowering perennial plant species is unknown in the wild. Human cultivation and use of saffron spans more than 3,500 years [1, 2] and spans cultures, continents, and civilizations. Saffron, a spice derived from the dried stigmas of the saffron crocus (Crocus sativus), has through history remained among the world's most costly substances.
With its bitter taste, hay-like fragrance, and slight metallic notes, the apocarotenoid-rich saffron has been used as a seasoning, fragrance, dye, and medicine. Crocus sativus, unknown as a wild plant, is considered to be a mutant that has derived from C. cartwrightianus. The cultivated clone was probably selected for its triploid vigour and extra long stigmas and has been maintained in cultivation for over 3000 years. The saffron crocus (Crocus sativus L.) is sterile and does not set viable seed. Therefore, the crop must be propagated by corm multiplication. The saffron crocus flowers in autumn shortly after planting, before, together with or after leaf appearance. The remainder of its growing season consists of initiation, filling up, and maturation of the daughter corms at the beginning of summer. Each corm only lasts a single season and is replaced by 1 to 10 cormlets, depending on the original size of the mother corm [3].
Corms are globular and depressed, up to 4.5 cm in diameter and covered with a tunic of parallel fibres. Corms are dormant during the summer and produce 5 to 11 erect, narrow, grass-like green leaves, up to

40 cm long, that emerge in autumn. Flowers are fragrant, up to 8 cm long, and usually pale lilac or mauve with darker coloured veins. The outstanding feature of the flower is its style, which divides into three brilliant red stigmas 25-30 mm long [3].

Saffron is native to the moderate environment, characterised by cool to cold winters, with autumn-winter spring rainfall, and warm dry summers with very little rainfall. It can withstand substantial frosts (-10°C), and can tolerate occasional snow in the winter [3].

Flower yield is highly dependent on corm density and corm size. Traditionally, saffron is grown on raised beds to allow good drainage and easy access for picking. Corms are planted out during their dormant period in summer. The best yields for flower and corm production are obtained by leaving a space of 2-3 cm between each corm in the furrow, with a planting depth of 8-10 cm. Optimal corm quantity per hectare is 13-15 tons, which is about 600-700 thousand corms with an average weight of 20-22 g each (45-48 corms/kg) [3].

Recommended planting depths for corms vary from 7.5-10 cm to 15-22 cm. Planting depth affects corm production; more buds sprout from shallow planted corms than from deep planted ones, resulting in more daughter corms. Corm size has a significant effect on the production of daughter corms and on the production of flowers and the yield of saffron. The larger the mother corm, the more daughter corms will be produced in the annual cycle, which increases the potential for higher yields in subsequent years. New saffron corms also grow above the old ones each season, so they creep towards the soil surface by 1-2 cm each year. Therefore, the crop needs to be lifted and replanted periodically. This occurs about every 4 years in Spain, but fields may last up to 12 years or more under non-irrigated conditions in Kashmir. Replanting is normally done when yields begin to fall due to overcrowding or damage to corms that are too close to the soil surface [3].

In traditional saffron culture, large amounts of farm yard manure were applied to the saffron fields before planting, and typically 20-30 tons per hectare are incorporated during cultivation. This material supplies nutrients, but its other major role is to improve soil moisture holding capacity and structure under nonirrigated conditions. Under traditional growing systems no further fertiliser was applied after corm planting. However, recent data suggest that at least some annual fertiliser applications are beneficial and a base dressing of 80 kg P/ha and 30 kg K/ha followed by a split application of 20 kg N/ha in autumn and again immediately after flowering is recommended [3].

Saffron flowers in the autumn, about 40 days after planting, and continues for 30-40 days, depending on the weather. The flowering period of each plant may last up to 15 days. Rain 10-15 days before flower picking results in excellent flowering and high production, whereas under drought conditions, small flowers with small stigmas can be expected. A cold period or a late planting can retard flowering [3].

Flowers are usually picked daily in the morning after the dew has evaporated but before flowers wither. The flower is cut at the base of the flower stem with a slight twisting movement or by cutting with the finger nail. Care is taken not to damage the leaves [3].

Following the separation of the stigmas from the flowers, it is essential to dry the flower heads immediately. Brightness of colour is aided by quick high temperature drying. Slow drying gives a poor quality product. A final dry matter close to 10% moisture is adequate for long-term storage [3].

The quality of saffron is dependent on its colouring power (crocin concentration), odour (safranal) and taste (picrocrocin). The best quality saffron has a high safranal content. Saffron is dry, glossy and greasy to the touch when freshly dried, turning dull and brittle with age. It is easily bleached if not stored in the dark, and also stores better under conditions of low temperature and low relative humidity. An International Standard for saffron is available (ISO3632-1:1993). Saffron in filaments is classified into four categories based on the content of floral waste and extraneous matter, with category 1 (extra) having a maximum of 0.5% floral waste and 0.1% extraneous matter. Category 1 has the highest bitterness (as expressed in the absorbance test for picrocrocine), and the highest colouring test (as expressed in the absorbance test for crocine). Safranal levels, also based on an absorbance test, have a range for all grade categories.

2. Description of the study area

Najaf is one of the sacred cities of Iraq. The Najaf area is located 30 km south of the ancient city of Babylon and 400 km north of the ancient Biblical city of Ur. It is located on the edge of western plateau of Iraq, at southwest of Baghdad the capital city of Iraq, with 160 km far from the capital. It is raised upon sea level with almost 70 meters, and is situated on the longitude of 19 degree and 44 minutes, as well as on the latitude of 31 degree and 59 minutes [4].

3. Indigenization of saffron plant project

This project is implemented in two stages: The first stage: Pioneering Agriculture: where is planting limited area and under inspection for the purpose of follow-up agricultural conditions and took the basic data for the stages of plant growth. The second stage: Production Agriculture: where is planting large areas and productive quantities. Ten stations have been elected in ten districts in Najaf city as shown in Figure1. The weather data for the city has been recorded during the year 2014 as shown in Figures 2-8.

4. Results

The cluster that have been planted in the experimental stations, shown in Figure 9. After four weeks, about 85% of planted clusters have been grown up, where the vegetative growth complete. Figure 10 shows the vegetative part of the saffron, grass like leaves. Within the four to six weeks, growing of the vegetative part of the plants, few of flowers have been witnessed in some of stations. Figure 11 shows the saffron crocus flowers. Within 2-4 days of flower growth, the red threads of the flower were harvested (picked) and dried then stored, as shown in Figure 12.

5. Recommendations

1- The preliminary results obtained from the first stage of the project, were very encouraging, therefore the second stage it is believed to be an encouraging and productive.

2- The availability of suitable climate with accepted range of minimum/maximum temperature degree and humidity in addition to suitable soil type, make the investment of large tracts of land in saffron planting, feasible.

3- In order to protect the saffron fields from the effects of extreme weather conditions, it would be appropriate that the saffron fields are under tall trees such as palm and citrus trees, as well as the corms should be cultivation about 15-20 cm deep in the soil.

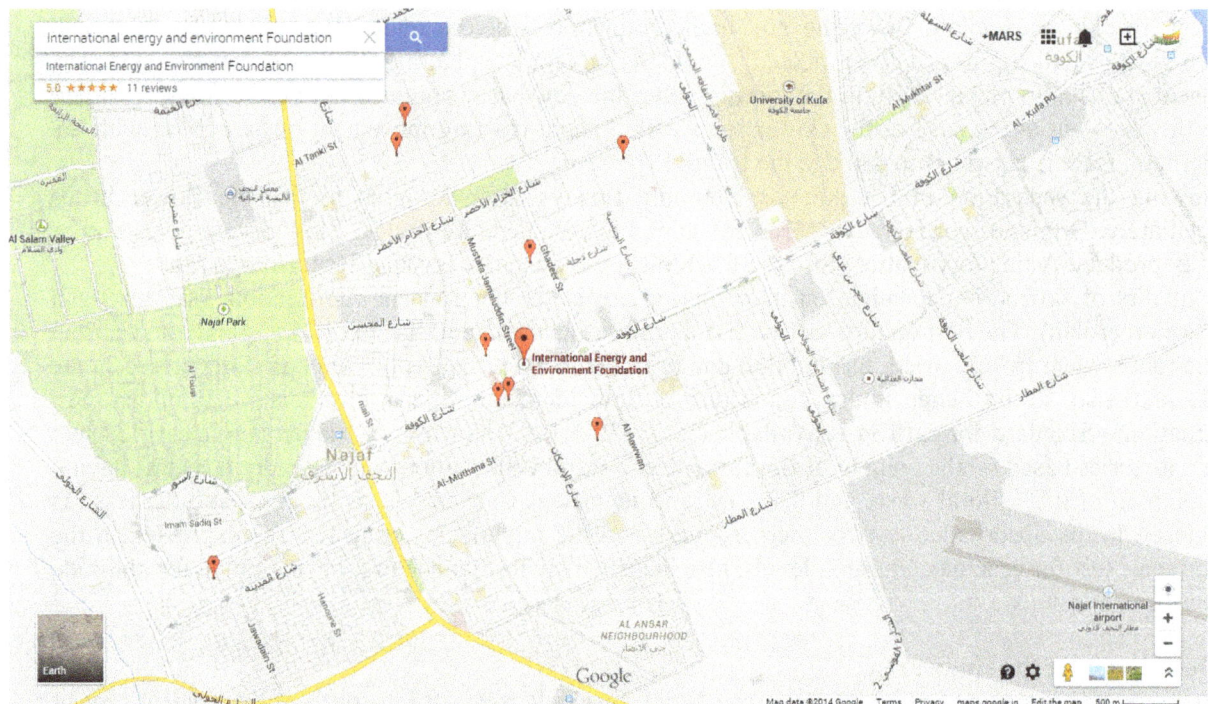

Figure 1. Satellite image of Najaf city explain the locations of the planting areas of Saffron [Google earth, 2014].

Air temperature (°C)

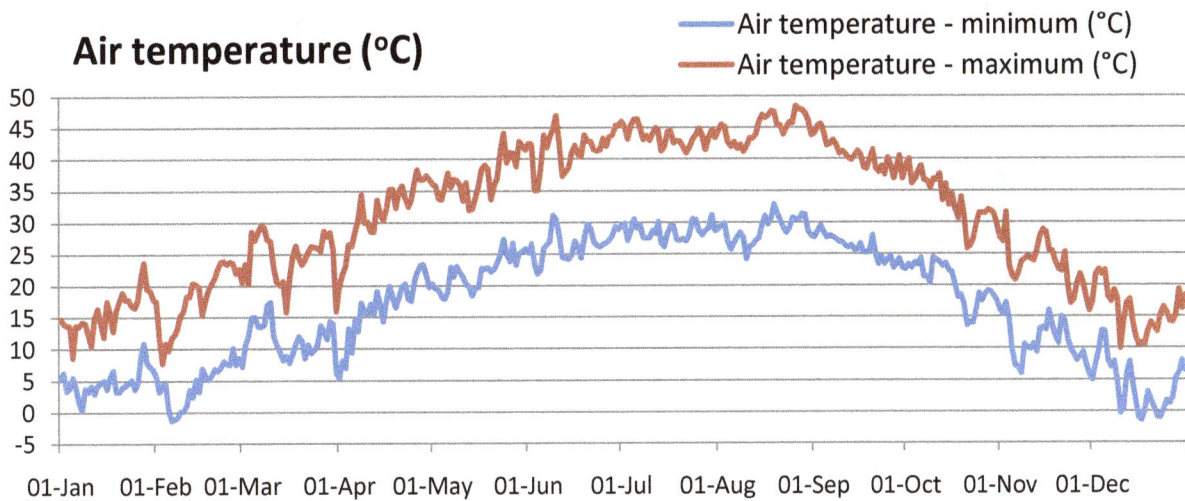

Figure 2. Maximum and minimum air temperature in Najaf city during 2014.

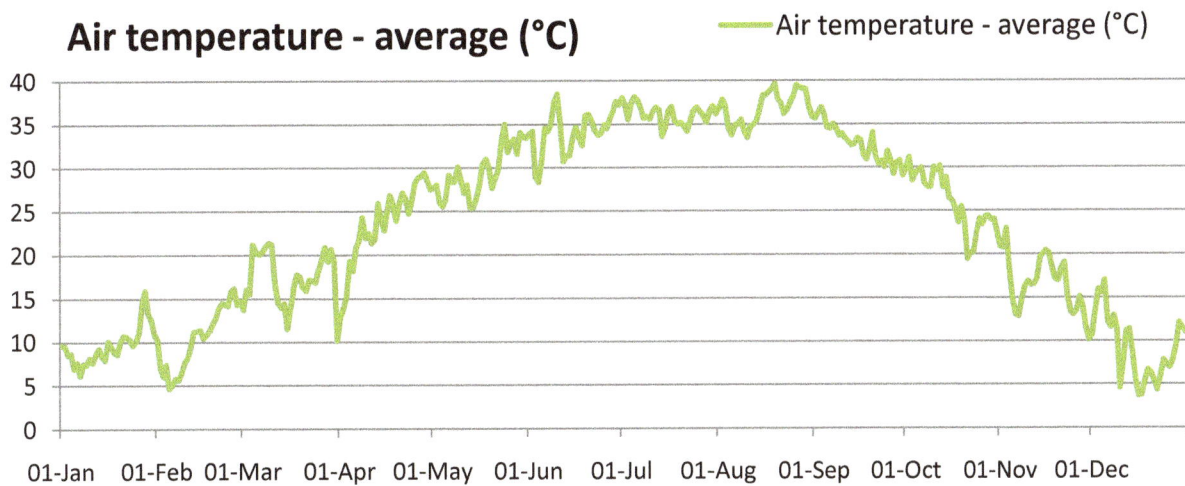

Air temperature - average (°C)

Figure 3. Average air temperature in Najaf city during 2014.

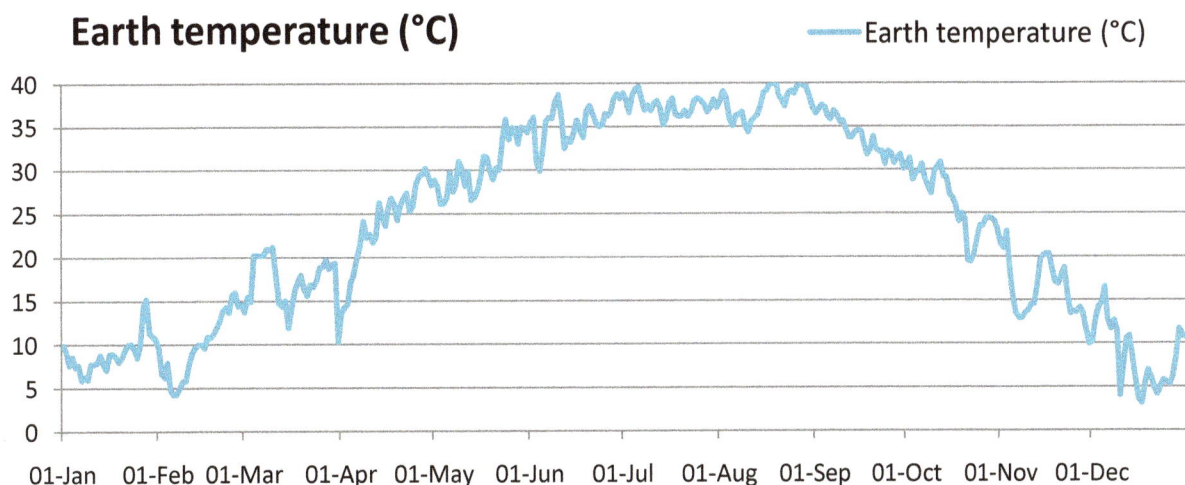

Earth temperature (°C)

Figure 4. Earth temperature in Najaf city during 2014.

Relative humidity (%)

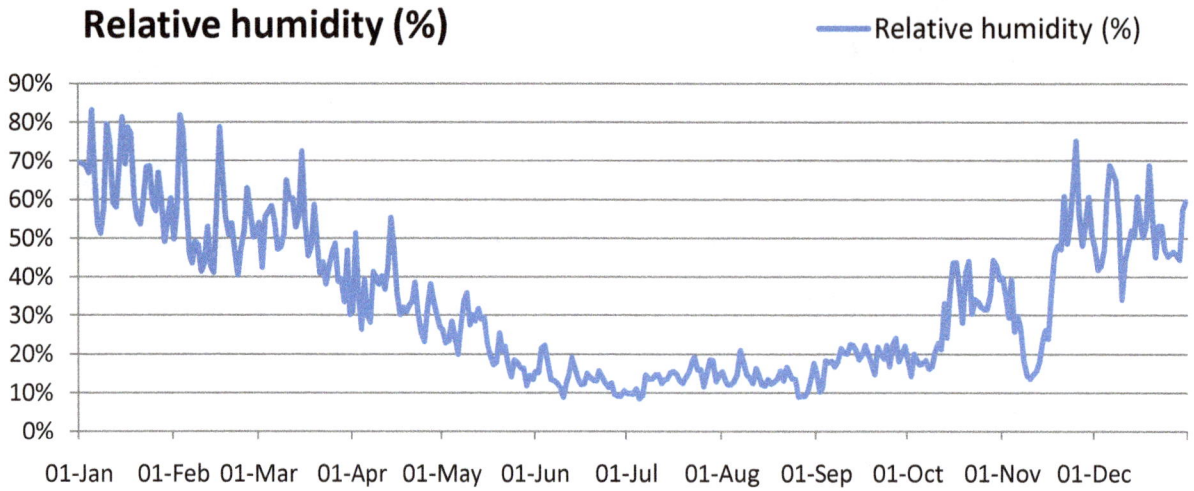

Figure 5. Relative humidity in Najaf city during 2014.

Wind speed (m/s)

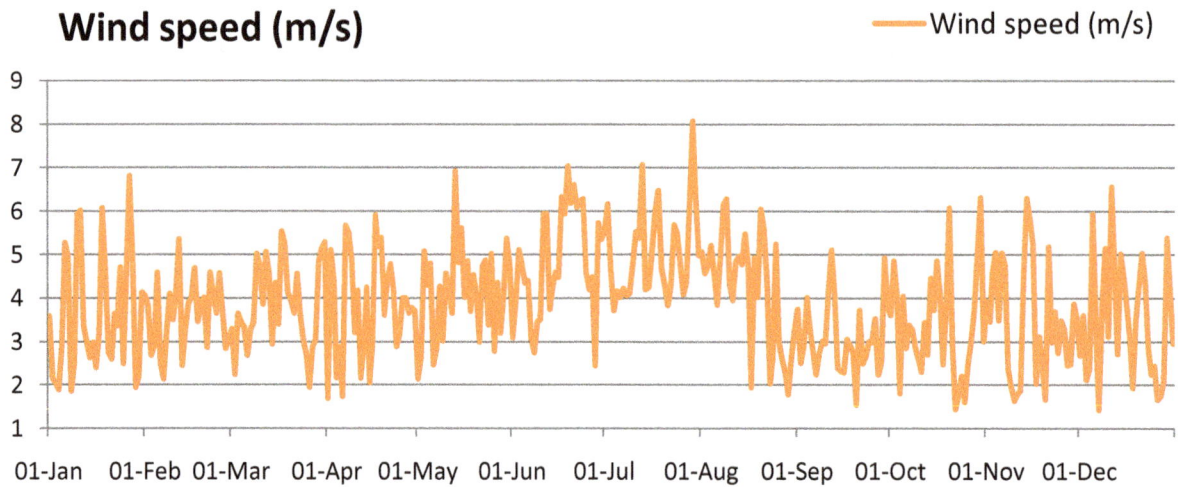

Figure 6. Wind speed in Najaf city during 2014.

Atmospheric pressure (kPa)

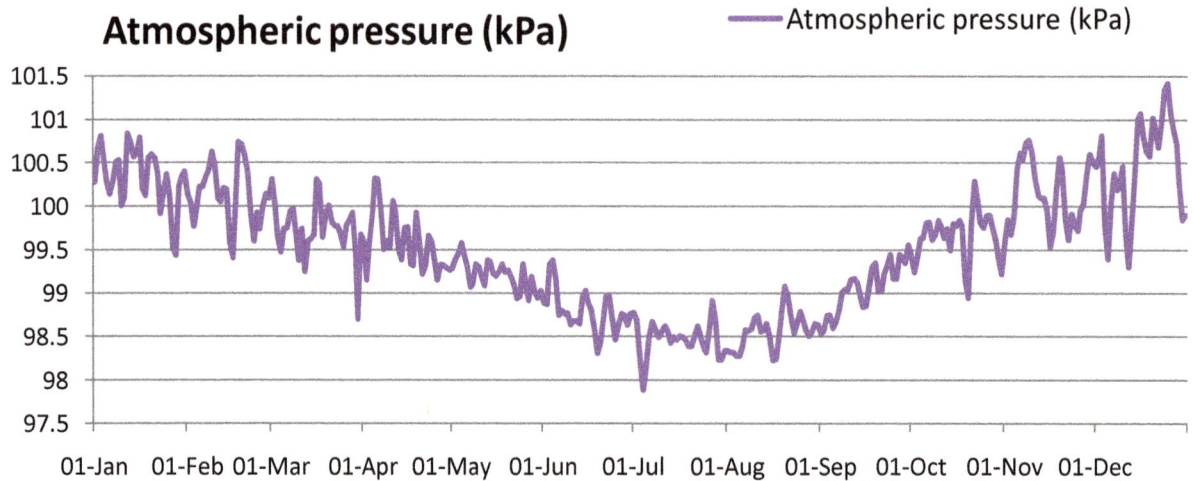

Figure 7. Atmospheric pressure in Najaf city during 2014.

Daily solar radiation - horizontal (kWh/m²/d)

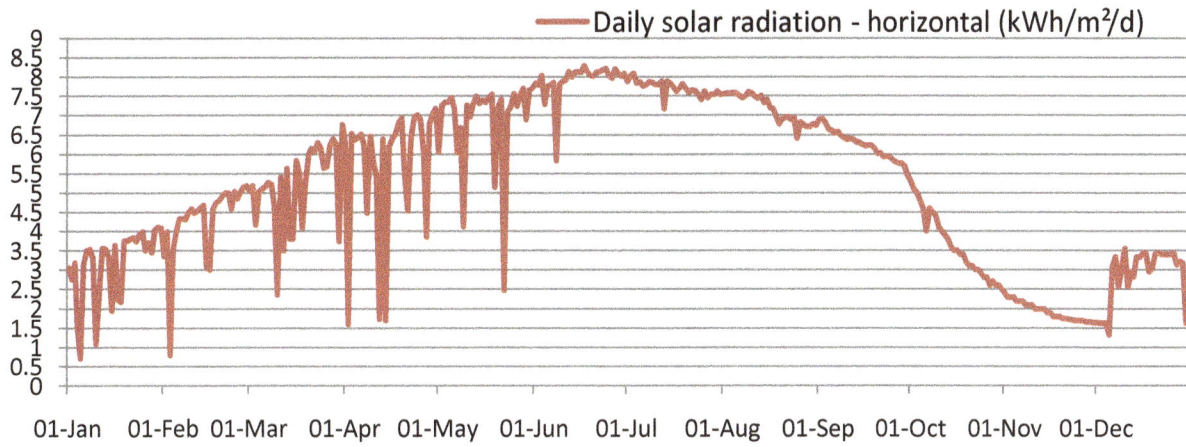

Figure 8. Daily solar radiation in Najaf city during 2014.

Figure 9. A cluster of new daughter corms.

Figure 10. The 'grass like' leaves of the saffron crocus during the autumn to spring vegetative stage.

Figure 11. continued.

Figure 11. Saffron crocus flowers in autumn either before, together with (as above), or after leaf production.

Figure 12. Saffron red threads.

6. Conclusion

In conclusion, this project is an important achievement, which is the first of its kind in Iraq in general and in the holy city of Najaf, in particular, where large tracts of land will be invested to cultivate this important and useful plant, and in addition to its commercial and medical benefits, there is an environmental benefits where it will contribute stabilize the topsoil and preservation of erosion and improve the climate in the adjacent areas.

Acknowledgements

This work is part of a R&D project conducted at the environment department in International Energy and Environment Foundation (IEEF), Iraq.

References

[1] Dalby A., Food in the Ancient World from A to Z, Routledge, 2003, ISBN 978-0-415-23259-3
[2] Moshe Negbi. Saffron: Crocus sativus L. Medicinal and Aromatic Plants - Industrial Profiles. ISBN:9780203303665, CRC Press, 2003.
[3] Growing saffron - the world's most expensive spice. New Zealand Institute for Crop & Food Research Ltd., A Crown Research Institute, 2003.
[4] Najaf heritage at risk., IEEF heritage group, ISBN: 978-1490932651, IEEF 2013.

13

Material waste in the China construction industry: Minimization strategies and benefits of recognition

Sulala M.Z.F. Al-Hamadani[1,2], **ZENG Xiao-lan**[1,2], **M.M.Mian**[1,2], **Zhongchuang Liu**[1,2]

[1] Three Gorges Reservoir Area's Ecology and Environment Key Laboratory of Ministry of Education, Chongqing University, Chongqing, 400045, China.
[2] National Centre for International Research of Low-carbon and Green Buildings, Chongqing University, Chongqing, 400045, China.

Abstract

Waste minimization strategies and the relative importance of benefits of material waste recognition were examined using a survey of construction companies operating in Chongqing city China. The results showed that a remarkable proportion of respondent companies have specific policies for minimizing construction waste. Amongst the strategies, minimizing waste at source of origin is practiced to a large degree by construction companies with specific waste minimization strategies. However, considerable quantities of construction waste are generated. These quantities need to be reused or recycled or combination of them. The study also revealed that recycling is not highly practiced because it needs a lot of capital and an area, except for those high scrap value recycling materials like steel, whereas other non-profitable will be sent to C&D landfills directly.

Respondents' perceptions towards the benefits of material waste recognition revealed that materials waste is primarily considered an environmental and financial problem and its minimization a cost cutting activity and protection of the environment. In contrast, the contractual benefits were considered less important by surveyed companies.

Keywords: Construction industry; Minimization strategies; Waste; 3R; Benefits of recognition; China.

1. Introduction

In the construction industry, it is well known that there is a relatively large volume of material being wasted due to a variety of reasons. The problem of material waste on construction sites is not an isolated issue and is of environmental concern. Therefore, waste minimization has become an important issue in the construction industry. Waste minimization has been defined as: any technique, process or activity which avoids, eliminates or reduces waste at its source or allows reuse or recycling of the waste [1]. There were three main waste minimization strategies used in construction projects: (1) reducing waste; (2) reusing materials; and (3) recycling waste. It has been concluded from many studies that minimization of waste at source must be prioritized when developing strategies for waste minimization. The other strategies are reusing and recycling waste which means putting the materials that are wasted into beneficial use. If the three strategies had not been applied, the generated waste will end up at landfill.

China has developed the same 3R principle. Similar to other countries, the hierarchy of waste management in China is emphasizing on reduce, reuse and recycle the waste and it is said to be the best approach in China [2]. Chinese government improves China's waste management system by way of changing the "China Waste hierarchy" into "The Danish Waste hierarchy", which means to reduce the landfill proportion and increase the proportion of waste reduce, reuse, and recycling (3R) as in Figure 1. The earlier of China's waste hierarchy model, the landfill occupied more than 80% for waste disposal and landfill is the least priority in the waste hierarchy [3]. This execution of the Danish model was successful in reaching a high recycling rate for construction and demolition waste [4]. The scarcity of landfill in China is the reason behind adopting the Danish Waste Model to minimize the usage of land for waste disposal by way of maximizing the waste reuse and recycling.

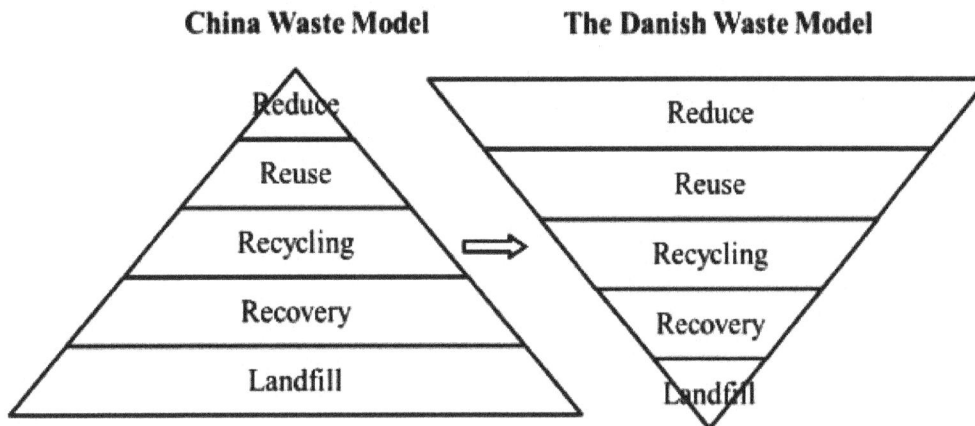

Figure 1. China's adoption for the Danish hierarchy model

Construction material waste recognition can provide financial, environmental and other benefits. These benefits can be appreciated over a short term or long-term period. But overall, cost benefits can be appreciated throughout the whole building process by carrying out an analysis of the life cycle costs. Environmental benefits, which are important to be considered due to the alarming situation of waste in China, where a total of 400 million tons of construction waste was generated in China each year [5], and the share of Chongqing city in 2008 was around 12 million cubic meters of C&D [6].

In this survey, the identification of the benefits of material waste recognition was obtained from literature reviews of both Refs [7, 8]. These are: It helps the contractors to know the real requirements of the project represented in materials, time and cost; It encourages companies and firms to decrease waste; It participates in project success and profit achievement; It keeps environment out of pollution; To get enough finance for a project; and It helps contractors in pricing bids.

Adopting other waste management and minimization strategies other than disposal in a landfill has become a pressing need for China to reduce construction and demolition waste and increase the lifespan of landfills.

This paper focuses on the following objectives:

- Identifying waste minimization strategies employed by construction companies in Chongqing city, China;
- Determining the relative importance of the benefits of material waste recognition; and
- Suggesting strategies to improve the current C&D waste management situation in Chongqing China.

2. Research methodology

To obtain the aims and objectives, the essential stages of methodology were performed in several stages. The waste minimization strategies and benefits of material waste recognition outlined in the questionnaire were identified from a review of related literature. The questionnaire contained three sections to reach the aim of this research, as follows: Profile of the respondent, waste minimization strategies and benefits of material waste recognition. Respondents were asked to indicate if their company had a specific policy for minimizing construction waste, and if they did, respondents were asked to indicate which of the strategies were utilized by their company to reduce waste generated on construction project sites. Respondents were asked to indicate if they utilized a combination of the listed strategies or any other strategies not included in the questionnaire.

The five-point Likert scale is used in the questionnaire to quantity opinions of respondents about the benefits of material waste recognition, with the scale ranging from strongly disagree to strongly agree. A total of 90 questionnaires were distributed through direct visits to the construction companies and sites. The target groups were randomly selected from industry practitioners in China Chongqing city. Finally a total of 80 valid responses were received, representing an overall response rate of 89 percent. The data were then analyzed by using statistical analysis techniques including ranking analysis for the benefits. These research activities were facilitated with the aid of SPSS 12.0 software. The process of data analysis is interpreted as follows. The relative importance of all identified benefits was examined by ranking them based on their mean values and standard deviation. In this research, only those benefits with mean values that are greater than 3.00 were recognized as important Benefits of material waste recognition.

Furthermore, five construction professionals and three researchers were interviewed to investigate strategies for construction waste minimization. The interviews were conducted in May 2014 and each lasted about 30 min. The selection of these professionals was based on their prominent experience in construction and demolition waste management (Table 1).

Table 1. Interviewees profiles in the interviews

Expert no	Position	Affiliation	Experience in C&D waste management (Years)
1	Professor	University	10
2	Professor	University	12
3	Associate professor	University	10
4	Project manager	Contractor	7
5	Project manager	Contractor	8
6	Site engineer	Contractor	5
7	Supervision engineer	Supervision company	6
8	Supervision engineer	Supervision company	9

3. Results and discussion

3.1 Respondents' profile

Respondents' occupation clarifies that the majority of respondents, 48% in this survey are Civil Engineers, followed by 14% Project Managers, 11% Construction Technicians, 9% Site Engineer, 6% Architects, 6% Cost Engineers, 5.0% Quantity Surveyors and 1% Foreman. So, a total of 62% of the respondents was a combination of Civil Engineers and Project Managers. It can be noticed that Civil Engineers and Project Managers play a major role in this research. This indicates that the majority of respondents are highly involved in the construction site problems.

The respondents' years of experience in descending manner as follows: 32% from 1 to 3 years, 21% from 5 to 10 years, 18% from 3 to 5 years, 13% more than 20 years, 10% from 10 to 15 years and 6% from 15 to 20 years. This means that most of the respondents have a good understanding about issues in connection with the construction.

3.2 Construction material waste minimization strategies

The results from the present survey show that a remarkable proportion of construction firms have a specific policy for minimizing construction waste. Table 2 indicates that 69.5% of the respondents' construction firms had specific policies for minimizing waste on construction project sites, while 30.5% did not have a specific policy for minimizing construction waste.

Table 2. Proportion of the construction firms with a specific policy for minimizing construction waste

Specific policy (Y/N)	Proportion of construction firms
Yes	69.5%
No	30.5%

The distribution of the waste minimization strategies adopted by respondents who had specific policies for minimizing construction waste is shown in Table 3. Only 25 (44.6%) of respondents consider the recycling waste only as a waste reduction strategy. It is interesting to note that the number of respondents increased from 25 (44.6%) to 45 (80.4%) in the recycling category when it is combined with minimizing waste at the source of origin. Minimizing waste at source of origin is widely practiced as a waste minimization strategy by construction firms with specific waste minimization policies. Similarly, only 30 (53.6%) of respondents in this category limited their waste minimization strategy to reusing waste only. It is also notable that the number of respondents raised from 30 (53.6%) to 42 (75.0%) when minimizing waste at the source of origin involved. Reusing waste strategy is applied by construction firms, since they use the generated materials in other projects. This strategy is applied in renovation and rehabilitation projects. 32 (57.1%) of the respondents agreed with minimizing waste at the source of origin only as a sole waste minimization strategy. However, the number of respondents increased dramatically to 53 (94%) when a combination of reusing and recycling waste practiced with minimizing waste at the source of origin. Other waste minimization measures given by respondents include: ordering just what is needed of material; take-back arrangement with suppliers; appointment of waste manager on site; using materials before expiry dates; use of more efficient construction equipment; accurate measurement of materials during batching; good construction management practices; encourage re-use of waste materials in projects; recycling of some waste materials on site; and use of low waste technology.

Table 3. Distribution of waste minimization strategies employed by respondents' firms with specific waste reduction plans

Waste reduction strategy	Frequency (Agree)	Percent	Frequency (Not agree)	Percent	Total
Combination of re-using waste, recycling waste and minimizing waste at the source of origin	53	94.6%	3	5.4%	56
Combination of recycling waste and minimizing waste at the source of origin	45	80.4%	11	19.6%	56
Combination of reusing waste and minimizing waste at the source of origin	42	75.0%	14	25.0%	56
Minimizing waste at the source of origin only	32	57.1%	24	42.9%	56
Reusing waste only	30	53.6%	26	46.4%	56
Recycling waste only	25	44.6%	31	55.4%	56

3.3 Benefits of material waste recognition

It can be seen clearly from Table 4 that all benefits receive a mean value of greater than 3.000, which implies that these benefits are all critical and important in terms of construction material waste recognition. The analysis on the benefits of material waste recognition shows that the environmental benefit is ranked first by respondents, yet the mean value of the second important benefit 'It encourages companies and firms to decrease waste' is 3.84 indicating the agreement of most respondents with the benefit of waste reduction. These findings are evidence of the increasing interest and awareness of the environmental issues in the Chinese construction industry.

The third and fourth important benefits in the ranking are 'To get enough finance for the project' and 'It participates in project success and profit achievement' with mean values of 3.73 and 3.68 respectively. This indicates that the financial benefits of material waste recognition are important aspect for Chinese construction practitioners. The importance of the financial aspect is totally obvious for the construction industry in Chongqing city, which its average growth rate of investment in construction activities is about 25.4% from 1997 to 2006 [9].

It has been also noticed from this study that 'It helps the contractors to know the real requirements of the project represented in materials, time and cost' and 'It helps contractors in pricing bids' are the lowest two important benefits of material waste recognition with mean values of 3.63 and 3.41 respectively. This shows that contractual benefits are considered relatively less important by the respondents of this survey.

Table 4. Ranking of the benefits of material waste recognition

The benefits of material waste recognition	Mean	Std. Deviation	anking
It keeps environment out of pollution	4.05	1.113	1
It encourages companies to decrease waste	3.84	1.037	2
To get enough finance for a project	3.73	1.125	3
It participates in project success and profit achievement	3.68	0.965	4
It helps the contractors to know the real requirements of the project represented in materials, time and cost	3.63	0.933	5
It helps contractors in pricing bids	3.41	0.990	6

3.4 Recommended strategies to improve waste minimization
Based on the interview discussions with construction professionals and researchers. Several strategies were suggested to improve the existing construction and demolition waste management in Chongqing City, China. These are: providing guidelines on effective waste management methods; enhancing on-site waste management plans to minimize the waste; adopting low-waste construction technologies and behavior; involving environmental protection in the design stage and implementing waste minimization's design; developing an appropriate waste landfilling charge scheme; and developing a mature market for trading recycled materials.

4. Conclusions
This paper has presented results from a survey on waste minimization strategies employed by construction firms and the relative importance of benefits of material waste recognition. The results of the survey indicate that a sizeable proportion of construction firms have specific policies for minimizing waste generated on construction project sites. Amongst the firms which do have a specific waste minimization policy, minimizing waste at source (either alone or in combination with other waste minimization strategies) was the most widely practiced waste minimization strategy. However, unlike recycling, it is perceived to be more effective only when undertaken in combination with other waste reduction strategies. This can be attributed to the fact that it is practically impossible to minimize waste at source to the point where waste generation is completely eliminated. Therefore, any waste that is generated after minimization at source would need to be reduced using another waste minimization strategy such as reusing or recycling. From the analysis on the benefits of material waste recognition, conclusions can be drawn that the environmental and financial benefits are considered strong drivers for material waste recognition, meanwhile the contractual benefits are perceived relatively less important by respondents. Also, it is interesting to conclude that environmental issues have climbed the ladder to become a priority in the China construction industry.

Acknowledgment
This work was supported by the Faculty of Urban Construction and Environmental Engineering, Chongqing University, Chongqing, 400045, P. R. China.

References
[1]   Poon C.S., Jaillon L.A. Guide for Minimizing Construction and Demolition Waste at the Design Stage. The Hong Kong polytechnic University, 2002.
[2]   Hoornweg D., Leader T.T, Lam P., Chaudhry M. Waste management in China: Issues and recommendations. East Asia Infrastructure Department, World Bank, Washington DC, 2005.
[3]   National Bereau of Statistics of China, "China Statistical Yearbook," 2009.
[4]   Danish Environmental Protection Agency, "Waste in Denmark," 2012.
[5]   Lu K.A. The actuality and integrate using of China's construction wastes. Journal of Construction Technology, 1999, 128(5): 44-45.
[6]   Chongqing Environmental Department [Internet]. Chongqing (CN): Evaluation of the construction and demolition waste landfills in Chongqing city. [Cited 2009 September 4] (in Chinese). Available from: http://jjckb.xinhuanet.com/jcsj/2009 06/04/content_161872.htm
[7]   Al-Moghany S.S. Managing and Minimizing Construction Waste in Gaza Strip. Palestine: Islamic University of Gaza, 2006.

[8]   Al-Hajj A., Hamani K. Material Waste in the UAE Construction Industry: Main Causes and Minimization Practices. J. Architectural Engineering and Design Management, 2011, 7(4): 221-235.

[9]   Chongqing Municipal Bureau of Statistics [CD-ROM]. Beijing: Chongqing Statistical Yearbook; 2003-2007.

14

Assessment of air pollution elements concentrations in Baghdad city from periods (May-December) 2010

Ahmed F. Hassoon

Department of Atmospheric Sciences, College of Sciences, Al-Mustansiriyah University, Baghdad, Iraq.

Abstract

Air pollution in developing countries has recently become a serious environmental problem, which needs more active air quality monitoring and analyses. To assess air quality characteristics over the city of Baghdad. temporal variations in CO, NO, NO_2, NO_X, O_3, SO_2 and PM_{10} Concentrations measured between May-December 2010 (245 days), at period from 8:00-16:00 daily hour from location called AL-Jadriyah station (44.1E -33.3N, 38.5m above sea level). From diurnal variability of these concentration, we see high daily values of CO and CH_4 3.25, 1.9 PPM at November while NO_X record 0.23 PPM at December that consider as highest daily value. While other pollutant concentration don't have large variation have 0.14-0.18 PPM. Particular matter at 10um (PM_{10}) have $1.6g/m^3$ at 21/7/2010. At winter season and specifically at December month, there is good relation between the hourly concentration of PM_{10} and other chemical pollutant concentration such as CO,SO_2, NO, NO_X this can be putting by correlation coefficient r =0.7-0.5. The monthly mean concentration of pollutant CO, NO_X, CH_4, NO, NO_2 recorded high value at August Month. While O_3 have large mean concentration in November, while PM_{10} have large monthly mean concentration at June and July months where there is most frequent dust-storm events. High concentration and its frequency distribution shifts towards large values concentrated at summer seasons June, July, Aug. about 0.25-1.25 PPM and have frequency percent about 82.6%. In winter the frequency distribution shifts towards large values of O_3 even above 61.2% in range 0.085-0.105 PPM. Regarding the frequency distribution of SO_2 all season where shifts towards lower except spring (May) 88.2% at range concentration 0.025-0.125 PPM. Nitrate oxide have different concentration an frequency at several season but NO_X have large frequency at summer, other pollutant concentration CH_4 and PM_{10} have high concentration frequency at this period.

Keywords: Air pollution; Air quality; Frequency of occurrence; Temporal variations.

1. Introduction

Air pollution is defined as the emission of particulate toxic elements in to the atmosphere by natural or anthropogenic sources [1]. These sources can be further differentiated in to either mobile or stationary sources [2]. Anthropogenic air pollution commenced with human s systematic use of fire. Its historical development has been characterized by steadily increasing amounts of total emissions. The inversion of new sources of pollution emission as well as the emission of pollutants that had not formerly been emitted by man-made sources. This development has had the greatest impact on the air quality of so-called Mega-cities (cities with over 10000000 inhabitants). Today the major sources of man –made air pollution are motorized street traffic (especially exhaust gases and tire abrasion). The burning of fuels, and larger factory emissions. Depending on the pollutant particles size, they can be carried for distances of several thousand miles. With decreasing diameter, they are able to infiltrate finer lung structures [3]. The world health organization (WHO) estimate 2.4 million fatalities due to air pollution each year [4]. Since the breathing of polluted air may have severe health effects such as asthma, or increased cardiovascular risks [5, 6]. Thus air pollution has presented one of the major environmental issues and is becoming a very important factor of the quality of life in urban areas, posing a risk both to human health and to the environment. According to the directive on ambient air quality assessment and management, ozone (O_3) nitric oxide, nitrogen dioxide (NO_2), sulfur dioxide (SO_2), carbon monoxide (CO) and particulate matter with diameter <10um (PM_{10}) are target species, due to their negative effects on human health and vegetation [7]. Ozone a secondary air pollutant has gained extensive attention in the literature due to its harmful effects in vegetation during the growing period. Emission of nitrogen oxides (NO_X = NO + NO_2), volatile organic compounds (VOCS) and sulfur compounds (including SO_2) can lead to complex series of chemical and physical transformations such as the formation of O_3 in urban and regional areas [8]. Meteorological conditions (temperature, relative humidity, wind speed, rainfall and atmospheric pressure) strongly influence the efficiency of photochemical processes leading to the ozone formation and destructions [9, 10]. The understanding of the O_3 behavior near surface layers is essential for a study of pollution oxidation processes in urban areas. Concentration of atmospheric trace gasses involved in forming O_3 and NO_X change rapidly accompanied by a change of wind speed and wind direction, temperature, humidity and solar radiation. All these factors play crucial role in production and destruction of O_3 [11]. Usually NO_2 in the atmosphere comes from two sources, either directly from emission sources (primary pollutant) or from chemical reactions in the atmosphere [12]. Nitrogen monoxide (NO), in turn, is converted to NO_2 by reactions with proxy radicals (RO_2) or O_3. Proxy radicals are produced mostly by the reactions of hydroxyl radical (OH) with reactive hydrocarbons and CO and photolysis of aldehydes which have both natural and anthropogenic origins.

Nitrogen dioxide is then photolysis in the atmosphere, and the released atomic oxygen combines with molecule O_2 to form O_3 [13]. High concentrations of CO generally occur in areas with heavy traffic and congestions. The point sources of CO emission also include industrial processes. Non transportation fuel combustion 2006) [12]. The present of sulphur dioxide in air is related to the fuel combustion and industrial processes. Primary pollutant (CO, SO_2) concentrations are usually higher in cold months (winter) than in hot months. Where the concentrations of the secondary pollutants (NO_2 and O_3) are higher in summer than in winter months [14]. Overall all these pollutant concentrated in crowed street where the United Nation estimated that over 600 million people worldwide in urban areas are exposed to dangerous levels of traffic–generated air pollutants [15]. Methane is invisible, odorless, and combustible gas present in trace concentrations in the atmosphere, it enters the atmosphere from both natural (30 percent) and anthropogenic (70 percent) sources. Methane is twenty three times more potent as a greenhouse gas than carbon dioxide. Both gases are targeted for emissions reduction in the Kyoto Protocol [16, 17]. Particulate matter (PM) in urban areas is mainly made up of metals, organic compounds, materials of biological origin and elemental carbon [18]. The particle core, which often forms bulk of the urban particulate matter, mainly comprises elemental carbon. Many organic compounds can lead to mutations and can be cancerous. The effects of air pollutions on health are very complex, after being inhaled, they affect human health severely damaging the lungs and respiratory system [19]. In present work analyzes the temporal variation of measurements CO, SO_2, NO, NO_2, NO_X, O_3, CH_4, PM_{10} concentrations were performed for the first time in may – December 2010 and the results are present in this paper to find simple evaluate the state of ambient air in urban area of Baghdad city middle – east of Iraq, and focusing mainly on the following five objectives:

1. Establishing baseline pollutant concentration levels at daytime, which could be used in the future to assess the effectiveness of any implemented emission control strategies at this period.
2. Comparing the observed pollutant concentrations levels to the corresponding would ambient air standards.
3. Revealing the role of wind speed and other atmospheric element, especially in hot and cold months, to the pollutant levels.
4. Examine the domain concentrations of these pollutants through the frequency of occurrence.
5. Test the relationship between several pollutant, this can be used in interpreted the origin of these elements.

It should be noted that such studies are lacking from this region and is the first that examines pollutant concentrations because there is not continues record through 24 hours of pollutant concentration in this city, especially in the previous years.

2. Data collection

Baghdad city located in the middle –east part of Iraq has a population of more than 8000000 (according to 2010 estimation) and is regarded as a urbanized area with several industries, a large number of automobile and huge number of small and large electric generator, spread in different parts it, specifically after 2003 where there luck in electric generated, all of these can contribute to the production of local pollutant. The climate of this city is semiarid to arid (specifically 2010 year), with a low annual average precipitation of 0.7mm occurring mainly in winter months (November to April). Monthly mean values of temperature, relative humidity (RH%), atmospheric pressure, and accumulated precipitation in Baghdad during the year of period study (May-December 2010) are plotted in Figure 1. The monthly mean temperature exhibits a clear annual pattern with low values in winter 13°C and high values 38°C in summer following the common pattern found in the northern mid-latitudes. RH% illustrates an inverse annual variation with larger values in winter 56% and very low values in summer below 20%, which are indicative of an arid environment. The atmospheric pressure is generally steady 1015 hpa from January to April, and then decreases significantly during summer and increases again in autumn, see Figure 1, The summer low pressure 999.5 hpa in July is attributed to the Indian thermal low that extends further to the west over the arid environment of Iraq and middle east as a consequence of the south Asian monsoon system.

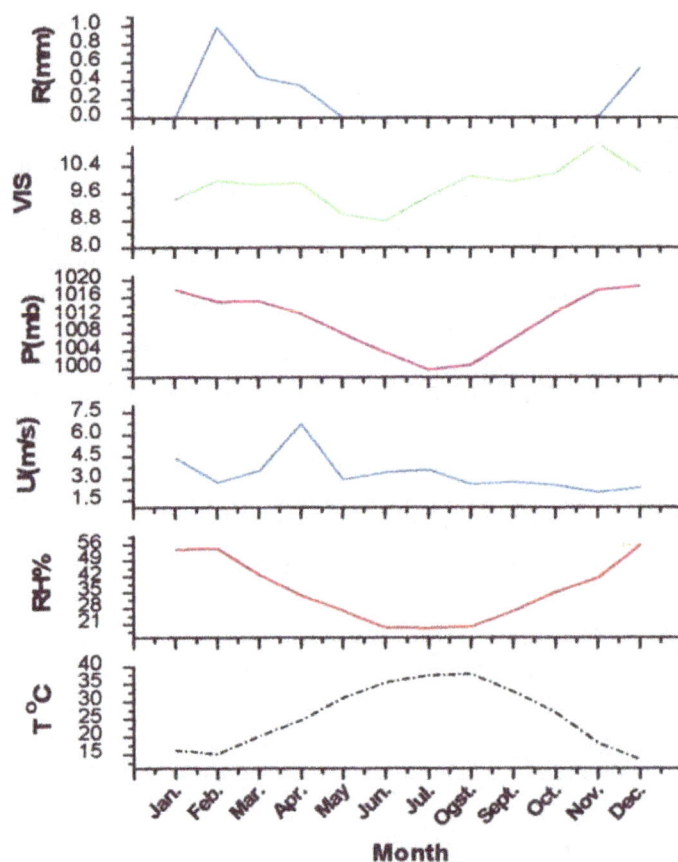

Figure 1. Monthly mean variation of meteorological variables in Baghdad, Iraq covering the period (January – December) 2010

In this study we taken the local recorded of daily hourly concentrations of SO_2, NO_X, NO, NO_2, CH_4, CO, PM_{10} and O_3 at near-surface level in city of Baghdad that have been systematically measured using environment pollution, AL-Jadriyah station (44.1 east, 33.3 north, 38.5m a.m.s.l where it located near residential site). The present measurements were carried out during the period May-December 2010 (total of 245 days). The measures of PM_{10} concentration were in milligrams per cubic meter, with a relatively high temporal resolution (1 hour) of recording. But the measurement of other chemical pollutions is recorded in ppm (part per millions) and the period of recorded is 30min, All center of the city, in a sparsely populated area without daily hour. The station is located at the outskirts center of city, in a sparsely populated area without any industries and direct influence by anthropogenic emissions. All the hourly measurement of this pollution was converted to daily average from which the monthly values and variations were obtained.

3. Result and discussion

3.1 Diurnal variability of pollutant concentrations

Daily mean pollutant concentration at period's daytime and from 2 May 2010 to 1 January 2011 are shown in Figure 2. We see that highest daily maximum values of the pollutant elements such as CO and CH_4 were 3.25, 1.9 ppm respectively this was at November 7/11/2010. On the other hand NO_X has highest concentration value at December 5/12/2010 about 0.23 ppm. But other pollutant element such as NO, NO_2, SO_2, O_3 have maximum values lived between (0.14 -0.18ppm), this refer to the changes of mean daily concentration of these element pollutant through period study is very low. Particular matter at diameter 10um (PM_{10}) make high mean daily value at 21/7/2010 about $1.6mg/m^3$.

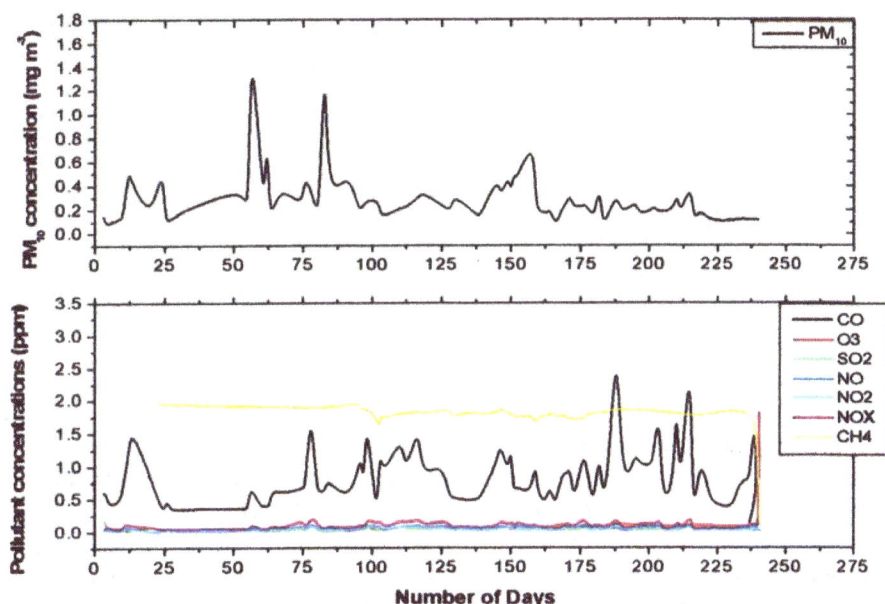

Figure 2. Daily pollutant concentrations at Baghdad during the period 1 May 2010 to 31 December 2010

This a threshold value of $1.62mg/m^3$ was chosen for the city of Baghdad in order to identify days with severe PM_{10} levels, while lowest daily value noted is at 2/11/2010. This mean that most highest value of PM_{10} is measured at summer season period (hot months) while lowest value is noted at winter season (cold days) this is parallel with domain atmospheric condition that is makes high frequency of dust storm mostly at summer season where its loaded by high concentration of PM_{10} because active wind through is period, days with severe PM_{10} concentrations are observed mainly in summer months on 21/7/2010, significant daily variability is observed especially for PM_{10} with several peaks and gaps attributed to the intensity of local emissions, regional meteorology, boundary layer dynamics, and long-range transported aerosols. This value can comparable in magnitude to those observed during an intense dust storm in Beijing [20]. This situation is different in the chemical pollutant element where we see that most of high concentration is domain in cold daily period, where wind speed is nearly calm. More explain to this circumstance we will take linear correlation coefficient between all the hourly concentration pollutant element at periods that represent winter, April, summer and autumn that extracted from period study, Table 1. At winter and specifically at December month, we see that there is a good values to correlation coefficient between PM_{10} and other chemical pollutant concentration such as CO, SO_2, NO, NO_X, PM_{10} levels in these seasons. This correlation coefficient values also founded between concentration of NO_X and other pollutant element such as SO_2, NO, NO_2 see Table 1 (winter period). Correlation coefficient between PM_{10} and chemical pollutant element is disappeared in other seasons period, April, summer, autumn period season see Table 1. But overall there is good correlation between NO_X, SO_2, NO, NO_2 in all seasonal daily periods and months, this may be due to the same references emission of this elements pollutant.

Table 1. Linear correlation coefficient among different pollutants in Baghdad urban area through the periods represents Winter, Spring, Summer and Autumn

Season		Pollutant concentrations							
		CO	O_3	SO_2	NO	NO_2	NO_X	CH_4	PM_{10}
Winter (Dec.)	CO	1	0.0556	0.5792	0.7780	0.8090	0.91541	0.4590	0.92497
	O_3		1	0.1169	-0.2685	0.30701	-0.07181	-0.28825	0.36641
	SO_2			1	0.25768	0.38534	0.59269	0.48941	0.81566
	NO				1	0.44406	0.92865	0.60292	0.65398
	NO_2					1	0.74423	0.12788	0.90042
	NO_X						1	0.54071	0.88528
	CH_4							1	-0.43833
	PM_{10}								1
		CO	O_3	SO_2	NO	NO_2	NO_X	CH_4	PM_{10}
Spring (April, May)	CO	1	-0.39861	0.25443	0.2544	0.60755	0.40598	0.23456	0.276
	O_3		1	-0.32433	-0.33192	-0.46832	-0.38853	0.04009	-0.16735
	SO_2			1	0.23768	0.31123	0.23752	0.70193	-0.24429
	NO				1	0.69567	0.87103	0.52626	-0.24098
	NO_2					1	0.78555	0.61677	-0.1137
	NO_X						1	0.55213	-0.22242
	CH_4							1	-0.35493
	PM_{10}								1
		CO	O_3	SO_2	NO	NO_2	NO_X	CH_4	PM_{10}
Summer (June, July, Aug.)	CO	1	0.10117	0.61214	0.57508	0.82317	0.69975	0.15713	-0.0434
	O_3		1	0.29566	-0.1311	0.14192	-0.06776	0.27855	-0.23331
	SO_2			1	0.57266	0.6731	0.64087	0.07754	-0.09658
	NO				1	0.68767	0.94615	-0.16564	0.02888
	NO_2					1	0.86232	0.00885	-0.26906
	NO_X						1	-0.14442	-0.08473
	CH_4							1	0.27479
	PM_{10}								1
		CO	O_3	SO_2	NO	NO_2	NO_X	CH_4	PM_{10}
Autumn (Sep., Oct., Nov.)	CO	1	0.10169	0.26392	0.39961	0.77345	0.60138	0.34074	0.26123
	O_3		1	0.00316	-0.25319	0.08903	-0.16759	0.06095	-0.23932
	SO_2			1	0.5709	0.52155	0.61913	0.15172	0.18205
	NO				1	0.49687	0.94048	-0.01066	0.07307
	NO_2					1	0.7571	0.19261	0.31387
	NO_X						1	0.05295	0.18013
	CH_4							1	0.1261
	PM_{10}								1

3.2 Monthly variability in pollutant concentration

Figure 3 illustrate the monthly variation of all pollutant of study that recordings during the period from May to December 2010. The vertical bars in this figure is refer to the standard deviation from the monthly mean and are indicative of the day-to-day variation. Thus, it is observed that the months with the highest pollutant concentration level also have largest standard deviations. This is occurs mainly in summer month specifically in August month where we noted the high mean concentration in pollutant concentration of CO, NO_X, CH_4, NO, NO_2 and SO_2 where it have 1.045±0.449, 0.1141±0.0425, 1.8144±0.0914, 0.07702±0.0295, 0.07018±0.0222 and 0.05333±0.0304 respectively this month period have high frequency wind speed events, in other hand have high amount of anthropogenic activity that is due to the emission of large amount of pollutant concentration from the vehicles and electric generated specially there is high needed to electricity at this period and the luck of the national electricity that supplied to the people of Baghdad city, Table 2. Although the wind speed is very active at this period or at this month from every year. On the other hand there is also large mean concentration in the November specifically at the ozone O_3 where its have largest mean value about 0.1024 ± 0.0163, CO have also large monthly mean and large standard deviation at this month 1.211 ± 0.7448, Table 2.

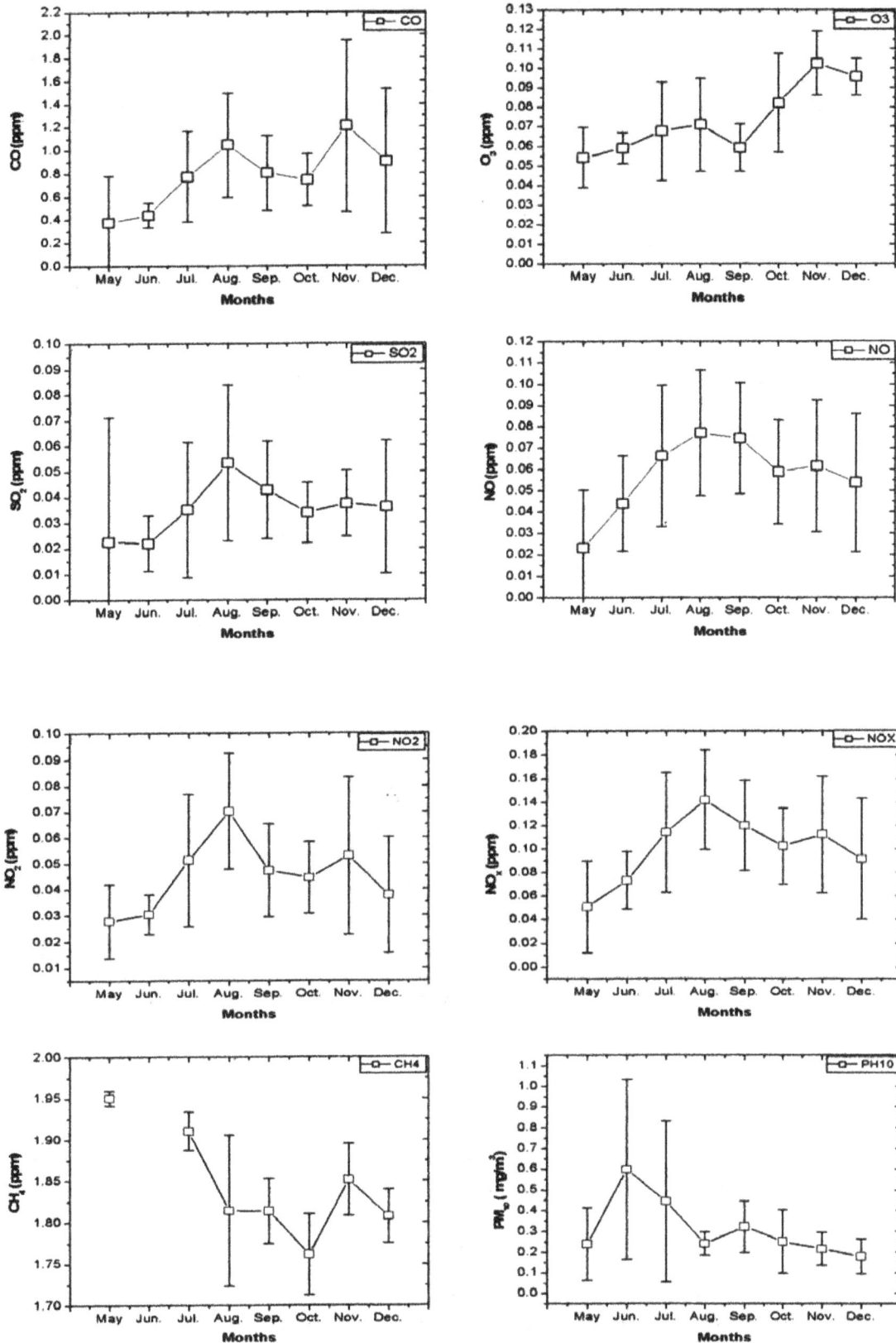

Figure 3. Monthly mean variation values of CO, O_3, SO_2, NO, NO_2, NO_X, CH_4 and PM_{10} at Baghdad during period May-December 2010

Table 2. Monthly mean, maximum and minimum pollutant concentration in Baghdad during period 1 May to 31 December 2010

		MONTHS							
		May	June	Jul.	Aug.	Sep.	Oct.	Nov.	Dec.
CO (ppm)	mean	0.37549	0.4392	0.77265	1.04562	0.8049	0.7425	1.211	0.9047
	min	0.29375	0.10619	0.39047	0.4497	0.3243	0.225	0.393	0.353
	max	1.735	0.64118	1.98163	1.8433	1.3949	1.183	3.230	2.8131
O_3 (ppm)	mean	0.05421	0.05903	0.06774	0.0709	0.0592	0.0821	0.102	0.0956
	min	0.01531	0.00797	0.02526	0.0238	0.01206	0.0253	0.0163	0.00939
	max	0.0965	0.07388	0.13413	0.1157	0.0796	0.1208	0.147	0.1099
SO_2 (ppm)	mean	0.02262	0.02198	0.03506	0.05336	0.04284	0.0338	0.0373	0.0360
	min	0.00975	0.01033	0.0063	0.0225	0.0163	0.0117	0.0128	0.0101
	max	0.18456	0.0432	0.10663	0.1133	0.0864	0.0613	0.068	0.1044
NO (ppm)	mean	0.02311	0.04385	0.0663	0.07702	0.0745	0.0587	0.0615	0.0537
	min	0.00318	0.02089	0.00843	0.023	0.0261	0.0243	0.0065	0.01
	max	0.0969	0.08515	0.13575	0.1251	0.1138	0.1204	0.120	0.1203
NO_2 (ppm)	mean	0.02778	0.03045	0.05125	0.07018	0.0473	0.0444	0.0529	0.0378
	min	0.01009	0.00762	0.02559	0.02229	0.0180	0.0138	0.023	0.0222
	max	0.0684	0.04413	0.1181	0.1135	0.0836	0.0765	0.143	0.1081
NO_X (ppm)	mean	0.05089	0.07342	0.1142	0.14186	0.1199	0.1024	0.112	0.0916
	min	0.01327	0.02435	0.04286	0.0425	0.0383	0.0323	0.0295	0.04
	max	0.1653	0.1133	0.2245	0.20975	0.1737	0.196	0.2048	0.228
CH_4 (ppm)	mean	1.95017		1.91048	1.81441	1.8139	1.7618	1.851	1.807
	min	0.00914	0	0.02337	0.09148	0.03942	0.0491	0.0433	0.0321
	max	1.966	0	1.93386	1.9547	1.8915	1.836	1.9057	1.867
PM_{10} (mg/m^3)	mean	0.23983	0.599	0.4452	0.2408	0.3218	0.2504	0.215	0.179
	min	0.089	0.165	0.1784	0.05567	0.1249	0.091	0.072	0.0839
	max	0.60017	1.4	1.62	0.3455	0.5056	0.814	0.3306	0.4012

The monthly mean recorded at the month of CH_4 is large at may and July but June don't have any monthly record. On the other hand the basic monthly of June and July have large monthly mean concentration about 0.599 ± 0.435 and 0.445 ± 0.387 mg/m^3. In this months is the periods with the most frequent dust-storm evens, one can therefore conclude that the intense dust-storms taking place on specific days during summer are predominantly responsible for large day-to-day variations at all PM concentrations. PM_{10} concentrations during winter months are significantly lower with 0.179 ± 0.0839 mg/m^3, this measured at December. The difference in the monthly mean variation between chemical air pollutant concentration and particular matter at diameter 10um suggests the differences in source regions for these pollutant components. Where the main anthropogenic source of PM_{10} and other pollutant element in Baghdad urban environment can be return to vehicle traffic, fossil-fuel combustions, industrial activities that release a large amount of near- surface anthropogenic components. In addition, the boundary layer mixing height is lower in winter and traps the pollutants near the ground as a result of temperature inversions. All the above explain the high concentration of chemical air pollutant at a cold daily period months, in contrast during summer months, thermal heating at the surface and the increase of mixing layer height favors buoyancy and the dilution of anthropogenic pollutant. On the other hand dust storm contribution to the total PM_{10} concentration is also expected to originate from Aeolian and traffic–driven resuspension. Some studies conduct in other urban environments, in Turkey [21] and in Athens [22] and other places in world found contrary to our results, in PM_{10}, that in winter both PM concentrations and concentration pollutant were high which was attributed to large use of fossil fuels in winter. Where monthly mean PM_{10} concentration in Athens have ranging from 60.3um/m^3 (January) to 88.9ug/m^3 (December). Its concluded that the city of Baghdad experiences much higher pollutant concentration levels. This is not only the case for summer element. This emphasizes the fact that air pollutant quality concentrations over Baghdad can be regarded as a real environment problem that possess a serious risk to quality of life and endangering human health.

3.3 Frequency of occurrence

Frequency of occurrence for the pollutant CO, O_3, SO_2 and NO for all period of study is depicted in Figure 4, we see that most concentration of CO is bounded between 0-1 ppm. Where frequency have 300-400 but in O_3 its between 0.05-0.1 PPM and concentrated in frequency between 200-300. SO_2 most of concentration of this pollutant is concentrated from 0-0.05PPM in NO the matter is different where we see that there is not peak frequency for this concentration its distributed at several concentration from 0-0.1 PPM at this period of study. This concentration can be change between several seasons, the red values in the Table 1 show the highest seasonal hourly frequency percent bands of the pollutant CO, O_3, SO_2, NO, NO_2, NO_X and PM_{10} respectively at each season to compared the situation at each season for example CO that consider important gas to poisons gases and its represent the measured of unhealthy of air quality. We see that most of highest concentration and its frequencies distribution shifts towards larger values concentrated at summer seasons June, July, and Aug. about 0.25-1.25 PPM and frequency percent 82.6%. While other season have smaller than concentration and frequencies at this period, see the red values in the Table 1. In winter the frequency distribution shifts towards large values, of O_3 even above 61.2% in the range concentration about 0.085-0.105 PPM. While in autumn highest percent frequency is 33.6% at 0.11-0.13 PPM, see the red values in the Table 1.

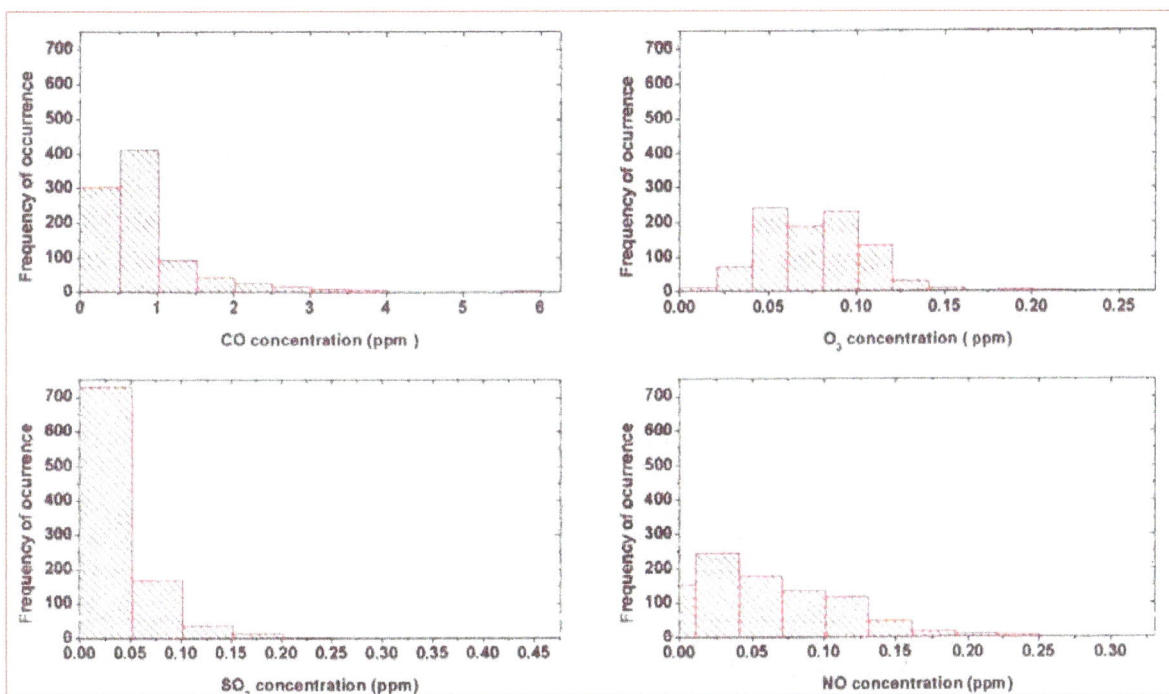

Figure 4. Frequency distribution of CO, O_3, SO_2, NO hourly concentration at all period of study in Baghdad (AL-Jadriyah station)

Regarding the frequency distribution of SO_2 all season where shifts towards lower except Spring (May) 88.2% at range concentration 0.025-0.125 PPM. On other hand concentration of nitrate oxide where have different concentration an frequency at several season. Overall NO and NO_2 were nearly similar at all season but NO_X have large frequency at summer other pollutant concentration CH_4 and PM_{10} also have summer month high concentration frequency about 69% at the concentration range 0.1-0.5mg/m^3.

4. Conclusion

Change in population, economic development, energy consumption, and technology can have consequences for air quality in Baghdad city and other megacities. Air pollution occurs as a result of a combination of several factors. Firstly, pollutants need to be emitted in to the atmosphere. Various pollutants may then undergo chemical reactions among themselves and with other species. The pollutants move as a result of transport by the wind, and they diffuse as a result of turbulence in air. The dispersion of the pollutants is a result of two inter-connected factors: local meteorology conditions, and local topography. Severe air pollution occurs when air pollutants are trapped within a region of the lower atmosphere close to their emission source. Megacities are particularly susceptible to air pollution because of the large quantity of pollutants that industry, motor vehicles and individuals will emit in to the air shed unless stringent control measures are in place. in Asia, cities such as Jakarta, manila and Bangkok are renowned for their poor air quality, in the Americas, Mexico city has the worst air quality of all the megacities whereas Los Angeles, which is the city in which photochemical smog was first identified, continues to have problems (Earth watch,1992). Air pollution also occurs in cities with lower populations than megacities, especially if there is substantial use of motor vehicles with unregulated emissions.

Temperature inversions are conditions in which the temperature of the atmosphere increases with altitude in contrast to the normal decrease with altitude. Strictly, a temperature inversion occurs when the potential temperature gradient is positive. When a temperature inversion occurs, cold air underlies warmer air. The greater density of the cold air inhibits vertical mixing. During a temperature inversion, air pollution released in to the atmosphere's lowest layer is trapped there and can be removed either by strong horizontal winds, or by strong irradiative heating. Temperature inversions may occur during the passage of a cold front or result from the inversion of sea air by a cooler onshore breeze. Overnight irradiative cooling of surface air often results in a nocturnal temperature inversion that is dissipated after sunrise by the warming of air near the ground. A more long-lived temperature inversion accompanies the dynamics of large high – pressure systems. Descending air near the center of the high-pressure system produces a warming (by adiabatic compression), causing air at middle altitudes to become warmer than the surface air, which is isolated within a boundary layer. Rising currents of cool air lose their buoyancy and are thereby inhibited from rising further when they reach the warmer. Less dense air in the upper layers of a temperature inversion. Because high-pressure systems often combine temperature inversion conditions and low wind speeds, their long residency over an industrial area usually results in episodes of severe smog.

References

[1]   Bernstein JA, Alexis N, Barnes C, Bernstein I.L., Bernstein J. A., NEL A., Peden D., Diaz-Sanchez D, Tarlo S.M., Williams P.B. Health Effect of air Pollution. Journal of allergy and clinical immunology, 114(5): 1116-1123,2004.

[2]   WHO: Air Quality and health-fact sheet No. 313, 2008.

[3]   WHO: Air Quality guideline – global update 2005, Europe WHO –Rof: world health organization – regional office for Europe, 9-28, 2006.

[4]   The Top 10 causes of death-WHO fact sheet No. 310.

[5]   Gauderman WJ, et. Al., Effect of exposure to traffic on lung development from 10 to 18 years of Age: ACohort study, 369(9561): 571-577, lancet 2007.

[6]   Hoffmann b, et. al. Residential exposure to traffic is associated with coronary Atherosclerosis, 116(5) : 489-496, circulation 2007.

[7]   Fleming, J, Stern, R. & Yamartino, R. J. A new air quality regime classification scheme for O3, NO_2, SO_2 and PM10 Observations Site, Atmospheric environment, 39, 2005.

[8]   National research Council, us Rethinking the ozone problem in urban and regional air pollution, Washington DC. National Academy Press. 1991.

[9]   Vukovich, F. M. & Sherewell, J. An examination of the relationship between certain Meteorological parameters and surface ozone variations in the Baltimore–Washington corridor, atmospheric environment, 37,971-981, 2003.

[10]   Markovic, D. M. & Markovic, A M. The relationship between some meteorological parameters and the tropospheric concentrations of ozone in the urban area of Belgrade. Journal of the Serbian chemical society, 70, 1478-1495, 2005.

[11]   Minoura, H., Some Characteristics of surface ozone concentrations observed in an urban atmosphere, atmospheric research, 51, 153-169, 1999.

[12]   Han, x. & Naeher, P.L. A review of traffic-related air pollution exposure assessment studies in the developing world, environment international, 32, 106-120, 2006.

[13]   Aneja, P. V., Kim, D. S. & Chameides, W. L. Trends and analysis of ambient NO, NOX, CO, and ozone concentrations in Raleigh, north Carolina, chemosphere, 34, 611-623, 1996.

[14]   Barrero, M, A., Grimalt, J., O., & Canton, L. Prediction of daily ozone concentrations and maxima in urban atmosphere, chemo metrics and intelligent laboratory systems, 80, 67-76, 2006.

[15]   Caciola, R. R., Sarva, M., & Pasola, R. Adverse respiratory effects and allergic susceptibility in relation to particulate air pollution : flirting with disasters. Allergy, 657-281, 2002.

[16]   Delong, Eward F. " resolving a methane mystery", Nature 407: 577-579, 2000.

[17]   Simpson, Sarah. "Methane fever", scientific American 282(2):24-27, 2000.

[18]   Tasic, M., et. al. Physic –Chemical characterization of PM10 and PM2.5 particles in the Belgrade urban area. act chemical Slovenia, 53, 401-405. 2006.

[19]   Pandey, J. S., Kumar, R. & Devotta, S. Health risks of NO_2, SPM and SO_2 in Delhi (India), atmospheric environment, 39, 6868-6874, 2005.

[20]   Zhao X, Zhuang G. Wang Z., Sun Y. Wang y. Yuan H., Variation of sources and mixing mechanism of mineral dust with pollution aerosols in a super dust storm – revealed by the two peaks of a super dust storm in Beijing, atoms res 84:265-279, 2007.

[21]   Akyuz M, Cabuk H., meterological variations of PM2.5/PM10 concentrations and particle-associated polycyclic aromatic hydrocarbons in the atmospheric environment of zonguldak turkey,j hazard mater 170: 13-21, 2009.

[22]   Chaloulakou a, et. al. Measurements of pm10 and pm2.5 particle concentrations in Athens Greece, atmos envir. 37: 649-660, 2003.

15

CFD modeling of dust dispersion through Najaf historic city centre

Maher A.R. Sadiq Al-Baghdadi

CFD Center, International Energy and Environment Foundation, Najaf, P.O.Box 39, Iraq.

Abstract

The aim of this project is to study the influences of the wind flow and dust particles dispersion through Najaf historic city centre. Two phase Computational Fluid Dynamics (CFD) model using a Reynolds Average Navier Stokes (RANS) equations has been used to simulate the wind flow and the transport and dispersion of the dust particles through the historic city centre. This work may provide useful insight to urban designers and planners interested in examining the variation of city breathability as a local dynamic morphological parameter with the local building packing density.

Keywords: Dust dispersion, Particle transport, CFD, Airflow through a city, Najaf historic city centre.

1. Introduction

The fluid dynamics of airflow through a city controls the transport and dispersion of airborne contaminants. It also controls environmental air quality, wind forces on buildings, and the ambient noise level due to the winds. These are urban aerodynamics problems primarily, not meteorology. The space scales are short, a few meters to at most a few kilometers. The average airflow, the dynamic fluctuations, and the building-scale turbulence are all closely coupled to the complicated geometry. Buildings create large "rooster-tail" wakes; they shed vortices dynamically and support complex recirculation zones; there are systematic fountain flows up the backs of tall buildings; and dust in the wind can move perpendicular to or even against the locally prevailing wind direction. In principle, meteorology provides the aerodynamic boundary conditions for an urban region and influences the airflow in a city but the weather can be treated as known over times of a few minutes to a fraction of an hour. Urban aerodynamics is driven by a deep, stratified urban boundary layer with significant wind fluctuations. We require time-dependent, Computational Fluid Dynamics (CFD) to predict accurately these unsteady, obstructed, buoyant flows and the dynamic contaminant plumes that they drive. In typical urban scenarios most particulate and gaseous contaminants behave similarly with respect to the overall transport and dispersion but the full physics of multi-group particle and droplet distributions are required for some problems [1].

2. Description of the study area

Najaf is one of the sacred cities of Iraq. The Najaf area is located 30 km south of the ancient city of Babylon and 400 km north of the ancient Biblical city of Ur. It is located on the edge of western plateau of Iraq, at southwest of Baghdad the capital city of Iraq, with 160 km far from the capital. It is raised upon sea level with almost 70 meters, and is situated on the longitude of 19 degree and 44 minutes, as well as on the latitude of 31 degree and 59 minutes [2].

3. CFD modeling

The wind environment over Najaf historic city centre is governed by the conservation laws of mass and momentum. The flow is assumed to be incompressible Newtonian fluid with constant density. Based on the above assumptions, the Reynolds average Navier–Stokes (RANS) equations are as [3, 4];

$$\rho \frac{\partial \mathbf{U}}{\partial t} + \rho \mathbf{U}.\nabla \mathbf{U} + \nabla.\overline{(\rho \mathbf{u'} \otimes \mathbf{u'})} = -\nabla P + \nabla.\mu(\nabla \mathbf{U} + (\nabla \mathbf{U})^T) + \mathbf{F} \tag{1}$$

$$\rho \nabla.\mathbf{U} = 0 \tag{2}$$

where $\mathbf{U}$ is the averaged velocity field and $\otimes$ is the outer vector product.
The turbulent viscosity is modeled using k-ε model as;

$$\mu_T = \rho C_\mu \frac{k^2}{\varepsilon} \tag{3}$$

where C_μ is a model constant, k is the turbulent kinetic energy, and ε is the turbulent dissipation rate.
The transport equation for k is;

$$\rho \frac{\partial k}{\partial t} + \rho \mathbf{u}.\nabla.\left(\left(\mu + \frac{\mu_T}{\sigma_k}\right)\nabla k\right) + P_k - \rho \varepsilon \tag{4}$$

where the production term is;

$$P_k = \mu_T \left(\nabla \mathbf{u}:(\nabla \mathbf{u} + (\nabla \mathbf{u})^T) - \frac{2}{3}(\nabla.\mathbf{u})^2\right) - \frac{2}{3}\rho k \nabla.\mathbf{u} \tag{5}$$

The transport equation for ε is;

$$\rho \frac{\partial \varepsilon}{\partial t} + \rho \mathbf{u}.\nabla \varepsilon = \nabla.\left(\left(\mu + \frac{\mu_T}{\sigma_\varepsilon}\right)\nabla \varepsilon\right) + C_{\varepsilon 1}\frac{\varepsilon}{k}P_k - C_{\varepsilon 2}\rho \frac{\varepsilon^2}{k} \tag{6}$$

where $C_{\varepsilon 1}$, $C_{\varepsilon 2}$, σ_ε, σ_k are model constant.
The particle momentum comes from Newton's second law;

$$\frac{d}{dt}\left(m_p \mathbf{V}\right) = \mathbf{F}_D + \mathbf{F}_g + \mathbf{F}_{ext} \tag{7}$$

where m_p is the particle mass [kg], $\mathbf{v}$ is the velocity of particle [m/s], $\mathbf{F}_D$ is the drag force [N], $\mathbf{F}_g$ is the gravitational force vector [N], $\mathbf{F}_{ext}$ is any other external force [N].
The drag force ($\mathbf{F}_D$) is defined as;

$$\mathbf{F}_D = \left(\frac{1}{\tau_p}\right)m_p(\mathbf{u} - \mathbf{v}) \tag{8}$$

where τ_p is the particle velocity response time [sec], $\mathbf{u}$ is the fluid velocity [m/s]
The fluid velocity used in the drag force in turbulent dispersion becomes;

$$\mathbf{u} = \mathbf{U} + \mathbf{u'} \tag{9}$$

where $\mathbf{U}$ is the mean velocity and $\mathbf{u'}$ is a turbulent fluctuation defined as;

$$\mathbf{u'} = \varphi\sqrt{\frac{2k}{3}} \tag{10}$$

where k is the turbulent kinetic energy, and φ is a normally distributed random number with zero mean and unit standard deviation.
The gravity force is given by;

$$\mathbf{F}_g = m_p \mathbf{g}\frac{(\rho_p - \rho)}{\rho_p} \tag{11}$$

Where ρ_a is the particle density [kg/m³], ρ is the density of the surrounding fluid [kg/m³], and $\mathbf{g}$ is the gravity vector.

4. Results

The resolved built area and the area of interest are depicted in Figure 1. Najaf historic city centre building geometry used in CFD model is shown in Figure 2. Figure 3. shows the flow direction and wind speeds in the historic city centre. Figure 4. Shows dust particles dispersion in the historic city centre. Finally, Figure 5. Shows the dust particles in the wind over the city. The source was an instantaneous release of a neutrally buoyant tracer dust particles at ground level in the historic city centre. The frames show relative tracer concentrations at 2, 3, 4, 5, 6 and 8 minutes after release.

Figure 1. Satellite image of Najaf historic city centre [Google earth, 2014].

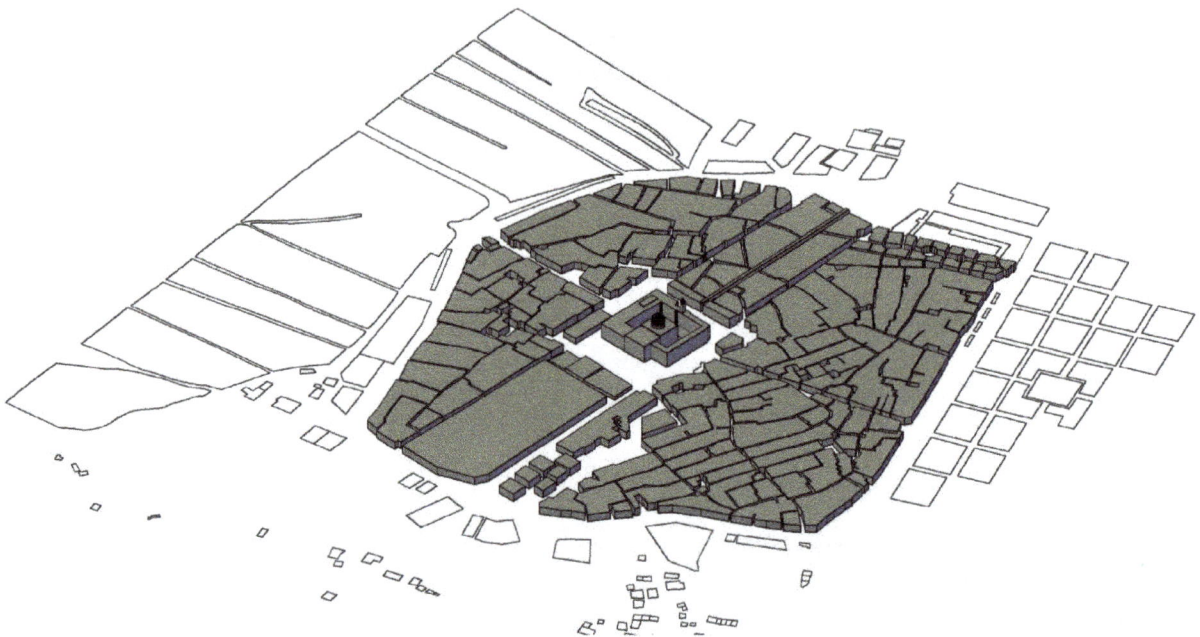

Figure 2. CAD model for a wind flow analysis of Najaf historic city centre.

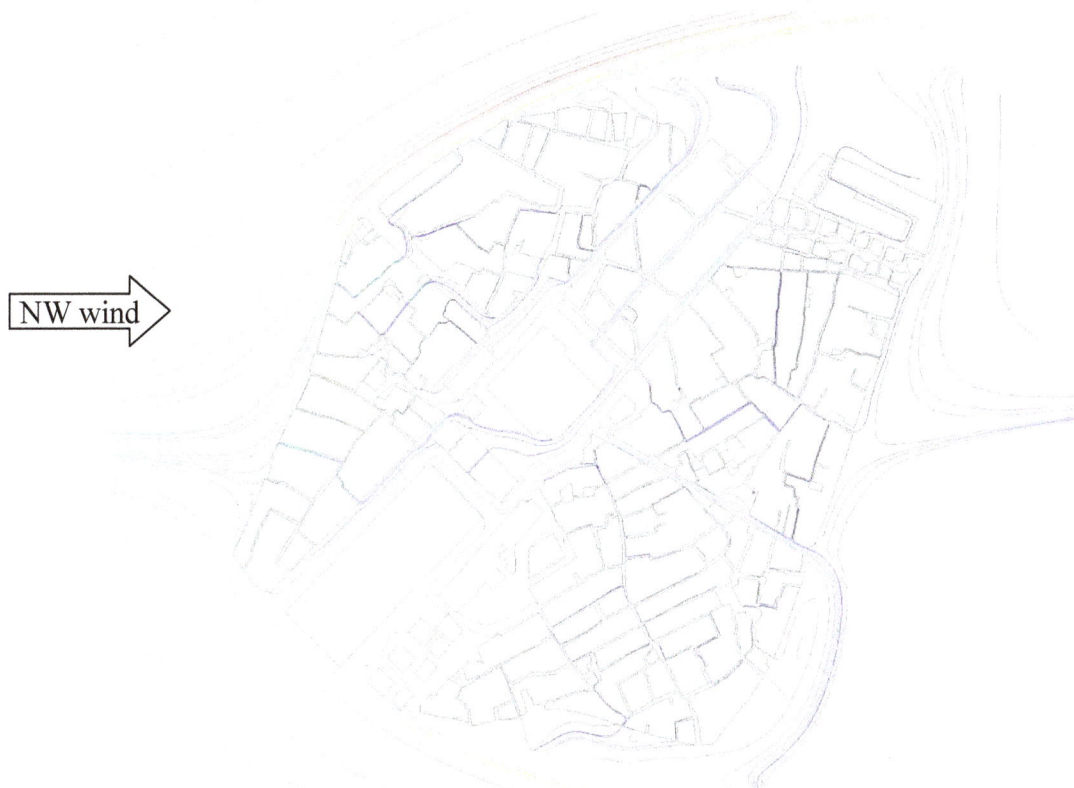

Figure 3. Plan view of streamlines showing flow direction and wind speeds in the historic city centre.

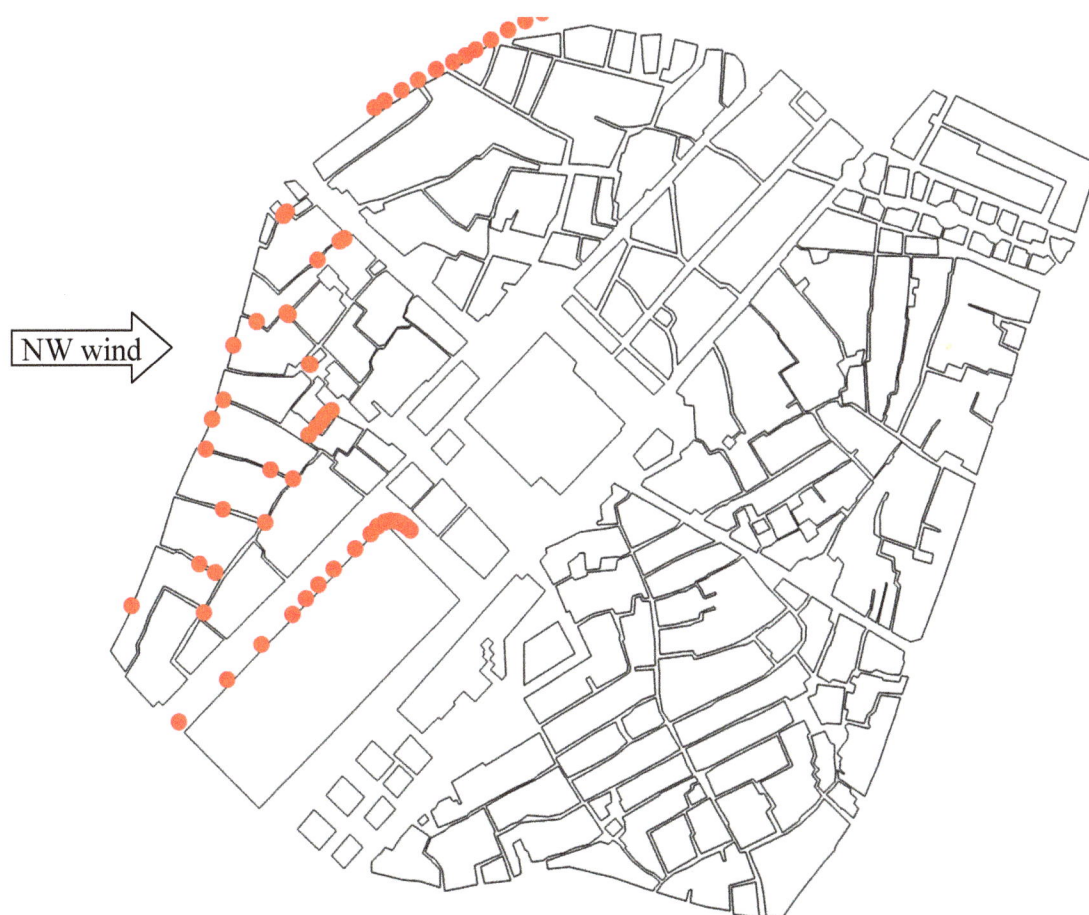

Figure 4. Dust particles dispersion through the historic city centre.

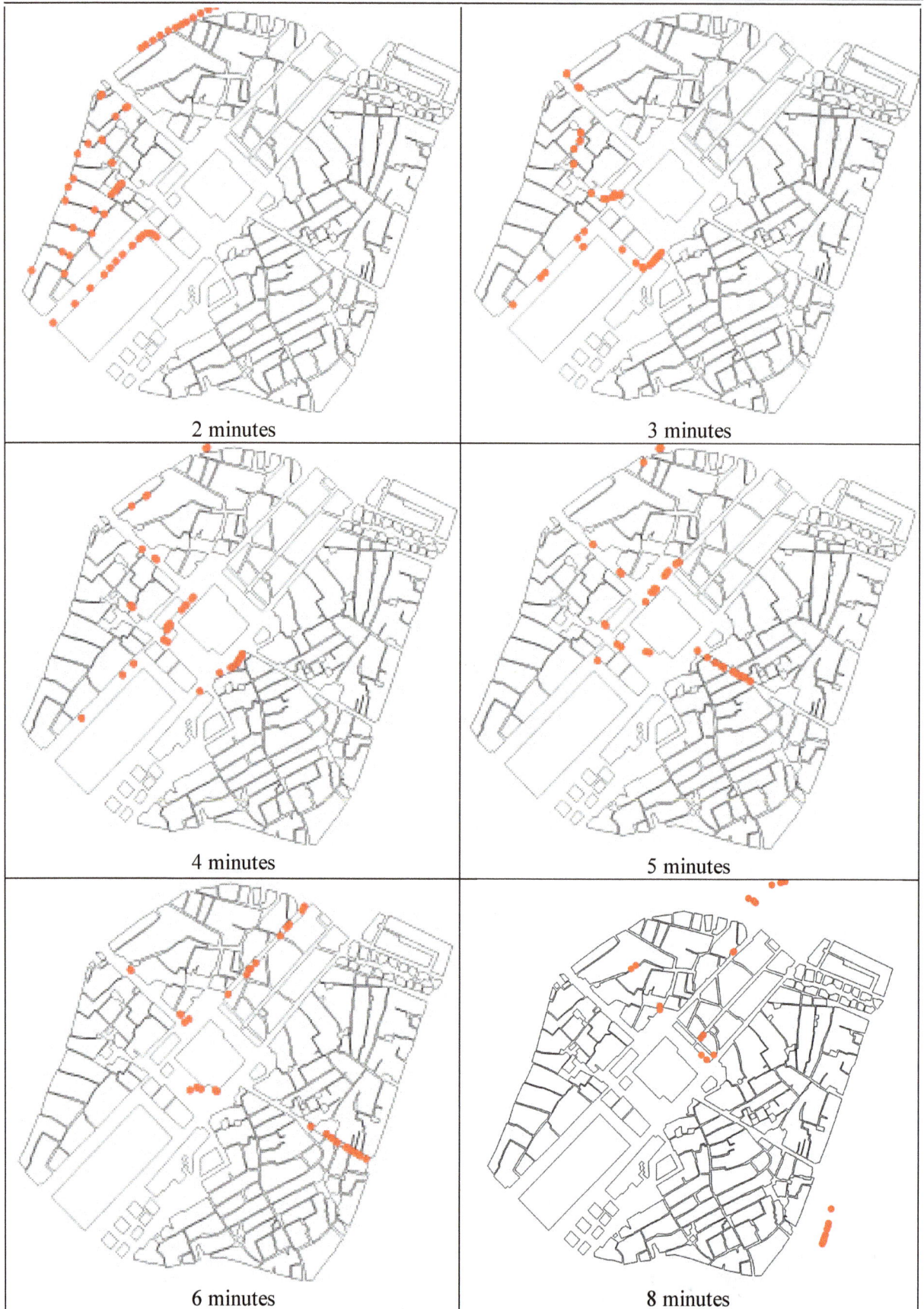

Figure 5. Visualizations of the dust particles in the wind through the city. The source was an instantaneous release of a neutrally buoyant tracer dust particles at ground level in the historic city centre. The frames show relative tracer concentrations at 2, 3, 4, 5, 6 and 8 minutes after release.

5. Conclusion

In conclusion, CFD benefits this project by cutting down the time and costs spent in parametric, case and design exploration studies. The wind flow and dust particles dispersion through a realistic complex urban geometry were examined using numerical results from a CFD simulation.

Acknowledgements

This work is part of an R&D project conducted at the CFD center in International Energy and Environment Foundation for advanced studies in Najaf historic city centre.

References

[1] Jay P. Boris. Dust in the Wind: Challenges for Urban Aerodynamics. American Institute of Aeronautics and Astronautics 2005, (AIAA-2005-5393).

[2] IEEF heritage group. Najaf heritage at risk. IEEF 2013, ISBN: 978-1490932651.

[3] Ioannis Panagiotou, Marina K.-A. Neophytou, David Hamlyn, Rex E. Britter. City breathability as quantified by the exchange velocity and its spatial variation in real inhomogeneous urban geometries: An example from central London urban area. Science of the Total Environment 2013, 442, pp.466–477.

[4] Priyadarsini Rajagopalan, Kee Chuan Lim, Elmira Jamei. Urban heat island and wind flow characteristics of a tropical city. Solar Energy 2014, 107, pp.159–170.

16

Experimental analysis of airtightness and estimation of building air infiltration using two different single zone air infiltration models

Tijo Joseph, Animesh Dutta

School of Engineering, University of Guelph, Guelph, Ontario, Canada.

Abstract

Building air leakage can contribute significantly to the energy consumption of a building. This paper presents the airtightness performance of a campus building located in Ontario, Canada. The air leakage rate through the building envelope was measured under stilted depressurization conditions following the ASTM E-779 standardized test method. With this test derived empirical leakage flow co-efficient and leakage flow exponent measures, the air infiltration rate for the building under varying wind and outside temperature conditions was calculated using two different single zone air infiltration models – the Lawrence Berkeley Laboratory model and the Alberta Air Infiltration model thus also allowing for a comparison of the results between the two mathematical models.

Keywords: Airtightness; Building air infiltration; Air leakage testing; Single zone infiltration modelling; LBL model; AIM-2 model.

1. Introduction

A building is associated with energy inputs throughout its lifetime but most significantly, during its use phase [1]. In Canada, buildings are estimated to be responsible for around 50% of the total energy used [2]. Importantly though, a substantial amount of energy consumption during a building's operational phase is lost as a waste stream owing to inherent equipment inefficiencies, inadequacies in the design and implementation of the building construction and energy systems, improper maintenance, and not least irresponsible human behavior. On the other hand, these factors can lead to a higher than best case scenario building energy demand. Building air leakage is an important pathway that can result in energy loss or contribute to a higher energy consumption. It is only pertinent therefore to qualify and quantify the air leakage in buildings.

Primarily, there are three means by which an air exchange can occur between the interior and exterior of a building envelope; by natural ventilation, by infiltration (or exfiltration) or by mechanical ventilation. Natural ventilation, typically intentional, allows for outdoor air ingress into the building through fenestration. Mechanical ventilation can facilitate the exchange of air across the envelope with the operation of an exhaust or supply fan. Unlike these two means, infiltration or exfiltration is unplanned for and represents the leakage of air through openings in the building fabric. While infiltration represents flow from the exterior of the building to the interior, exfiltration stands for the reverse. In this paper, infiltration may be used to mean both infiltration and exfiltration.

Various studies have reported the impact of air leakage on building energy demand. In a parametric computer simulation study of residential building prototypes in the United States (USA), the authors estimated infiltration as one of the two major components of heat loss from buildings [3]. In terms of heat gain, the same study estimated the contribution from infiltration as around 16% [3]. In a study based in Finland, infiltration was reported to contribute to 15 to 30% of the energy consumed for space heating purposes [4]. A thirteen country European study attributed around 53% of space heating as going into leakage loss [5]. In a study of commercial buildings in the USA, the contribution of infiltration to energy demand was reported as ranging from 10% to 42% with the higher impact observed in cold climates[6]. In yet another study on commercial building prototypes, the results indicated that infiltration was one of the biggest contributors to heating loads [7].

Evidently, infiltration or exfiltration can impact and is a major source of the heating and to a lesser extent the cooling load requirements of a building [6, 8-10]. In the case of cooling loads though, infiltration under the right circumstances can offset some of the load requirement [9]. Whether free cooling provided by infiltration can be availed of is a matter of local climatic conditions [9]. For example, in regions where the humidity is very high, moisture laden air infiltration can result in the build-up of moisture levels within the building [9]. Ultimately, this may warrant interior dehumidification.

Air infiltration under conducive conditions can also indirectly result in damage to components of the building envelope. Air laden with moisture can seep into wall assemblies or into the interior of the building thereby contributing to potential moisture related issues like mold or rot formation [11, 12]. The contribution of infiltration to the ventilation requirements of building occupants and also to indoor air quality(IAQ) is yet another factor that underscores the relevance of airtightness studies for buildings [8, 10, 12]. In general though, it is desirable to reduce air infiltration in buildings as it can significantly contribute to the energy consumption of a building [13]. This narrates the importance of studying the air leakage characteristics of buildings. This paper presents the results from an experimental analysis of the airtightness of a campus building and further, also estimates the air infiltration rate for the building under varying climatic conditions.

2. Building details

The campus building under investigation is a historic two storied building with a look-out basement and was built circa 1882. The building is located in the Canadian province of Ontario (Figure 1). It has a gross floor area of 3478 square feet (sq.ft.) and a net floor area of 3090 sq.ft and can be classed as similar to a residential type building. The building height is around 7 meters (m) and the base shape is rectangular with a length to breadth sizing of 10.3m by 14.8m. The total conditioned volume for the building is around 596 cubic meters (m^3). The building is oriented in the north-east direction with a site altitude of 338m, and latitude and longitude references as 43° 31'53.03" and 80°13'34.17" respectively. The building is not serviced by any air conditioning or ventilation equipment and thus air ducts are not present.

Figure 1. Campus building located in Ontario, Canada

3. Pathways, drivers &characteristics of air leakage in buildings

There are various elements that shape the airtightness of a building and this includes the building foundation type, ventilation scheme, number of floors, how complex the building envelope is and not least the construction quality [14, 15]. The air leakage pathways for a building can be in the form of cracks in the building envelope, holes in the components of the building fabric or gaps in the interface between building elements. Typically, air leakage can occur through the walls, at the intersection between the wall and the floor or ceiling, at interfaces to the attic, basement or crawlspace and also at spots where non-effectively sealed off penetrations run through the air barrier [14, 16]. Air leakage can also occur around and through fenestration, around electrical switches and power outlets, through the fireplace chimney, and also through the heating, ventilation and air-conditioning (HVAC) ducts [14]. Table 1 provides an estimate of the dispersion of air leakage across different elements of a building as cited in the ASHRAE handbook of fundamentals [17].

Table 1. Air leakage distribution across building elements [17]

Component description	% Distribution range
Walls	18% − 50%
Ceiling	3% to 30%
HVAC	3% − 28%
Fenestration	6% - 22%
Fireplace	0% - 30%
Exhaust vents in conditioned spaces	2% - 12%

Building air leakage is fundamentally driven by a pressure or temperature differential or, a combination of both, between the building indoor and outdoor environments. In any case, this leads to a resultant pressure differential across the building envelope and is a result of one or a combination of the following occurrences: wind blowing onto a building facade, the stack effect phenomenon or the operation of in-house mechanical devices such as an exhaust fan [11]. Exhaust fans for example, in the process of pulling out and exhausting interior air, can result in depressurizing the building interior and consequently pulling in outside air into the building. Stack effect is the result of a temperature differential between the exterior and interior of the building and is more pronounced with building height and a higher temperature difference [11]. Stack effect is also the major contributor of infiltration in buildings [18].

Generally, the flow characteristics of air infiltration, that is, whether laminar or turbulent, the dispersion and physical magnitude of the leakage paths and the pressure differential between the interior and exterior of the building influences the air leakage rate through the building envelope [19]. The infiltration rate depends on the weather conditions and is a variable subject to daily and seasonal variations throughout the year [15, 20].Under stack effect, buoyant warm air in the interior of the building rises and can escape from leakage areas located at the top level of the building [11]. This simultaneously draws in cold air through leakage points at the base level. This is the case during winter. However in summer, the stack effect operates in reverse but is much less pronounced than is the case in winter. In the reverse scenario, cold air drops to the lower level of the building and thereby draws out this generally conditioned indoor air through the leakage areas at the base level [11]. Typically with stack effect, a neutral pressure plane develops where the switch occurs from exfiltration to infiltration over the building height.

Wind induced infiltration develops as wind blows over a building thus working up pressure on the windward side. Wind is thus pushed in through leakage areas on the respective windward facade and at the same time results in air flowing out from the leeward end [12]. Wind is variable in nature both in terms of velocity and direction and thus so is the infiltration behavior induced by wind. In general, based on average wind speeds of 10 to 15mph, the average annual wind induced pressure differential is reported as ranging from 10 to 14Pascal(Pa) [12]. The local terrain features and the building shape also influences wind induced air leakage [11].

4. Airtightness measurement methodology

A mass balance approach to airflow across the building envelope along with details of the leak path characteristics derived using approximate methods, forms the basis of estimating the total air leakage of a building [15, 19]. There are different standards that define the measurement methodology for ascertaining the airtightness of a building. This includes the CGSB 149.10-M86 'Determination of the Airtightness of Building Envelopes by the Fan Depressurization Method' standard, the ASTM E-779 'Standard test method for Determining Air Leakage Rate by Fan Pressurization' standard or the ISO 9972 'Thermal Insulation - Determination of Building Airtightness - Fan Pressurization Method' standard.

The ASTM standard E-779,selected as the reference standard for this investigation, defines a standardized test methodology for measuring the air leakage rate through a building envelope under controlled depressurization conditions [8]. A fan induced depressurization is carried out typically using a blower door assembly to move air through the building envelope. The blower door unit has a variable speed controlled calibrated fan. When the fan is run under depressurization mode thus blowing air out of the building, the resultant is an artificial pressure differential between the building interior and exterior [18].

Depressurization is applied under single zoning to the building envelope at different relatively high pressures ranging between 10 to 60Pa and the corresponding steady-state airflow through the building envelope, required to maintain a stable pressure differential, is logged [16]. An airflow and pressure differential measuring system provides the required measurement data. The high test pressures serve to mitigate the effect of wind or stack effect [18]. The relationship between the depressurization values and the corresponding logged airflow rates forms the basis of deriving a leakage flow exponent (n) and leakage flow coefficient (C) which can then be used to estimate the airtightness or effective leakage area(ELA) of the building [8, 20].

The ASTM standard E-779 also defines certain limiting conditions above which depressurization testing will not yield good results. The first condition is to ascertain if the stack effect is too pronounced and is based on checking if the product of interior-exterior temperature differential in absolute terms and the building height is greater than $200m^{\circ}C$. The second recommendation is pertaining to desirable outside environmental conditions; which is wind speeds of 0 to 2m/s and temperatures ranging between 5 to $35^{\circ}C$.

In this study, a Minneapolis Blower Door^{TM}(Model 3) system along with a DG-700 pressure and flow gauge was used to conduct the airtightness test. The blower unit and the pressure and flow gauge are depicted in Figure 2. After set-up procedures were completed including zeroing for building baseline pressure correction, the building was subjected, as a single zone, to a multi-point depressurization test with the test procedures following the ASTM E779 standard. Air leakage data, as cubic feet per minute (CFM), was logged for varying pressures subject to a ten seconds time averaging function available with the DG 700 unit. Interior and exterior temperatures were logged during the testing. Local wind speed was also noted using a vane anemometer to ensure test conditions conformed to the ASTM E779 standard. The reference standard conditions of temperature as $20^{\circ}C$ and air density as $1.2041kg/m^{3}$ applied.

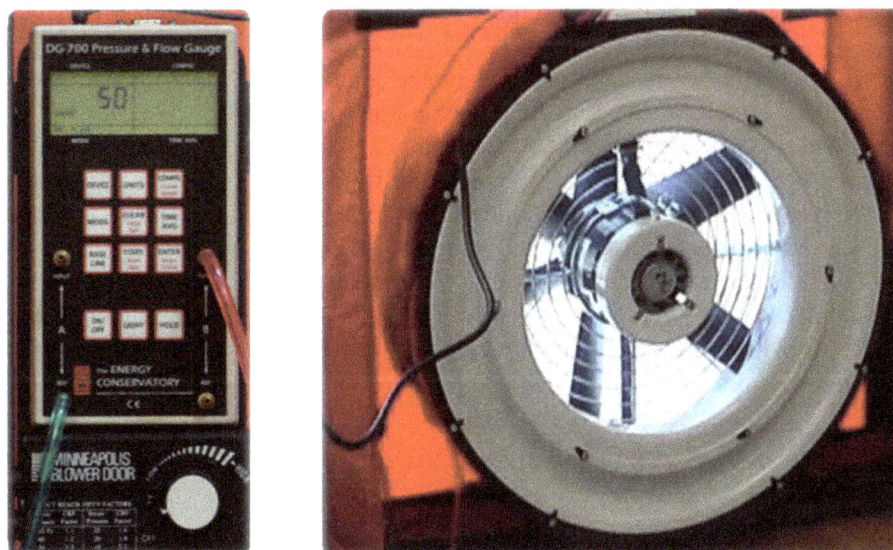

Figure 2. Minneapolis model 3 blower unit and a DG-700 digital gauge

An empirical relationship which takes the form of a power law function, $Q = C\Delta P^n$, generally describes the infiltration flow across the building envelope [15, 20, 21, 22]. Data obtained from the blower door testing can be reduced to this empirical relationship between airflow (Q) and the corresponding pressure differential (ΔP). It is relevant to note that the blower door data indicates the leakage rate under stilted status and does not represent infiltration rate as yet [15, 20]. The power law function can be used to forecast the infiltration under normal conditions [15, 20, 21]. The flow exponent n and the flow coefficient C are empirical measures representing the building leakage characteristics and are specific to the building. These empirical values stand for the leakage sizing, leakage path and airflow regime present in the building [15, 20].

5. Estimating air infiltration under varying climatic conditions

Based on the empirical C & n measures and or the estimated leakage area provided by the depressurization test, there are different approaches in estimating the building air infiltration rate under typical local environmental conditions [10]. For buildings modeled as a single zone, one of these approaches is to use a single zone superposition model [19]. In this paper, two such air infiltration models, namely the Alberta air infiltration (AIM-2) model and the Lawrence Berkeley Laboratory (LBL) model, are used to estimate the air infiltration rate for the studied building under varying wind and outside temperature conditions. This also allows for a comparative assessment of the results from the two mathematical infiltration models.

The AIM-2 model singly works out the airflow associated with wind and likewise due to stack effect and then calculates the combined effect using a superposition equation [23]. The model is based on the $Q = C\Delta P^n$ empirical relationship and in addition, accounts for the flue outlet of the building as a separate leakage area. However, for this study, the flue outlet parameters required in the model are not considered as the building under investigation does not have any flue outlets. The C & n values are required inputs to the model. Another model input is based on defining the leakage distribution in the building envelope in terms of fractions of the total leakage. A uniform leakage distribution is assumed in this case. The wind speed, another required input, is considered at eaves height of the building [23]. The wind shelter factor which represents the local shielding effect around the building is assigned the shelter type 'light local shielding' based on the observed site topography [23].

The LBL model describes the infiltration in a building taking into account the leakage characteristics established through blower door diagnostics and then accounts for infiltration driving forces due to wind and temperature climatic conditions [15, 20]. One of the required parameters for the LBL model is ELA. ELA stands for the area of a nozzle shaped hole which has a discharge coefficient of 1 and which, on an equivalent basis, can result in the same infiltration rate as is the case for the building under a 4Pa pressure differential [15, 20, 21]. As with the previous model, a uniform leakage distribution is assumed. A generalized shielding coefficient parameter based on the local shielding present is also a model input and for this case, a 'light local shielding with few obstructions' category value is assigned [15, 20]. A 'terrain class III' is selected as terrain classification towards assigning the terrain class parameters in the model [15, 20].

Two climatic aspects for the site, which are required inputs for both the superposition air infiltration models, were also examined. One is the average daily temperature range observed locally and the other is the local wind speed trend. Historical averages climate data (1971 to 2000) from the Guelph Arboretum weather station gives the average daily temperature as ranging from lows of -11°C to highs of 26°C [24]. For the wind trend, the wind rose diagram for the site was generated based on typical metrological year (TMY) and modelled weather data using Autodesk® Vasari's wind rose tool. The overlaid wind rose diagrams on the building site for two separate periods - November to the following March (winter) and May to September (summer)is given in Figure 3 and provides information on the frequency and speed of wind at the site. From the wind rose diagram, it is evident the site experiences wind speeds as high as +7m/s and chiefly from a west direction.

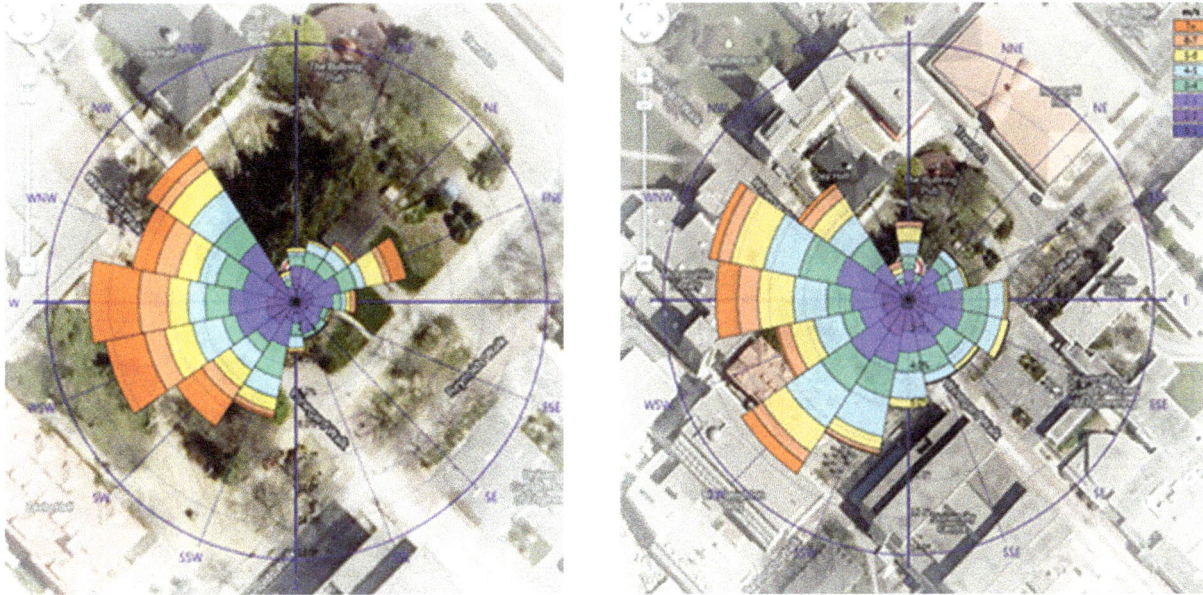

Figure 3. Wind rose diagrams for the building site over winter and summer periods [Courtesy – *Autodesk® Vasari*]

6. Airtightness field measurement data and air change per hour standards

In order to ensure a standard basis for comparison, or put another way, for the fact that buildings having a larger wall, floor or roof area have more potential for leakage spots, air leakage results are typically normalized in terms of volume or area [25]. Air change per hour (ACH) is one such measure and represents the frequency of air replacement within the building on a per hour basis [18]. The corresponding ACH under 50Pa is termed ACH50. Table 2 is a summary of the observed range of air change rate values at 50Pa from field measurements carried out in Canada covering residential buildings.

Table 2. ACH50 field measurement data for Canada

Sample Size	Range for ACH50 (/hr)		Reference
	Lower	Upper	
222	0.4	11	[26]
47	0.13	2.6	[26]
100	1.876	4.653	[25]

A review of airtightness field measurements carried out in Europe and North America for residential buildings reports a wide range from 0.5 to 84 hr^{-1} @ ACH50 [14, 25, 27-31]. In a survey study of airtightness measurements carried out on buildings in Canada, the authors reported a typical average value of 4.4hr^{-1} @ ACH50 for residential buildings and a range of 1.1 to 1.6hr^{-1} @ ACH50 for commercial buildings [32]. As far as standards, there are various airtightness standards across different countries. These standards also differ in the quoted maximum permissible air change rate. The R-2000 standard in Canada is set at 1.5hr^{-1} @ ACH50 [33]. Examining standards in Europe gives a range of 0.5 to 6hr^{-1} @ ACH50 as the standard limit for allowable air change rate [27].

7. Results and discussion

The measurement data from the blower door test, corrected for baseline pressure, site altitude and air density using a proprietary software (TECTITE^{TM}) provided by the blower door manufacturer, is presented in Figure 4.

Based on this data, and using the calculation methodology provided in the ASTM E779 standard, the following measures are obtained:

Flow coefficient, C = 0.246 m^3/(sPa^n)

Flow exponent, n = 0.6

95% confidence limit for n = 0.012 & 95% confidence limit for ln(C) = 0.044

Effective leakage area, ELA = 0.215m^2

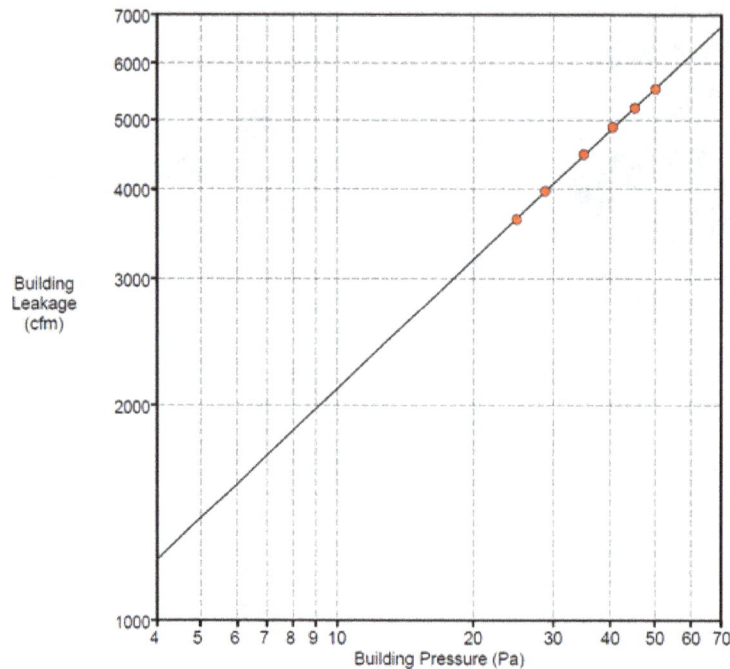

Figure 4. Depressurization testing– CFM versus building pressure (Courtesy – *TECTITE*^{TM}*software*)

Guided by information provided in Chan et al., 2013 [34],the obtained n value of 0.6 indicates that the leakage sites in the building are constituted predominantly of "specific openings". The ELA value of 0.215m^2 represents the cumulative sizing of the leakage areas in the building. To set this in better perspective, the area of a window in the main section of the building is around 1.46m^2. The TECTITE^{TM} software generated ACH50 value for the building based on the blower door test data is 15.5hr^{-1}. Taking into consideration the limit set by the R-2000 standard for permissible air change rate (1.5hr^{-1}@ ACH50), this obtained building ACH50 value does not meet the standard and also establishes that the building is leaky. Based on a building leakage characterization table (covering leakage class A to J corresponding to ACH50 values of 1hr^{-1} and >27hr^{-1} respectively) provided in Sherman, 1995 [20], the campus building can be categorized as of leakage class G. According to EN ISO 13790, if the ACH50 is greater than 10 under natural infiltration conditions, which is the case here, the building would be categorized as having a low envelope tightness level [35].

Based on the site wind speed range as depicted in the wind rose plots and the potential daily average maximum temperature differential between the exterior temperature and expected interior set point temperature of the building, the AIM-2 infiltration model is used to obtain the cumulative effect airflow rate (Q, derived from superposition of the stack related Q_s and wind related Q_w airflow rates) and the corresponding ACH values. The variation of these obtained values versus wind speed and temperature differential are presented as plots in Figures 5 to 8. As observed from Figure 6, with the temperature differential assumed to be holding at a constant of 17°K, the estimated ACH value increases from around 1ACH to 2.75ACH corresponding to a wind speed change from 1m/s to 8m/s. On the other hand, with the wind speed assumed to be holding at a constant of 2m/s, the estimated ACH value drops from a high of 1.6ACH to a low of around 0.6ACH corresponding to temperature differentials of 37°K and 2°K respectively. Additionally, the ACH values are also estimated using the LBL model. There is a difference in the airflow rate and ACH estimations made by the two mathematical models and this is summarized in Tables 3 and 4. The ACH estimation difference between the two models range as high as 31%.

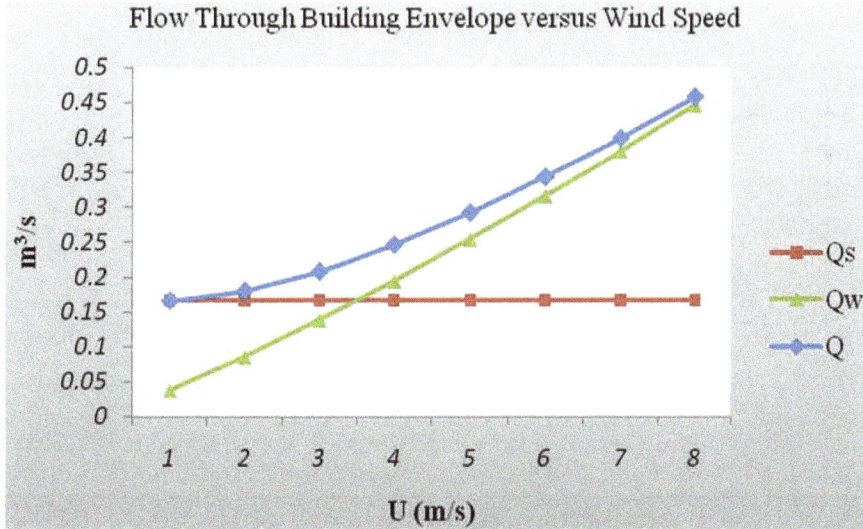

Figure 5. Stackrelated (Q_s), wind related (Q_w)&cumulative effect (Q) airflow rate versus wind speed

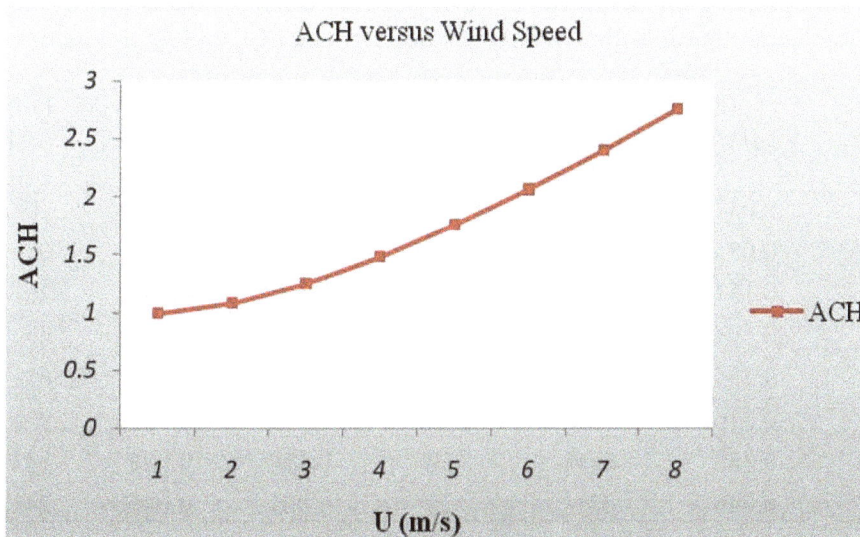

Figure 6. ACH versus wind speed

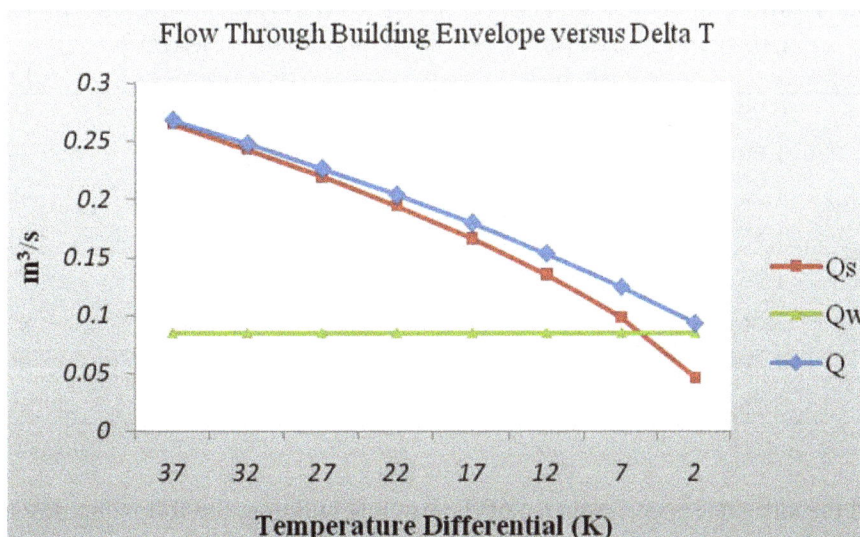

Figure 7. Stack related (Q_s), wind related (Q_w) & cumulative effect (Q) airflow rate versus temperature differential

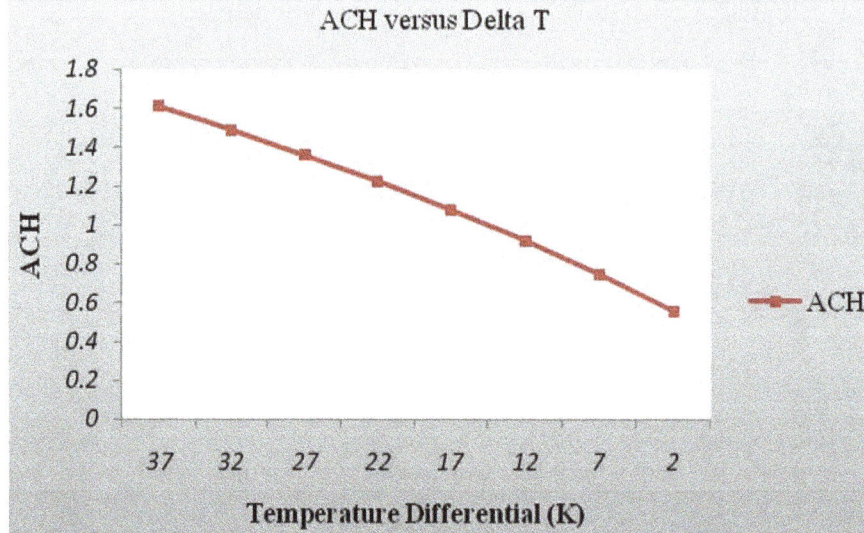

Figure 8. ACH versus temperature differential

Table 3. Difference in estimation of airflow rates between LBL and AIM-2 models

v' (m/s)	Q_w (m³/s) (*LBL*)	Q_w (m³/s) (*AIM*)	Difference - AIM over LBL %	ΔT (K)	Q_s (m³/s) (*LBL*)	Q_s (m³/s) (*AIM*)	Difference - AIM over LBL %
1	0.038	0.037	2%	37	0.262	0.265	-1%
2	0.075	0.085	-12%	32	0.243	0.243	0%
3	0.113	0.137	-22%	27	0.224	0.220	2%
4	0.151	0.194	-29%	22	0.202	0.194	4%
5	0.188	0.254	-35%	17	0.177	0.166	6%
6	0.226	0.316	-40%	12	0.149	0.135	9%
7	0.264	0.380	-44%	7	0.114	0.098	14%
8	0.301	0.446	-48%	2	0.061	0.046	24%

Table 4. Difference in estimation of ACH between LBL and AIM-2 models

v' (m/s)	ACH (hr⁻¹) (*LBL*)	ACH (hr⁻¹) (*AIM*)	Difference - AIM over LBL %	ΔT (K)	ACH (hr⁻¹) (*LBL*)	ACH (hr⁻¹) (*AIM*)	Difference - AIM over LBL %
1	1.09	0.99	9%	37	1.63	1.61	1%
2	1.16	1.08	7%	32	1.53	1.49	3%
3	1.26	1.24	1%	27	1.41	1.36	4%
4	1.40	1.47	-6%	22	1.29	1.22	5%
5	1.55	1.75	-13%	17	1.16	1.08	7%
6	1.72	2.06	-20%	12	1.00	0.92	8%
7	1.91	2.40	-26%	7	0.82	0.74	9%
8	2.10	2.75	-31%	2	0.58	0.55	4%

8. Conclusions

Having investigated the airtightness performance of the campus building, the following describe the key conclusions and follow-up recommendation:

- The reported building ACH50 value of 15.5hr⁻¹, when compared to the ACH50 range observed from field measurements for residential buildings across Canada (0.13hr⁻¹ to 11hr⁻¹), is a clear outlier. The building ACH50 value also far exceeds the R-2000 standard permissible air change rate of $1.5hr^{-1}$ at ACH50.

- The estimated ELA for the building is $0.215m^2$ and the nature of the leakage sites is assessed as 'specific openings'.

- Even with accounting for potential uncertainties in the test results including with flow or pressure recordings (the DG 700 specification sheet reports a flow accuracy of +/-3%) or with normalizing the test data to the conditioned volume, the campus building can conclusively be categorized as having a low envelope tightness level. This effectively translates to energy loss for the building during the heating season.

- The ACH range exhibited against varying wind speed and interior – exterior temperature differentials typical at the building site, is estimated using two different single zone air infiltration models, the AIM-2 model and the LBL model, to also allow for model intercomparison. The resultant plots indicate an ACH ranging from $1hr^{-1}$ to $2.75hr^{-1}$ against wind speeds in the range of 1 to 8m/s and a range of $0.5hr^{-1}$ to $1.6hr^{-1}$ corresponding to a temperature differential range of $2^{\circ}K$ to $37^{\circ}K$. There is an ACH estimation difference between the two models and this ranges as high as 31% in the case of varying wind speed conditions, but is relatively narrower registering a high of 10% in the case of varying temperature differentials.

- The air infiltration modeling result is based on assumptions made with regards to the leakage distribution in the building. Identification of the leakage pathways in the building can allow for a revised assignment of the leakage distribution so as to reflect the actual case.

- In light of this airtightness study which classes the investigated building as leaky, it is recommended, as an energy conservation measure, to further investigate the building envelope using methods like smoke tracing and or infrared thermography towards identifying the air leakage pathways and applying retro-fit measures in addressing these leakage sites.

References

[1] Perez-Lombard, L., Ortiz, J., & Pout, C. (2008). A review on buildings energy consumption information. Energy and Buildings, 40(3), 394-398.

[2] Natural Resources Canada's Office of Energy Efficiency. (2011).Comprehensive Energy Use Database, 1990 to 2011. Retrieved April18, 2014, from

http://oee.nrcan.gc.ca/corporate/statistics/neud/dpa/comprehensive_tables/list.cfm?attr=0

[3] Huang, J., Hanford, J., & Yang, F. (1999). LBNL-44636 Residential heating and cooling loads component analysis. California: Building Technologies Department, Environmental Energy Technologies Division, Lawrence Berkeley National Laboratory, University of California.

[4] Jokisalo, J., Kurnitski, J., Korpi, M., Kalamees, T., & Vinha, J. (2009). Building leakage, infiltration, and energy performance analyses for Finnish detached houses. Building and Environment, 44(2), 377-387.

[5] Orme, M. (2001). Estimates of the energy impact of ventilation and associated financial expenditures. Energy and Buildings, 33(3), 199-205.

[6] Emmerich, S. J., McDowell, T. P., & Anis, W. (2005). Investigation of the impact of commercial building envelope airtightness on HVAC energy use. US Department of Commerce, Technology Administration, National Institute of Standards and Technology.

[7] Huang, Y. J., & Zhang, H. (1999). Commercial heating and cooling loads component analysis. Energy Analysis Program, 7.

[8] ASTM Standard E779, 2003, "Standard Test Method for Determining Air Leakage Rate by Fan Pressurization," ASTM International, West Conshohocken, PA, 2003, DOI: 10.1520/E0779-03, www.astm.org.

[9] Rasouli, M., Ge, G., Simonson, C. J., & Besant, R. W. (2012). Uncertainties in energy and economic performance of HVAC systems and energy recovery ventilators due to uncertainties in building and HVAC parameters. Applied Thermal Engineering, 50(1), 732-742.

[10] Wang, W., Beausoleil-Morrison, I., & Reardon, J. (2009). Evaluation of the Alberta air infiltration model using measurements and inter-model comparisons. Building and Environment, 44(2), 309-318.

[11] Building Energy Codes. (2011). Building Energy Code Resource Guide: Air Leakage Guide. Catalogue No. PNNL-SA-82900. U.S.A.: U.S. Department of Energy.

[12]   Anis, W., & Wiss, F. A. I. A. (2006). Air barrier systems in buildings. Whole Building Design Guide, National Institute of Building Science.

[13]   Alfano, F. R., Dell'Isola, M., Ficco, G., & Tassini, F. (2012). Experimental analysis of air tightness in Mediterranean buildings using the fan pressurization method. Building and Environment, 53, 16-25.

[14]   Kalamees, T. (2007). Air tightness and air leakages of new lightweight single-family detached houses in Estonia. Building and Environment, 42(6), 2369-2377.

[15]   Sherman, M. H. (1987). Estimation of infiltration from leakage and climate indicators. Energy and Buildings, 10(1), 81-86.

[16]   Sherman, M. H., & Chan, W. R. (2004). Building airtightness: research and practice. LBNL report, 53356.

[17]   ASHRAE. Infiltration and ventilation requirements, Ch. 27 of ASHRAE handbook of fundamentals. Atlanta: American Society of Heating, Refrigerating and Air Conditioning Engineers; 2009.

[18]   Minneapolis Blower Door® Operation Manual for Model 3 and Model 4 Systems. Energy Conservatory.

[19]   Awbi, H. B. (2003). Ventilation of buildings (2nd ed.). New York: Spon Press.

[20]   Sherman, M. (1995). The Use of Blower-Door Data. Indoor Air, 5(3), 215-224.

[21]   Persily, A. K. (1998). Airtightness of commercial and institutional buildings: blowing holes in the myth of tight buildings. Building and Fire Research Laboratory, National Institute of Standards and Technology.

[22]   The Energy Conservatory. (2014). Automated Blower Door Systems and Accessories. Retrieved April 18, 2014, from http://www.energyconservatory.com/products/automated-blower-door-systems-and-accessories

[23]   Walker, I. S., & Wilson, D. J. (1998). Field validation of algebraic equations for stack and wind driven air infiltration calculations. HVAC&R Research, 4(2), 119-139.

[24]   Government of Canada. Calculation of the 1981 to 2010 Climate Normals for Canada. Retrieved April 18, 2014, from http://climate.weather.gc.ca/climate_normals/normals_documentation_e.html

[25]   Harris, J. (2009). Air Leakage in Ontario Housing. Retrieved April 18, 2014, from http://www.mah.gov.on.ca/Asset8296.aspx?method=1

[26]   Hamlin T. (1997). Airtightness and Energy Efficiency of New Conventional & R-2000 Housing in Canada. Ottawa, ON: Canada Centre for Mineral and Energy Technology, Natural Resources Canada.

[27]   Pan, W. (2010). Relationships between air-tightness and its influencing factors of post-2006 new-build dwellings in the UK. Building and Environment, 45(11), 2387-2399.

[28]   Persily, A., Musser, A., Emmerich, S.J. (2010). Modeled infiltration rate distributions for US housing. Indoor Air, 20(6), 473-485.

[29]   Blomsterberg, Å., Carlsson, T., Svensson, C., & Kronvall, J. (1999). Air flows in dwellings-simulations and measurements. Energy and buildings, 30(1), 87-95.

[30]   Jokisalo, J., Kalamees, T., Kurnitski, J., Eskola, L., Jokiranta, K., & Vinha, J. (2008). A comparison of measured and simulated air pressure conditions of a detached house in a cold climate. Journal of Building Physics, 32(1), 67-89.

[31]   Juodis, E. (2000). Energy saving and airtightness of blocks of flats in Lithuania. Indoor and Built Environment, 9(3-4), 143-147.

[32]   Fennell, H. C., & Haehnel, J. (2005). Setting airtightness standards. ASHRAE journal, 47(9), 26.

[33]   Natural Resources Canada. (2012). More information on the R-2000 Standard. Retrieved April 18, 2014, from https://www.nrcan.gc.ca/energy/efficiency/housing/new-homes/5089

[34]   Chan, W. R., Joh, J., & Sherman, M. H. (2013). Analysis of air leakage measurements of US houses. Energy and Buildings, 66, 616-625.

[35]   Sfakianaki, A., Pavlou, K., Santamouris, M., Livada, I., Assimakopoulos, M. N., Mantas, P., & Christakopoulos, A. (2008). Air tightness measurements of residential houses in Athens, Greece. Building and Environment, 43(4), 398-405.

17

The spatial distribution of dust sources in Iraq by using satellite images

Kamal H.Lateef[1], Azhaar K.Mishaal[1], Ahmed M.Abud[2]

[1] Ministry of Science and Technology- Renewable Energy Directorate, Iraq.
[2] Ministry of Science and Technology- Environment and Water Directorate, Iraq.

Abstract

Dust storms phenomenon occurs in the most regions of Iraq during the year, this paper is study this phenomenon by using the technique of satellite images, it has been used satellite images (Meteosat-9) with the sensor (SEVERI) and selected different dates of dust storms in 2012, geographic information system programs (ERDAS-GIS) has been used to discrimination the regions that cause this phenomena within the study area to prepare the images to read the real geographic coordinates and determines the regions that caused the occurrence of the dust storms represented by geographical location (Lon/Lat) and making Iraqi map describes these regions for year 2012 and compared with maps for previous years.

Keywords: Dust sources; Spatial distribution; ERDAS; Satellite images.

1. Introduction

A dust storm or sand storm is a meteorological phenomenon common in arid and semi-arid regions, dust storms arise when a gust front or other strong wind blows loose sand and dirt from a dry surface. Particles are transported by saltation and suspension, a process that moves soil from one place and deposits it in another , Dust storms, one type of dust event are in most cases the result of turbulent wind [1], which raise large quantities of dust from desert surfaces and reduce visibility to less than 1km. This dust reaches concentrations in excess of 6000 µg/m3 in severe events [2]. Dust storms cause a great variety of environmental impacts. Tropsopheric aerosols, including dust, are an important component of the earth's climate system and modify climate through their direct radiative effects of scattering and absorption [3], through indirect radiative effects via their influence on clouds microphysics [4], and by their role in processes of atmospheric chemistry [5].

According to the WMO (World Meteorological Organization) protocol, Dust events are classified according to visibility into the categories of:

(1) Dust-in-Suspension: widespread dust in suspension not raised at or near the station at the time of observation; visibility is usually not greater than 10km;

(2) Blowing Dust: raised dust or sand at the time of observation, reducing visibility to 1 to 10km;

(3) Dust Storm: strong winds lift large quantities of dust particles, reducing visibility to between 200 and 1000m; and

(4) Severe Dust Storm: very strong winds lift large quantities of dust particles, reducing visibility to less than 200m the frequency of all dust events is [6]:

In the year 2011, GERIVANI submitted a paper can be help to find the impact of geological units on the wind erosion for finding dust storm sources in regions of western parts of Iran [7]. The researchers in reference [8] have calculated the dust storm velocity by determining the front pattern for the storm which are found that the velocity value is (37.62) km/h. The researchers in reference [9] found that the most important reason of the occurrence of dust storms in Iraq is the passage of a low-pressure system over Iran , the carry cool air from that region towards warmer region or warmer air of regions like eastern Syria and Iraq.

2. Materials and methods

2.1 The study area

Iraq is located in south-west of Asia between (29-37 N), (39-48 E), thus it occupies the northeast corner of the Arab world, Iraq overlooking across the south coast of the Arabian Gulf for distance of 60km, bordered by Kuwait and Saudi Arabia to the south and west, Jordan and Syria to the northwest, Turkey to the north and Iran to the east. Iraq area is 435 052 km^2 (Figure 1).

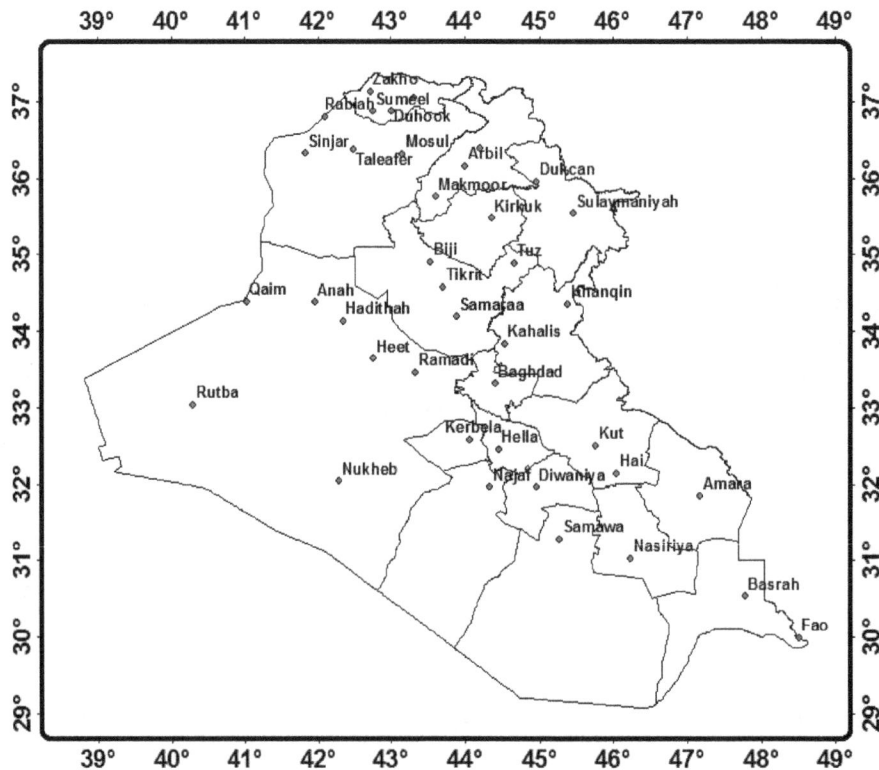

Figure 1. The area study

2.2 Data

In this study it has been used a remote sensing techniques and geographic information systems programs (GIS - ERDAS).Using sensor images (SEVERI) borne on the satellite Meteosat-9 with spatial resolution (3km) for different dates of year 2012 included (10/3) - (17.3) - (7-6) - (18.6) - (5/7) and time between (4 - 630 UTC). Add data meteorological of wind speed.

Satellite images have different dates and different months included seasons spring and summer, these images have a high and various wind speeds to discrimination the regions that cause the emission of dust particles and thus the occurrence of this phenomenon within the study area. These images represent the beginning of dust storms (start point reigns of dust) because it is in the early hours of the day. (Figure 2) shows the selected images [10].

It has been used meteorological data for different meteorological stations for the study area, these data represented by wind speed [11]. Figure 3 illustrated the average of wind speed (m/sec) during study period (January - late July).

10-3-2012

17-3-2012(530)

17-3-2012(6)

7-6-2012

18-6-2012

5-7-2012

Figure 2. The satellite images [10]

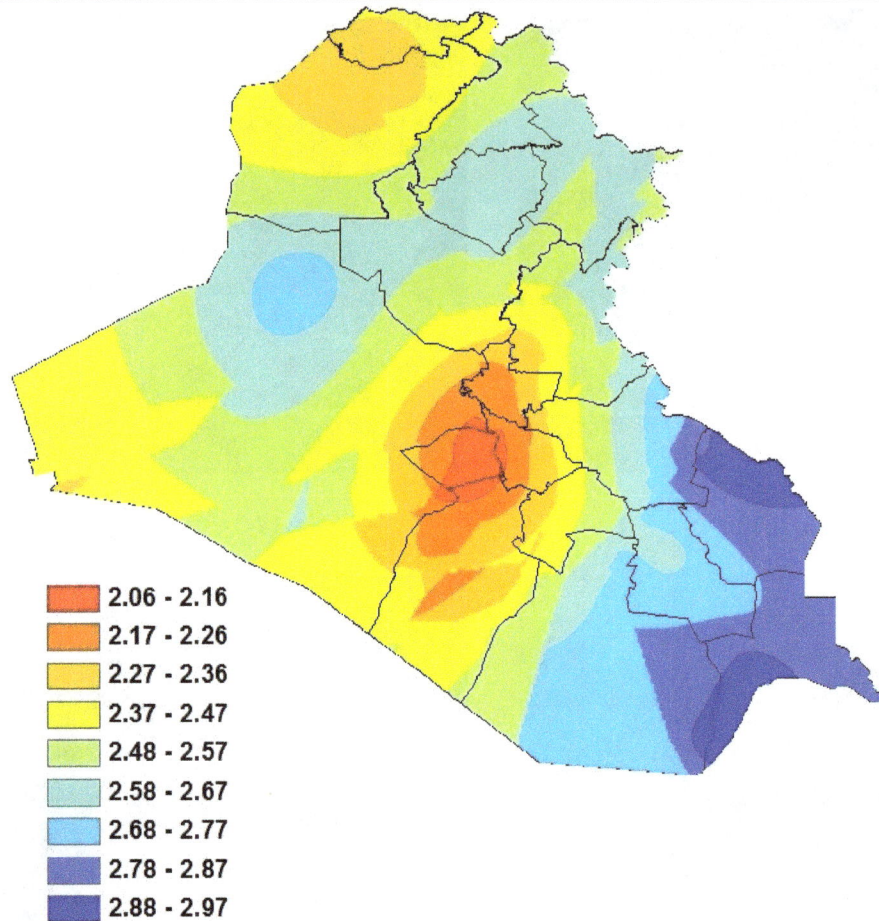

![]	2.06 - 2.16
![]	2.17 - 2.26
![]	2.27 - 2.36
![]	2.37 - 2.47
![]	2.48 - 2.57
![]	2.58 - 2.67
![]	2.68 - 2.77
![]	2.78 - 2.87
![]	2.88 - 2.97

Figure 3. Average of wind speed [10]

3. Results and discussion

3.1 Geometric correction

Satellite images usually contain the distortions of engineering for several reasons such as: mile line scanning, speed of the satellite and the Earth's rotation and therefore cannot be relied upon to produce a correct dimensions map. The geometric correction process is necessary to convert satellite images from a grid of pixels to images read real coordinates and thus determines the accurate location; Figure 4 shows the satellite images after geometrically corrected.

3.2 Determining regions that cause the emission of dust

Geographic location of the regions that emit particles of dust and thus cause the phenomenon of dust storm in the study area (Iraq) has been determined, the determination of these points haves been identified depending on the usage the style of visual interpretation through very important known elements (Tone) and (Texture).

Figure 5 identifies regions inside and outside the study area, so it will suffice by regions that cause dust phenomenon within the study area only, as show in Figure 6.

It can determine the geographical location of these points or regions that cause the phenomenon of dust. See Table 1.

It can make map for the study area (Iraq) that represents the regions cause the dust storms, as shown in Figure 7.

The researchers (Walter M. and Wilkerson) [12] made a map of dust sources regions in Iraq and Syria that direct impact on Iran before 1991, in addition, in 2005 the researchers (Jalali and Davoudi) [13] sketched a map of the regions of dust sources in Iraq and Syria, and they sketched other map in 2008, as shown in Figure 8.

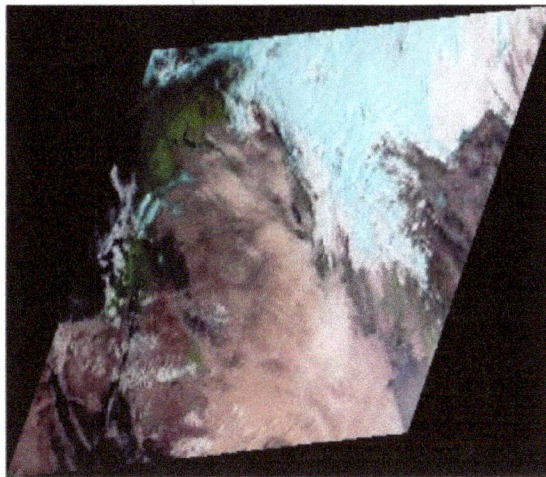

10-3-2012   17-3-2012(530)

17-3-2012(6)   7-6-2012

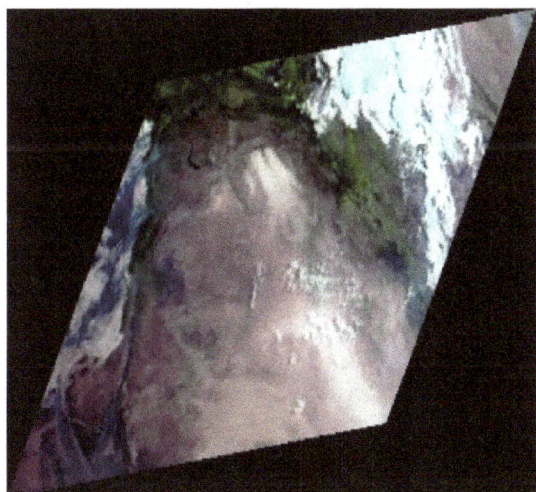

18-6-2012   5-7-2012

Figure 4. Corrected satellite images

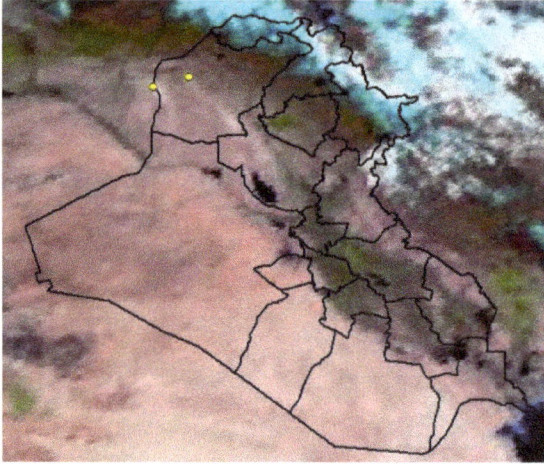

10-3-2012

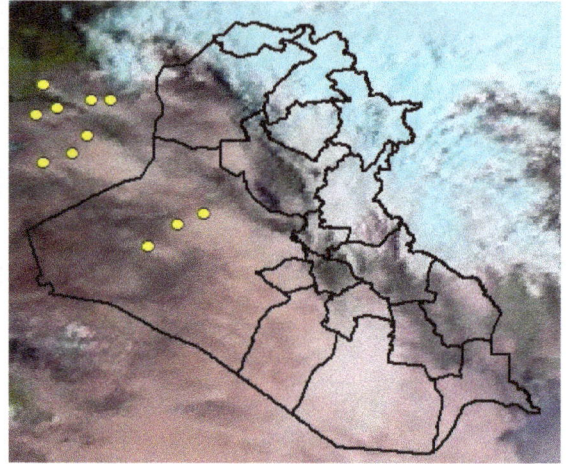

17-3-2012(530)

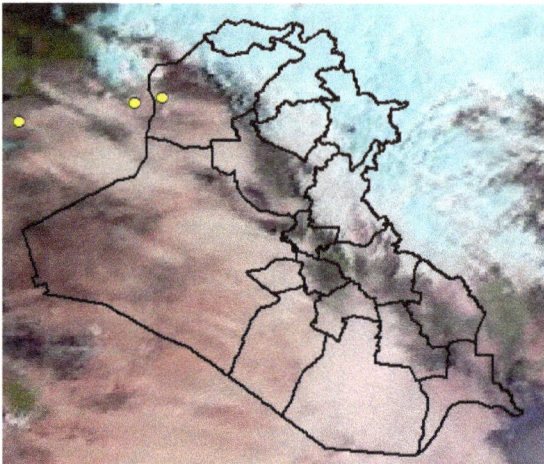

17-3-2012(6)

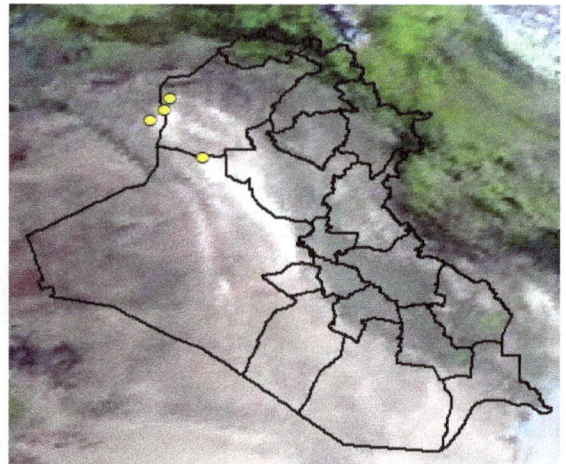

7-6-2012

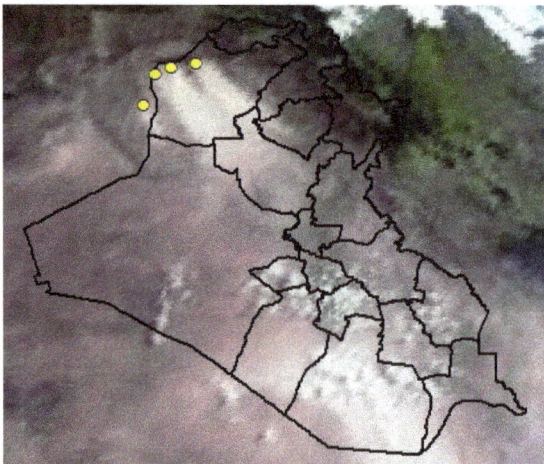

18-6-2012

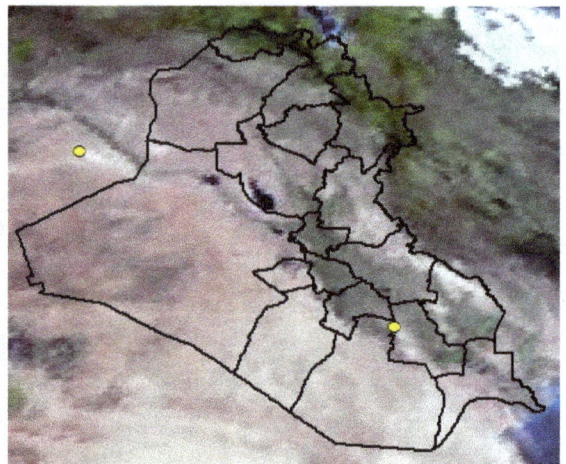

5-7-2012

Figure 5. The regions that cause dust particles emission

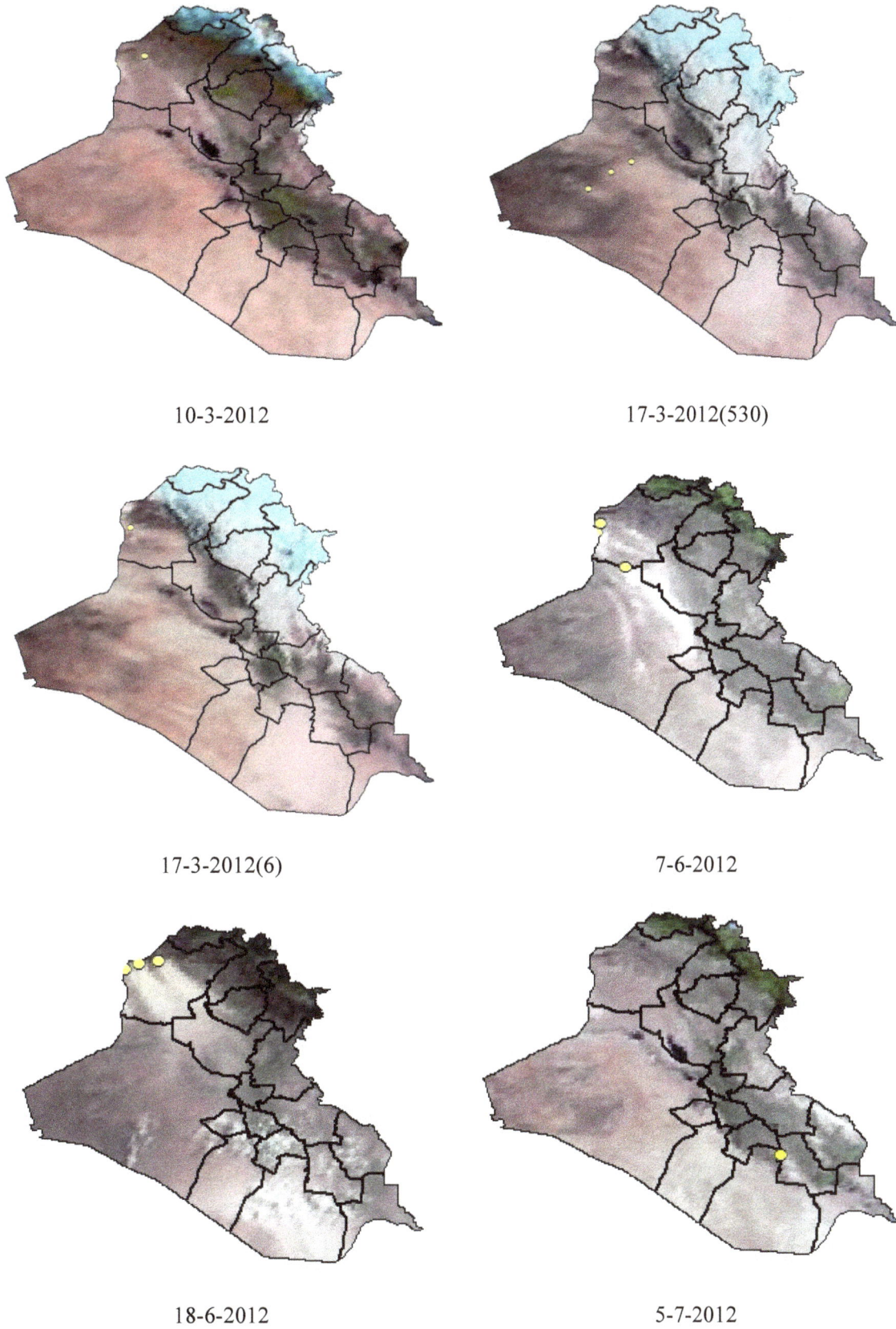

10-3-2012

17-3-2012(530)

17-3-2012(6)

7-6-2012

18-6-2012

5-7-2012

Figure 6. The regions that cause dust particles emission in study area

Table 1. The geographical coordinates for the determined regions in satellite images

point	date	Time (UTC)	longitude	latitude
1	10/3/2012	630	41.934	36.201
2	17/3/2012	530	41.134	33.081
3	17/3/2012	530	41.683	33.484
4	17/3/2012	530	42.182	33.710
5	17/3/2012	6	41.521	35.919
6	7/6/2012	4	41.355	35.881
7	7/6//2012	4	41.459	36.123
8	7/6/2012	4	42.081	34.948
9	18/6/2012	4	42.150	36.535
10	18/6/2012	4	41.666	36.463
11	18/6/2012	4	41.344	36.320
12	5/7/2012	530	45.822	31.503

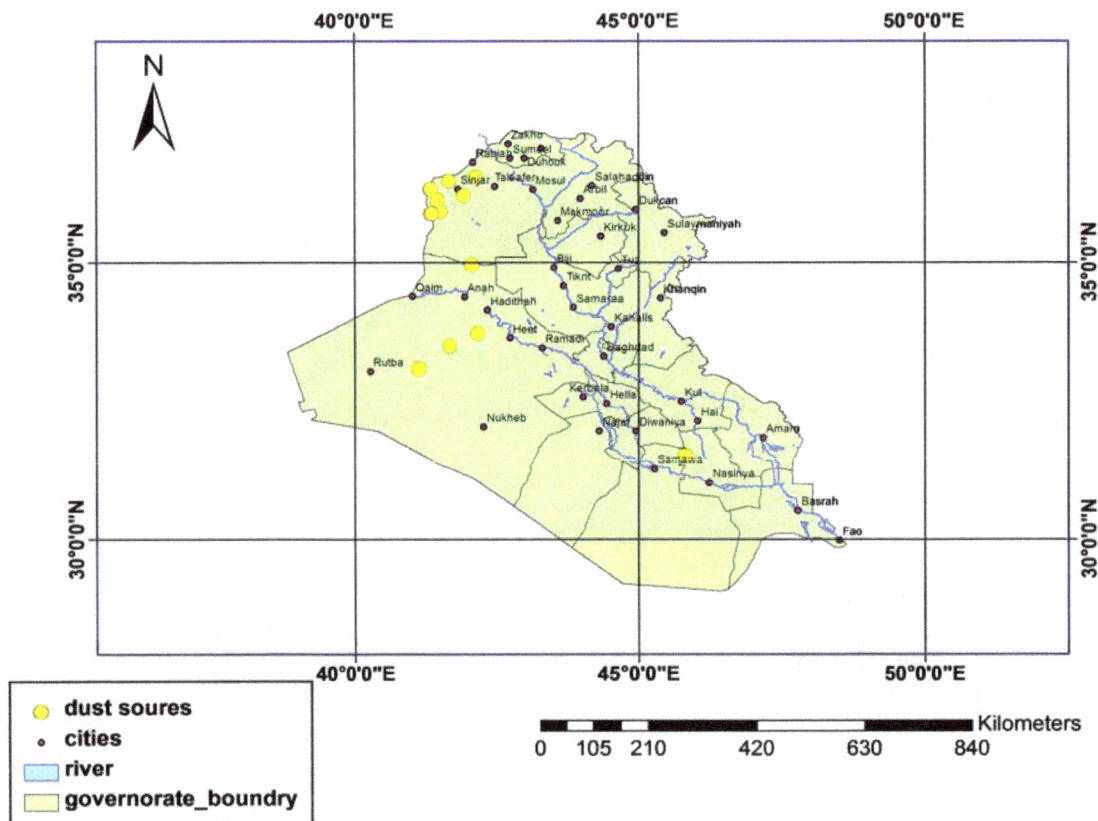

Figure 7. The spatial distribution for the regions that cause the occurrence of dust phenomenon in Iraq for year 2012

When comparing the map in 2012 with maps in years (1991-2005-2008) results showed that the northwestern region and the western region are the sources of the emission of dust particles, these reigns represented in the governorates of Al-Anbar and Ninwah.

Through the study of climate data observed in Ninewah city, it shows that the general trend of temperatures increase, heat waves are increasing, the number of hot days increase in the summer, decreases the amount of rainfall, especially during the last two decades, bad distribution of rainfall during the rainy season, late rainy season and increases the frequency of drought cycles and severity in the region. 2008 was severe drought, severity reached (-2.48) according to the Standard Precipitation Index (SPI). Sandstorms significantly increase in repetition and intensity which is reached unfamiliar number (32) storms and 71 days of thick dust [14].

1991

2005

2008

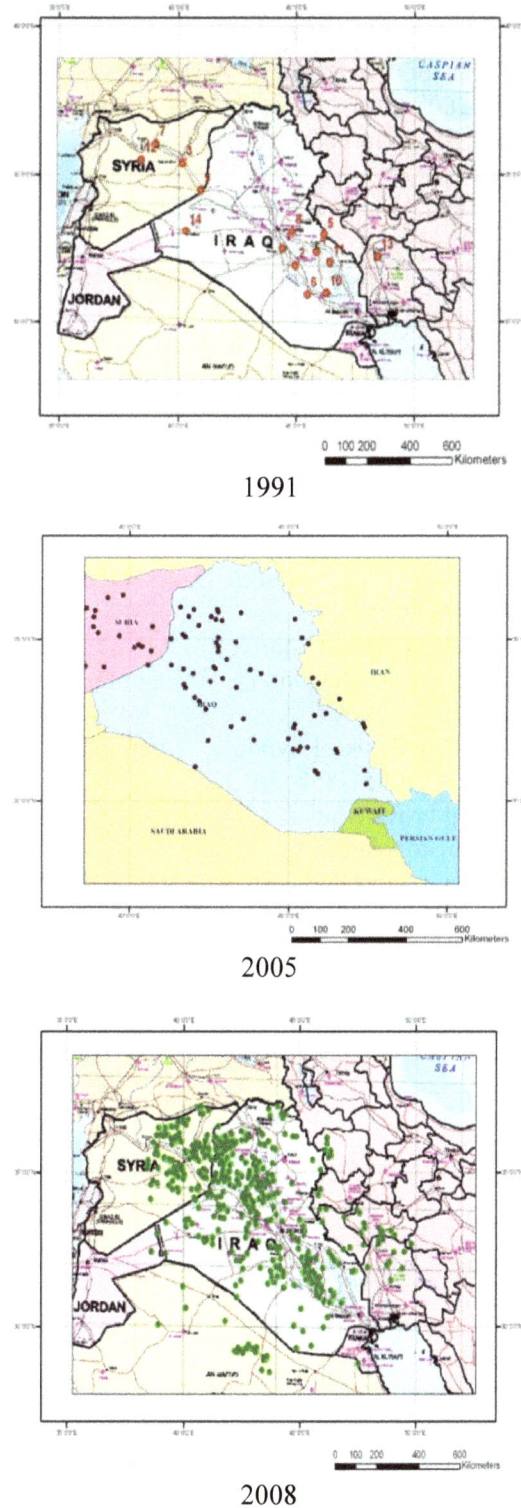

Figure 8. The spatial distribution for the regions that cause the occurrence of dust phenomenon for different years [12, 13]

4. Conclusions

1. It has been made a map for the study area appears regions that cause the phenomenon of dust storms.
2. The north-western region and neighboring to the Syrian border (eastern Syria) represents of the most important regions that cause the emission of dust particles and the occurrence of the phenomenon of dust in various forms (dust rising – suspense dust - dust storms).
3. The satellite images of the satellite Meteosat-9 give us a good possibility to observe this phenomenon from the beginning (early hours of the day) and determine the emission regions of dust particles.

References

[1]   Miller, S.D., Kuciauskas, A.P., Liu, M., Ji, Q., Reid,J.S., Breed, D.W., Walker, A.L. & Al Mandoos,A. Haboob dust storms of the southern Arabian peninsula. Journal of Geophysical Research- Atmosphere. 113, 26, 2008.

[2]   Song, Z., Wang, J. & Wang, S. Quantitative classification of northeast Asian dust events. J.Geophys. Res., 112, 8, 2007.

[3]   Tegen, I., A. A. Lacis, and I. Fung. The influence on climate forcing of mineral aerosols from disturbed soils. Nature 380(6573): 419–22, 1996.

[4]   Rosenfield, J. E., D. B. Considine, P. E. Meade, J. T. Bacmeister, C.H. Jackman, and M. R. Schoeberl. Stratospheric effects of Mount Pinatubo aerosol studied with a coupled two-dimensional model. Journal of Geophysical Research 102 (D3): 3649–70, 1997.

[5]   Schwartz, S. E., R. Wagener, & S. Nemesure. Microphysical and compositional influences on shortwave radiative forcing of climate by sulfate aerosols. Abstracts of Papers of the American Chemical Society 209: 2-ENVR Part 1, 1995.

[6]   Shao,Y & Dong,C.H. A review on East Asian dust storm climate,modelling and monitoring. Journal of Global and Planetary Change. 52, 1–22, 2006.

[7]   Gerivani,H., Lashkaripour,G, R,.Ghafoori, M&Jalali,N. The Source of dust storm in Iran: A case study based on geological information and Rainfull Data .Journal of Earth and Environmental Sciences, Vol 6, No 1M, P 297-308, 2011.

[8]   Kahdom, A. &Abdul Ali. Monitoring 3rd of March 2011 Dust Storm in Iraq Using Meteosat 9 Images. Journal of Iraqi Science. Proceeding of the 1st Conference on Dust Storms and their environmental effects, 2012.

[9]   Al-Jumaily, K.J& Ibrahim.M, K. Analysis of synoptic situation for dust storms in Iraq. Journal homepage. Volume 4, Issue 5, 2013 pp.851-858, 2013.

[10]   http://www.icare.univ-lille1.fr/msg/browse/index

[11]   http://classic.wunderground.com/history/index

[12]   Walter ,M. & Wilkerson D.Dust and Sand Forecasting in Iraq and Adjoining Countries. Technical Report, Air Weather Service, Scott AFB (AWS/XTX), IL 62225-5008, 1991.

[13]   Jalali ,N. & Davoudi M.H. Inspecting the origins and causes of the dust storms in the Southwest and West parts of Iran and the regions affected. Internal reports of Soil Conservation and Watershed Management Research Institute (SCWMRI), Iran, 2008.

[14]   Belaal,A,A & Bader,H,H.Change of Climate and Hydrology in Ninwaa. Journal of Damascus University, Vol 28,N1,P53-65,2012.

18

Assessment of anaerobic co-digestion of agro wastes for biogas recovery: A bench scale application to date palm wastes

Zainab Ziad Ismail, Ali Raad Talib

Department of Environmental Engineering, Baghdad University, Baghdad, Iraq.

Abstract

Anaerobic digestion is a technology widely used for treatment of organic waste to enhance biogas recovery. In this study, recycling of date palm wastes (DPWs) was examined as a source for biogas production. The effects of inoculum addition, pretreatment of substrate, and temperature on the biogas production were investigated in batch mode digesters. Results revealed that the effect of inoculum addition was more significant than alkaline pretreatment of raw waste materials. The biogas recovery from inoculated DPWs exceeds its production from DPWs without inoculation by approximately 140% at mesophilic conditions. Whereby, the increase of biogas recovery from pretreated DPWs was 52% higher than its production from untreated DPWs at mesophilic conditions. The thermophilic conditions improved the biogas yield by approximately 23%. The kinetic of bio-digestion process was well described by modified Gompertz model and the experimental and predicted values of biogas production were fitted well with correlation coefficient values > 0.96 suggesting favorable conditions of the process.

Keywords: Biogas; Anaerobic Co-digestion; Agriculture waste; Date palm wastes; Methane recovery.

1. Introduction

Renewable energy is a socially and politically defined category of energy sources. Among the different forms of renewable sources, biomass is undoubtedly one of the most promising [1]. About 16% of global final energy consumption comes from renewable resources, with 10% of all energy from traditional biomass, mainly used for heating, and 3.4% from hydroelectricity. When biomass is burnt or digested, the emitted CO_2 is recycled into the atmosphere, so not adding to atmospheric CO_2 concentration over the lifetime of the biomass growth [2]. Anaerobic digestion has been, and continues to be, one of the most widely used processed for the stabilization of biosolid waste, such as from the agro and municipal waste to industrial waste. The widespread use of this technology stems from its potential advantages including, the production of energy of methane, a reduction of 30–50% of waste volume requiring ultimate disposal, and a rate of pathogen destruction-particularly in the thermophilic process. The stabilized biomass can also be utilized as an excellent soil conditioner after appropriate treatment [3]. The composition of biogas varies depending upon the types and relative contents of different raw materials, as well as upon the different conditions and fermenting phases. The quality of biogas generated by organic waste materials does not remain constant but varies with the period of digestion [4]. Several studies have been reported about the co-digestion of lignocellulosic waste materials and agro wastes for biogas production. Rincón et al. [5] studied the methanogenic stage of a two-stage anaerobic digestion process treating two-phase olive oil mill solid residue (OMSR) at mesophilic temperature (35°C). A methane yield of 0.268 ± 0.003 L CH_4 at standard temperature and pressure conditions (STP) g^{-1} COD eliminated was achieved.

Jaafar [6] verified the possibility of using a special type of Iraqi date fruit named Zahdi (normally used for syrup production) as a resource for biogas production at thermophilic digestion with activated sludge as inoculum. Methane was produced with a yield of 570 mL/ VS of substrate. Addition of 1% yeast extract solution as nutrient increased methane yield by 5.9%. Marňóna et al. [7] studied the production of biogas co-digestion of cattle manure with food waste and sewage sludge mesophilic and thermophilic conditions using continuously stirred-tank reactors. Maximum obtained value was 603 LCH_4/kg VS feed for the co-digestion of a mixture of 70% manure, 20% food waste and 10% sewage sludge at 36°C. Lower methane yields were obtained when operating at 55°C. Kafle & Kim [8] evaluated the performance of anaerobic digesters using a mixture of apple waste (AW) and swine manure (SM). This mixture improved the biogas yield by approximately 16% and 48% at mesophilic and thermophilic temperatures, respectively, compared to the use of SM only, but no significant difference was found in the methane yield. Tampio et al. [9] compared the anaerobic digestion of autoclaved and untreated source segregated food waste (FW) over 473 days in semi-continuously fed mesophilic reactors with trace elements supplementation. Methane yields were 5–10% higher for untreated FW than autoclaved FW. However, none of the previously reported studies have dealt with the date palm wastes. The date palm *Phoenix dactylifera* has played an important role in the day-to-day life of the people for the last 7000 years. Today worldwide production, utilization and industrialization of dates are continuously increasing since date fruits have earned great importance in human nutrition owing to their rich content of essential nutrients. Tons of date palm wastes are discarded daily either as an agricultural by product of no economic wastes or by the date processing industries without proper waste management leading to environmental problems. Thus, there is an urgent need to find suitable applications for this waste [10]. Current study, aimed to assess for the first time the biogas production and recovery from the anaerobic co-digestion of date palm wastes.

2. Materials and methods
2.1 Materials
The date palm wastes (DPWs) used in this study involved mixed petiole, rachis, fronds, and leaflet waste materials resulted from the tapping and trimming processes of the date palm trees. This type of solid waste materials is abundantly available in Iraq without proper management and application.

The average measured values of total solids (TS), volatile solids (VS), and pH for the mixed date palm wastes samples were found to be 45.91 ± 2.57, 41.42 ± 1.04, and 7 ± 0.2, respectively. Cattle manure which is known to be rich in methanogenic anaerobic bacteria was used to inoculate the digesters. Cattle manure was freshly collected from a local slaughter house, prepared as slurry, and then added to the digesters as a supplementary material to enrich the bacterial activity and enhance the anaerobic co-digestion process.

2.2 Pretreatment of wastes materials
The pretreatment of the collected date palm wastes, was carried out to facilitate the hydrolysis of cellulose component existing in the substrate. Cellulose and lignin has a highly crystalline structure due to the presence of an extensive hydrogen bond and inter-chain in the cellulose structure. After cleaning manually the collected DPWs samples to remove dirt and dust, the cleaned materials were crushed, and sieved to different particle sizes. Chemical pretreatment included the addition of $Ca(OH)_2$ to the sieved DPWs at concentrations ranged from 0.1 to 0.2g $Ca(OH)_2$/g TS of waste was carried out then the mixtures were autoclaved at 121°C for 20 min. The calcium will precipitate and removed as $CaCO_3$ by flushing the autoclaved mix with CO_2 [11]. Inoculum slurry was prepared by mixing 50g of cattle manure with 400 mL distilled water and was manually homogenized with glass rod.

2.3 Digesters set up
As the main objective of this study was the anaerobic co-digestion of date palm wastes (DPWs) for biogas production, four bench-scale digesters operated in batch mode were set up in duplicate as given in Table 1. The digesters were of 500-mL Pyrex borosilicate heatproof code glass bottles. The components of each digester were maintained at 1:10 which is equivalent to 40 g solid waste material: 400 mL (inoculum slurry or distilled water). Each digester was tightly plugged with rubber stopper contains 2 holes each of 4mm diameter through which a piece of glass tube was submersed into the digester and the

other end of the glass tube was connected with rubber tube for the produced biogas transfer to the gas measuring apparatus. The rubber stoppers were tightly wrapped with parafilm to prevent any release of the produced gas. Digesters were immersed in a thermostatic water bath to maintain the required temperature conditions. Manual shaking of digesters were performed daily to insure that substrate molecules and bacterial come into close. Sodium bicarbonate ($NaHCO_3$) was used for pH adjustment and phenolphthalein was used for coloring water in the displacement bottle. The digesters were flushed with nitrogen for 10 min to provide anaerobic environment conditions.

Table 1. Digesters with waste setup material and temperature condition

Digester No.	Waste materials mix in digester	Temperature condition
1	Pretreated waste inoculated with cattle manure	Mesophilic (38°C)
2	Pretreated waste with distilled water	
3	Untreated waste inoculated with cattle manure	
4	Pretreated waste inoculated with cattle manure	Thermophilic (55°C)

2.4 Methods of analysis

The measurement of total solids (TS) and volatile solids (VS) were carried out in triplicate according to the procedure outlined in the *standard methods* [12]. pH was measured using pH meter (Model: WTW, Inolab 720). The recovered biogas was measured by three approaches; the manometer which is a simple apparatus consisted of glass U-tube shape with 10mm internal diameter filled with potassium hydroxide solution. The U-tube hitched with tap to adjust the level of solution with atmospheric pressure after CO_2 removal. The tube was provided with two ports, one for a biogas injection, and the other for gas outlet after removal of CO_2. The released gas was fractioned in a percentages (i.e. methane and CO_2 percentages) using the 4% potassium hydroxide. All measurements were carried out at room temperature and atmospheric pressure. The volume of gases was recalculated for standard temperature and pressure (STP: 0°C and 1 bar) according to Hansen et al. [13]. The other gas measuring approach is the water displacement method in which the gases were first passed through an airtight washing bottle containing 1 molar sodium hydroxide solution in order to eliminate the carbon dioxide. Then the remaining methane passed to a 500-ml glass container; displacing the water which overflowed into a measuring cylinder. The volume of displaced colored water represents the volume of produced methane. Gas chromatography was used to determine the major components of the produced biogas.

2.5 Soil conditioning with digestate

To examine the overall validity of the selected treatment approach, the digestate resulted from the anaerobic digestion process was used for soil conditioning. Cress seeds were selected for this test. The seeds were planted in a digestate-conditioned soil contained in suitable pots. The pots were irrigated and observed on a daily basis for a period of one week.

3. Results and discussion

In order to determine the best conditions for maximum biogas production from DPWs material, the effect of key parameters including inoculum addition, chemical pretreatment of the digestive waste materials, and temperature were carefully considered in this study.

3.1 Effect of inoculum addition

This part of work was carried out to study the effect of inoculum on biogas production. The biogas production in digesters No. 1 and 2 for pretreated DPWs with inoculum and pretreated DPWs without inoculum respectively was monitored for 117 day. The profiles of biogas production are given in Figures 1-3. Results of the specific biogas production revealed that the use of inoculum improved the co-digestion process and anaerobic biodegradation of waste materials (Table 2). The increase of biogas production associated with the inoculum addition is significantly related to the increase of active microorganism since the cattle manure is a rich source for bacteria. However, the existence of cellulose digestive bacteria could be another potential assumption for the increase of biogas generation rates. This type of bacteria is capable to attack the tight association between lignin and cellulose bond. These results are in a good agreement with the previously outlined findings for biogas production from anaerobic digestion of cattle manure as a substrate [14]. They found out that rumen fluid inoculum increased the biogas production rate two to three times compared to the substrate without rumen fluid.

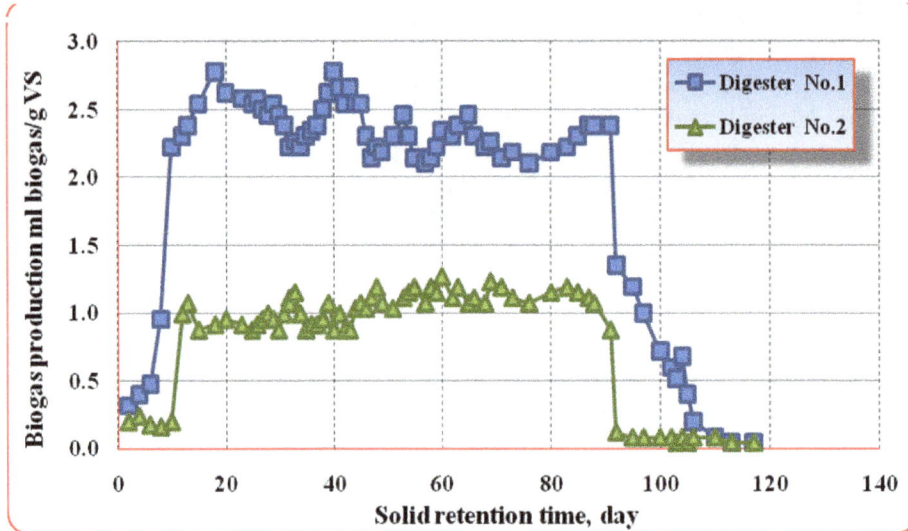

Figure 1. Biogas production profile for digesters No.1 and 2

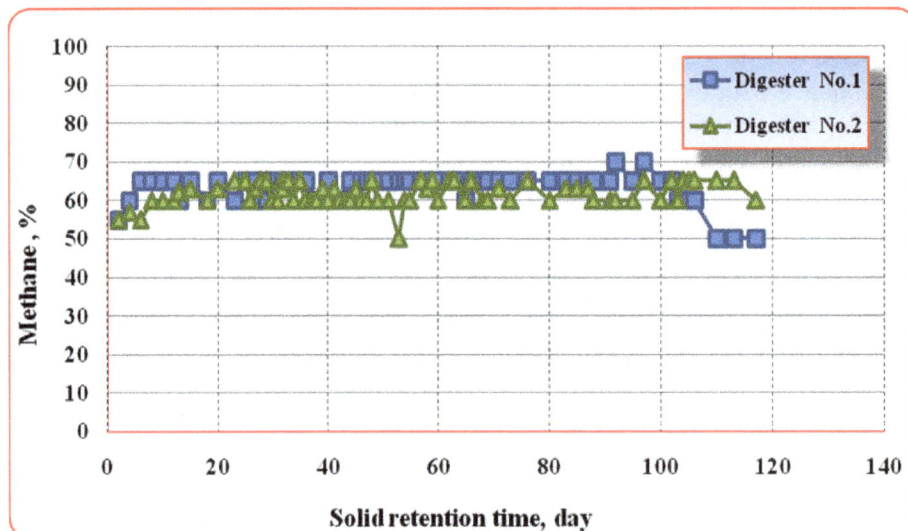

Figure 2. Percentages of CH_4 production for digesters No.1 and 2

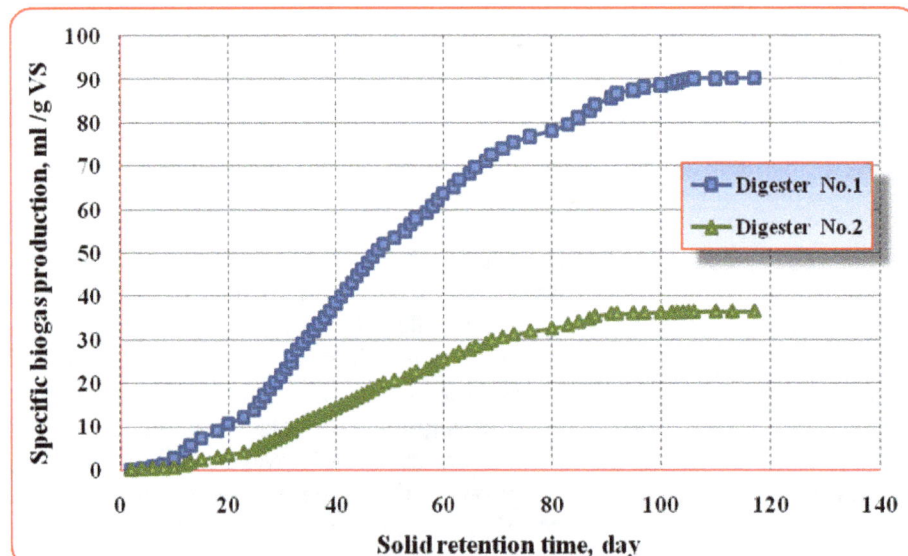

Figure 3. Specific cumulative biogas production profiles for digesters No. 1 and 2

Table 2. Effect of inoculum addition on biogas production

Digester No.	Inoculum	Maximum specific biogas production (mL/g VS)	Maximum specific CH$_4$ production (mL/g VS)	Biogas Increase (%)
1	Applicable	141.667 ± 8.1	90.381	
2	NA*	59.103 ± 2.4	36.493	139.7

Not applicable

3.2 Effect of chemical treatment

This section of work was devoted to investigate the effect of chemical pretreatment of DPWs on biogas production. The profiles of biogas production in digesters No. 1 and 3 for pretreated inoculated DPWs and untreated inoculated DPWs, respectively are given in Figures 4-6. These profiles indicate that the effect of alkaline pretreatment of DPWs was significant with respect to the enhancement of co-digestion process and the subsequent biogas production (Table 3). However, anaerobic digestion of lignocellulosic materials is a challenge because of the complex, rigid, and fibrous structure of these matters which under anaerobic conditions poorly degrades. Abdulkarim [15] reported that the addition of alkaline buffer based on total solid contents increased the biodegradability of the organic fraction of solid waste.

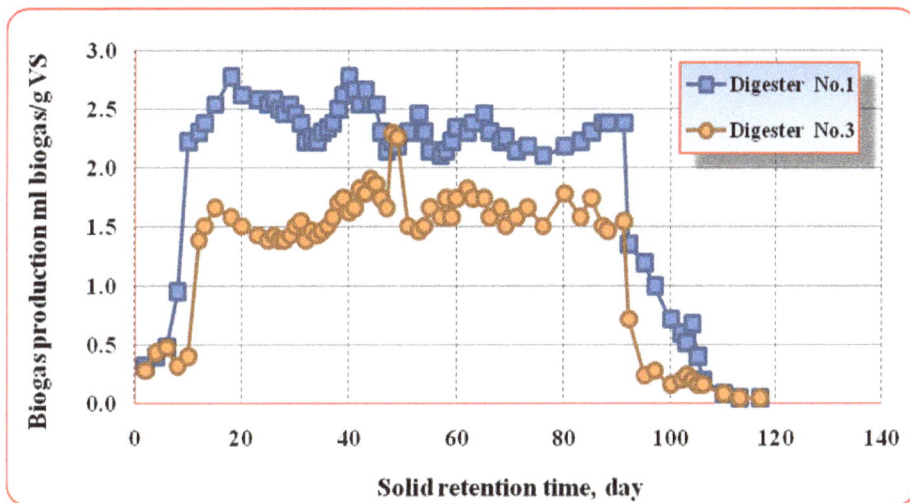

Figure 4. Biogas production profile for digesters No.1 and 3

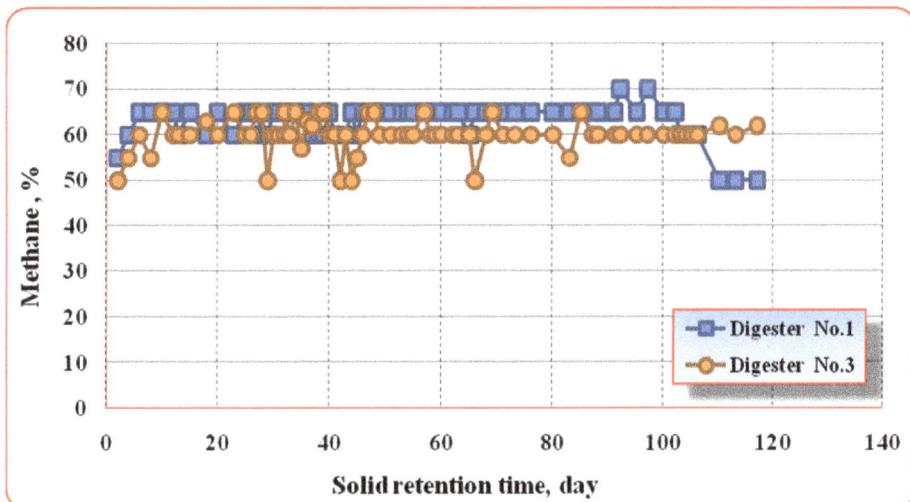

Figure 5. Percentages of CH$_4$ production digesters No.1 and 3

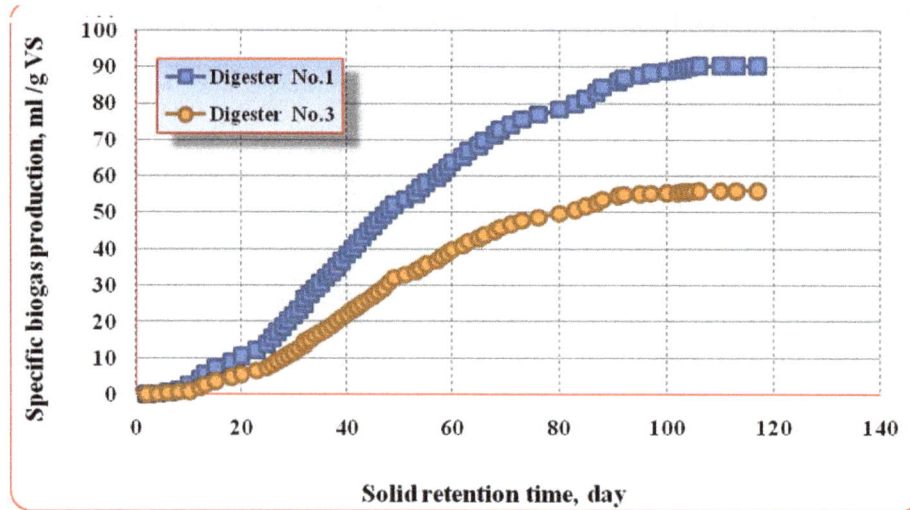

Figure 6. Specific and cumulative CH$_4$ production profiles for digesters No. 1 and 3

Table 3. Effect of pretreatment process of DPWs on biogas production

Digester No.	Pretreatment	Maximum specific biogas production (mL/g VS)	Maximum specific CH$_4$ Production (mL/g VS)	Biogas increase (%)
1	Applicable	141.667 ± 8.1	90.381	51.92
3	NA*	93.254 ± 4.2	56.107	

Not applicable

3.3 Influence of temperature

Results revealed a significant effect of temperature on biogas production. This is due to the fact that temperature is a very important operational parameter in anaerobic digestion processes. As given in Figure 7, the biogas recovery at thermophilic conditions was relatively higher than at mesophilic conditions. Table 4 summarizes the effect of temperature condition on the specific biogas production during 90 days-period observation indicating that biogas production at thermophilic conditions exceeds its production at mesophilic conditions by 92%. In conclusion, biogas yield with respect to methane content produced at thermophilic conditions is more favorable than its quality produced at mesophilic temperature range in this study. These observations are in a good agreement with the previously reported data regarding the biogas production at mesophilic and thermophilic conditions. Vindis et al. [16] reported a decrease in solid retention time and increase in biogas production from anaerobic digestion of maize silage under thermophilic conditions. Achu & Liu [17] realized higher biogas productivity under thermophilic conditions.

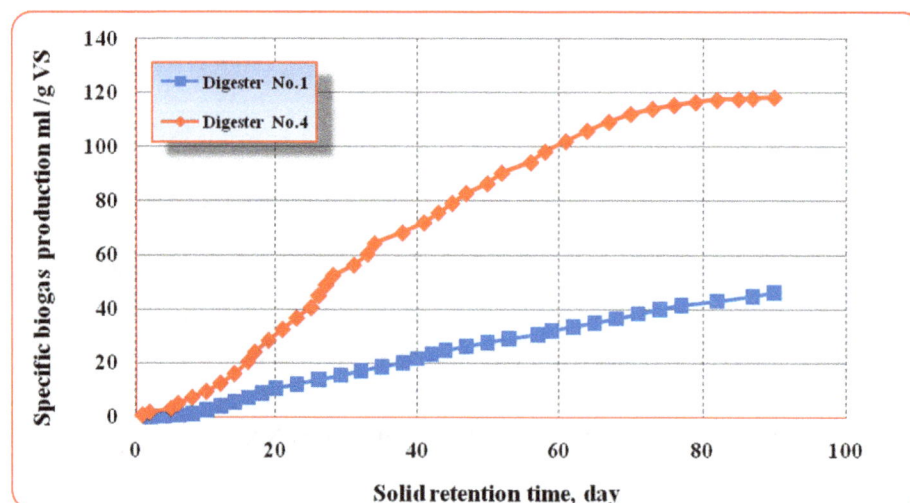

Figure 7. Specific and cumulative biogas production profiles in digesters No. 1 and 4

Table 4. Effect of temperature on the specific biogas production from pretreated inoculated DPWs

Digester No.	Temperature condition	Specific biogas production (mL/g VS)	Specific CH_4 production (mL/g VS)
1	Mesophilic	134.880	85.966
4	Thermophilic	166.468	118.389

3.4 Kinetic model

Biogas production rate in batch condition is corresponding to specific growth rate of methanogenic bacteria in the bio-digester. Accordingly, the predicted biogas production rate will obey Modified Gompertz Model [18] as follows:

$$G_{(t)}=G_0.\exp\{-\exp[((R_{max}.e)/G_0)(\lambda-t)+1]\} \tag{1}$$

where: $G_{(t)}$ = the cumulative biogas yield at a digestion time (mL/g VS), G_0 = the biogas potential of the substrate (mL/g VS), R_{max} = maximum methane production rate (mL/g VS-d), λ = lag phase (day) t = time (day), e = exp (1) = 2.7183.

A nonlinear least-square regression analysis was performed using SPSS [IBM SPSS statistics 18 (2009)] to determine λ, Rmax, and the predicted biogas and methane yield (Table 5). Plots of the measured and predicted values of biogas production are given in Figures 8-10. It is well observed that the predicted values of biogas production using modified Gompertz model is well fitted with the measured values. Results of this section are in a good agreement with the previously outlined findings. Kafle et al. [8] reported that the measured values of biogas produced from the bio-digestion of fish waste are well fitted with the predicted values using modified Gompertz model. Budiyono et al. [14] proved that the measured values of biogas produced from the digestion of cattle manure in batch mode are well fitted with the predicted data obtained by modified Gompertz model.

Table 5. Results of a kinetic study using Gompertz model at mesophilic conditions after 90 days

Digester No.	$G_{(t)}$ exp. (mL CH_4/g VS)	Gompertz model parameters				R^2
		λ (day)	R_{max}. (mL CH_4/g VS)	G_0 (mL CH_4/g VS)	$G_{(t)}$ model (mL CH_4/g VS)	
1	85.967	15.407	1.597	90.381	83.860	0.985
2	35.969	19.172	0.687	36.493	34.060	0.979
3	54.604	18.181	1.069	56.107	52.530	0.986

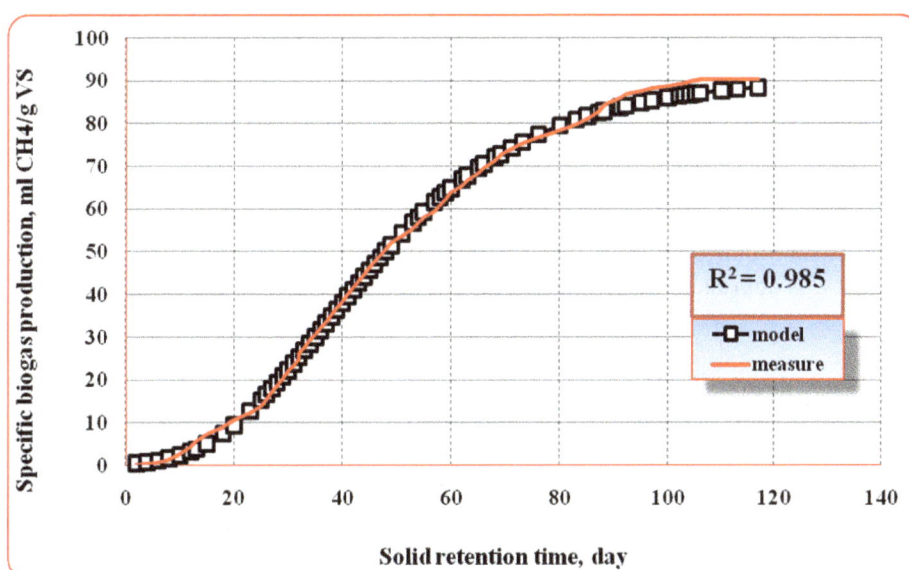

Figure 8. Measured and predicted results for biogas production from pretreated inoculated DPW

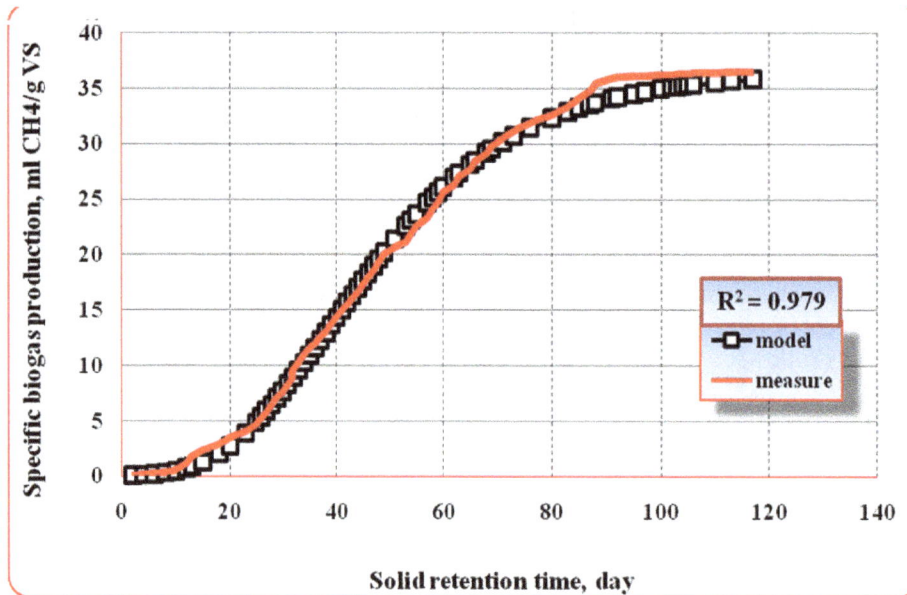

Figure 9. Measured and predicted results for biogas production from pretreated DPW

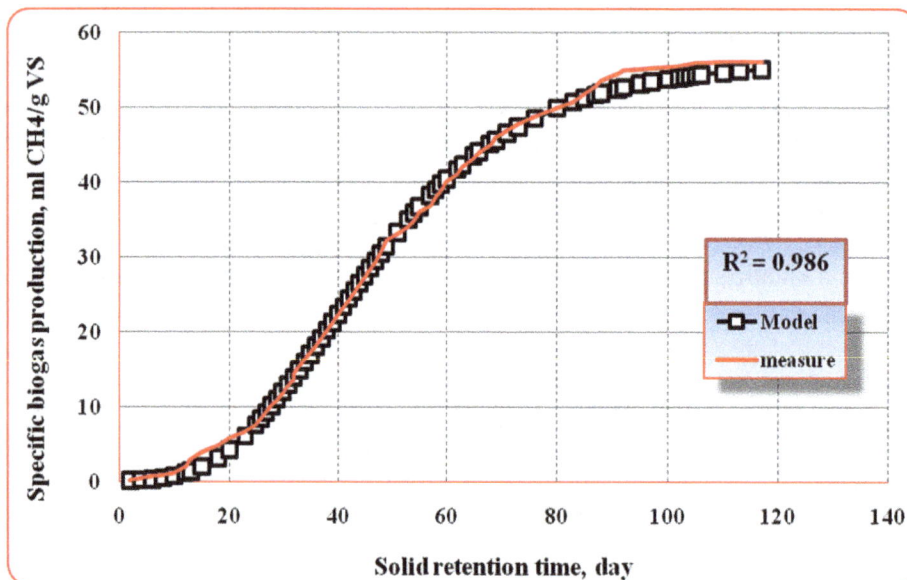

Figure 10. Measured and predicted results for biogas production from untreated inoculated DPW

3.5 Soil fertilization with residual digestates

The results of this part of work demonstrated that the selected process is a potential approach to treat the digestate resulted from the digestion process of DPWs. Figure 11 presents the growth progress of cress seeds after one week observation period. A healthy favorable growth of the planted Cress seeds was observed indicating that this approach is potential method to treat residues of digestive process.

4. Conclusion

This study was devoted to investigate the potential of anaerobic co-digestion for biogas production using abundantly available date palm waste materials of no economic value as the substrate. The experimental work demonstrated that the volume of produced biogas significantly affected by inoculum addition, pretreatment of waste materials, temperature conditions. The ultimate biogas yield from co-digesting of inoculated DPWs was estimated to be 141.667 ± 8.1 mL/g VS, whereby without inoculation it was 59.103 ± 2.4 mL/g VS. Maximum biogas production from co-digestion of alkaline pretreated DPWs was estimated to be 141.667 ± 8.1 mL/g VS, whereby, it was 93.254 ± 4.2 mL/g VS for untreated DPWs. The kinetic of bio-digestion process was well described by Modified Gompertz Model and the experimental and predicted values of biogas production were fitted well with correlation coefficient values > 0.96 suggesting favorable conditions of the process.

Figure 11. Growth observations for the planted cress seeds after one week, pot (A) is for non-conditioned soil, and pot (B) for digestate-conditioned soil

Acknowledgment

The authors are grateful to the staff of Sanitary Laboratory, Civil Engineering Department at Baghdad University for their technical support.

References

[1] Messineo A, Volpe R, Marvuglia A. Ligno-cellulosic biomass exploitation for power generation: A case study in Sicily. Energy 45, 613-625 (2012).

[2] Twidell J, Weir T. Renewable Energy Resources. 2nd edition, New York: Taylor & Francis (2006).

[3] Converti A, Del Borghi A, Zilli M, Arnì S, Del Borghi M. Anaerobic digestion of the vegetable fraction of municipal refuses: mesophilic versus thermophilic conditions. Bioprocess Eng. 21, 371–376 (1999).

[4] Abdel-Hadi MA. A simple apparatus for biogas quality determination. Misr. J. Agric. Eng. 25, 1055- 1066 (2008).

[5] Rincón B, Borja R, Martín MA, Martín A. Evaluation of the methanogenic step of a two-stage anaerobic digestion process of acidified olive mill solid residue from a previous hydrolytic–acidogenic step. Waste Manage. 29, 2566-2573 (2009).

[6] Jaffer KA. Biogas production by anaerobic digestion of date palm pulp waste. Al-Khwarizmi Eng. J. 6, 14-20 (2010).

[7] Marňóna E, Castrillón L, Quiroga G, Fernández-Nava Y, Gómez L, Garcìa MM. Co-digestion of cattle manure with food waste and sludge to increase biogas production. Waste Manage. 32, 1821-1825 (2012).

[8] Kafle GK, Kim SH. Anaerobic treatment of apple waste with swine manure for biogas production: Batch and continuous operation. Appl. Energy 103, 61-72 (2013).

[9] Tampio E, Ervasti S, Paavola T, Hearen, S, Banks, C, Rintala, J. Anaerobic digestion of autoclaved and untreated food waste. Waste Manage. 34, 370-377 (2014).

[10] Chandrasekaran M, Bahkali AH. Valorization of date palm (Phoenix dactylifera) fruit processing by-products and wastes using bioprocess technology – Review. Saudi J. Biol. Sci. 20, 105-120 (2013).

[11] Forgács G. Biogas production from citrus wastes and chicken feather: pretreatment and co-digestion. PhD Thesis, Borås University (2012).

[12] American Public Health Association (APHA). Standard methods of the examination of water and wastewater, Washington, DC (1998).

[13] Hansen TL, Schmidt JE, Angelidaki I, Marca E, Jansen J, Mosbaek H, Christensen TH. Method for determination of methane potentials of solid organic waste. Waste Manage. 24, 393–400 (2004).

[14] Budiyono B, Widiasa IN, Johari S, Sunarso S. The kinetic of biogas production rate from cattle manure in batch mode. Int. J. Chem. Biol. Eng. 3, 39 – 44 (2010).

[15] Abdulkarim BI, Evuti AM. Effect of buffer (NaHCO3) and waste type in high solid thermophilic anaerobic digestion. Int. J. Chem.Tech. Res. 2, 980-984 (2010).

[16] Vindis P, Mursec B, Janzekovic M, Cus F. The impact of mesophilic and thermophilic anaerobic digestion on biogas production. J. Ach. Mater. Manuf. Eng. 36, 192-198 (2009).

[17] Achu NI, Liu J. Effects of solid retention time on anaerobic digestion of dewatered-sewage sludge in mesophilic and thermophilic conditions. Renew. Energy 35, 2200-2206 (2010).

[18] Nopharatana AP, Pullammanappallil CW, Clarke P. Kinetics and dynamic modeling of batch anaerobic digestion of municipal solid waste in a stirred reactor. Waste Manage. 27, 595-603 (2007).

19

Comparative analyses of closed-landfill Methane (CH₄) and Carbon dioxide (CO₂) concentrations

Nwachukwu Arthur Nwachukwu

Williamson Research Centre for Molecular Environmental Sciences, School of Earth, Atmospheric and Environmental Science, University of Manchester, M13 9PL, UK.

Abstract

The time series data obtained from in-borehole measurement of CH_4 and CO_2 from a landfill site in Manchester, UK are given. Analysis reveals that they were variable for the period under investigation. There is a significant positive correlation between ground CH_4/CO_2 concentrations and their monitoring time. During this period, CH_4 concentration has increased from 0.5% to 62.7%. Similarly, CO_2 concentration has increased from 0.6% to 35.5%. Both gases have positive correlation coefficients of 0.5671 and 0.6653 respectively with time horizon of June – September 2011. Also, the two gases exhibit positive correlation coefficient of 0.9205 with each other; indicating that emission of CH_4 creates potential for emission of CO_2 and vice versa.

Keywords: Greenhouse gas; Global warming potential; Climate mitigation policies; Explosive mixture; Asphyxiation; Risk prediction; Closed landfill; Gasclam; Environmental controls.

1. Introduction

Landfills generate significant amounts of various gases during their active life and for a period of time after their closure [1]. Methane and carbon dioxide are the two major gases mostly generated in landfill emissions [2-6]. Volatile organic compounds (VOCs) are equally generated [7], however; they are usually in trace concentrations [8]. Landfill gas is produced by the decomposition of organic content of waste such as food, garden, wood and paper waste [9]. They are emitted into the atmosphere and can also travel long distances in the porous space of the soil medium. Their migration into the indoor or ambient environment is either by vapour intrusion, vapour emission or vapour release [8].

Globally, methane emissions from landfill accounts for about 30 teragrams per year or 6% of the total global methane emissions [10, 11]. In UK alone, landfill accounted for about 46% of the total methane emissions during 1996 [12]. The global emission of carbon dioxide from the soil respiration ranges from 68×10^{15} gcyear^{-1} [13] to 75×10^{15} gcyear^{-1} [14]; the magnitude of which depends on the activities of the belowground microbial community and root respiration [15]. Moreover, CH_4 and CO_2 are produced from several other sources containing biodegradable organic materials [16]. They can be trapped in materials such as coal and peat, and be released during activities like mining and pilling respectively [16] into the atmosphere.

Methane is regarded as the second most important anthropogenic greenhouse gas in the atmosphere next to carbon dioxide [12, 17]. Its global warming potential for a time horizon of 100 years is 25, which makes it an attractive target for climate mitigation policies [17]. It has a net life time of about 10 years

[17]. Methane is highly explosive at concentrations of approximately 5-15% by volume in air [9] and can accumulate to dangerous levels virtually undetected. It can also act as an asphyxiant and in particular circumstances it may be toxic.

Carbon dioxide on the other hand presents similar hazards to that of methane. It also poses an asphyxiation hazard when it collects in an enclosed space by displacing the existing air and creating an oxygen deficient environment [18]. CO_2 also causes adverse health effects, unconsciousness or even death at relatively low concentrations (at approximately 5% by volume in air) [9].

Methane and carbon dioxide are two main types of greenhouse gases with widely different warming potential. Though the concentration of CH_4 in the atmosphere is lower than CO_2 but it has 22 times the warming potential of CO_2 on a 100-yr time scale, therefore, it may have significant impacts on global climate change [19]. The present CO_2 concentration in the atmosphere is 384.8 ppm while the present CH_4 concentration is 1.74 to 1.86 ppm [20]. The annual increasing rate of the concentration of CO_2 and CH_4 in the atmosphere is 0.5% and 0.8%, respectively [21].

Given the fact that landfill soil is majorly made up of CH_4 and CO_2, there is a requirement to determine whether a change in one of the gases would bring about a concomitant change in the other and also how both correlate with time. This is particularly pertinent, since any variation to and from the soil would help in controlling atmospheric greenhouse effect. This paper uses time series data to establish these relationships.

2. Materials and method

The datasets analysed in this work were obtained with the help of an in-borehole ground-gas monitor, Gasclam (Ion Science, UK). This instrumentation has the capability to monitor continuously and simultaneously various ground-gases (CH_4, CO_2, CO, O_2, H_2S, and VOCs) and their environmental controls (temperature, barometric pressure, borehole pressure and soil water depth) on hourly sampling basis unmanned for up to three months. It logs long term, real trend information, allowing informed decision to be made on accurate, reliable data – a revolution in gas management and prediction. It measures the gases with the aid of the sensors incorporated into it. Its sampling frequency can be set and is variable from two minutes, to once daily. Data is downloaded to a PC or viewed remotely using the optional GPRS telemetry system.

The instrument was installed in a Landfill site in Manchester, UK. The gas monitors were set sampling on hourly basis and left in-situ to ensure a continuous monitoring of the ground-gases. By doing this, it gives one time series behaviour of the individual gases and their controls allowing room for prediction of their risk. This paper, however, does not look into their risk prediction, but investigates the relationship between two of the gases (that is, CH_4 and CO_2) in landfill soil.

3. Results

The time series data of methane and carbon dioxide concentrations collected for the months of June, July, August and September in 2011 are as shown in Figures 1-4 respectively. These figures illustrate the hourly concentrations of methane and carbon dioxide for the stated months. Figures 5-8 illustrate the relationships between methane and carbon dioxide for the respective months above. Figure 9 is the graph of methane and carbon dioxide against time for the entire monitoring period whilst Figure 10 displays the graph of methane against carbon dioxide concentrations for the same time horizon.

4. Discussions

Figures 1-4 show that methane and carbon dioxide concentrations are all variable. Figure 1 shows initial methane concentration to be 18.5%. It then rose to 25% and remained fairly constant for the next 300 hours and then dropped to 0%. From 0%, it rapidly increased to 60.9% and remained fairly constant for 211 hours before going down to 0% again. Carbon dioxide followed the same trend, however; its concentration is much lower than that of methane.

Figure 2 shows very high variability of methane and carbon dioxide concentrations with initial methane concentration as 51.7%. It then rose to 60% within 11 hours and remained fairly constant for the next 41 hours before going down to 20.3%. It then quickly rose to 60.9% and remained there until after 200 hours and then dropped to 10%. Carbon dioxide had initial and final concentration of 26.7% and 29.1% respectively but peaked at 34.1%. Once again, carbon dioxide shows the same trend with methane.

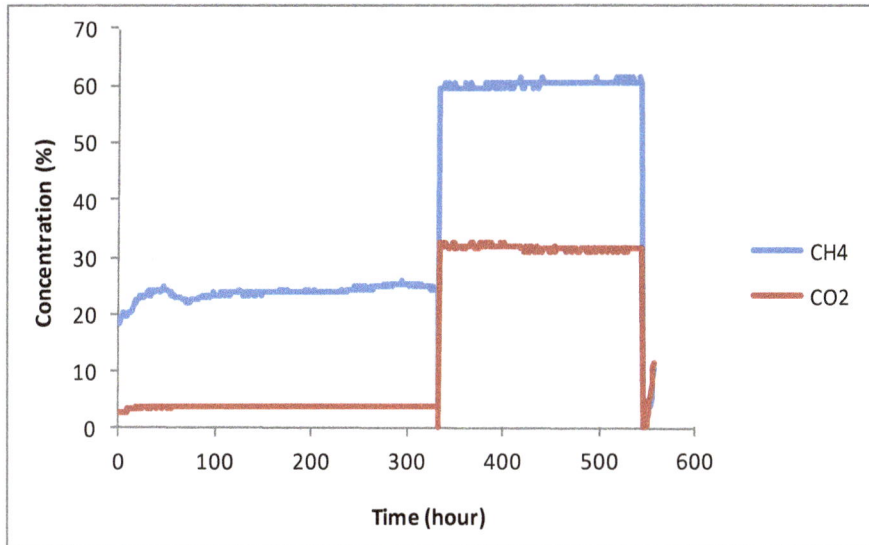

Figure 1. Hourly concentrations of methane and carbon dioxide (June, 2011)

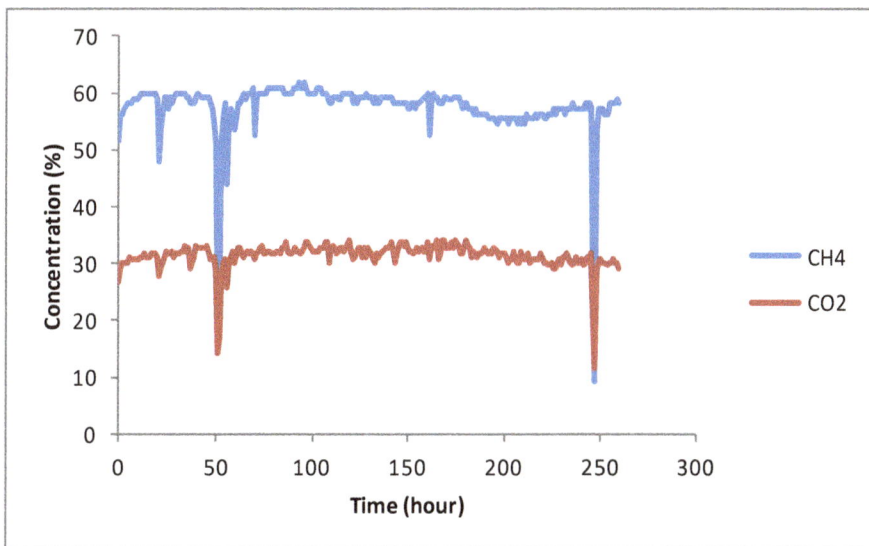

Figure 2. Hourly concentration of methane and carbon dioxide (July, 2011)

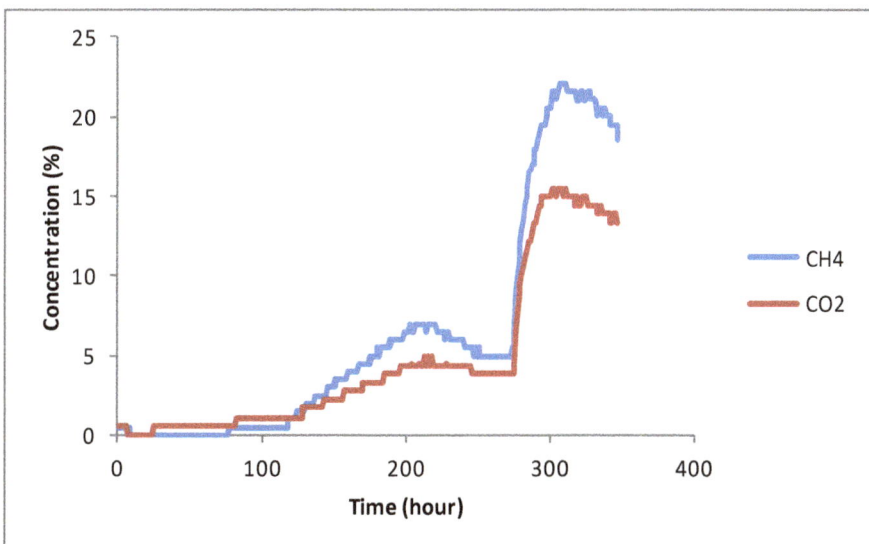

Figure 3. Hourly concentration of methane and carbon dioxide (August, 2011)

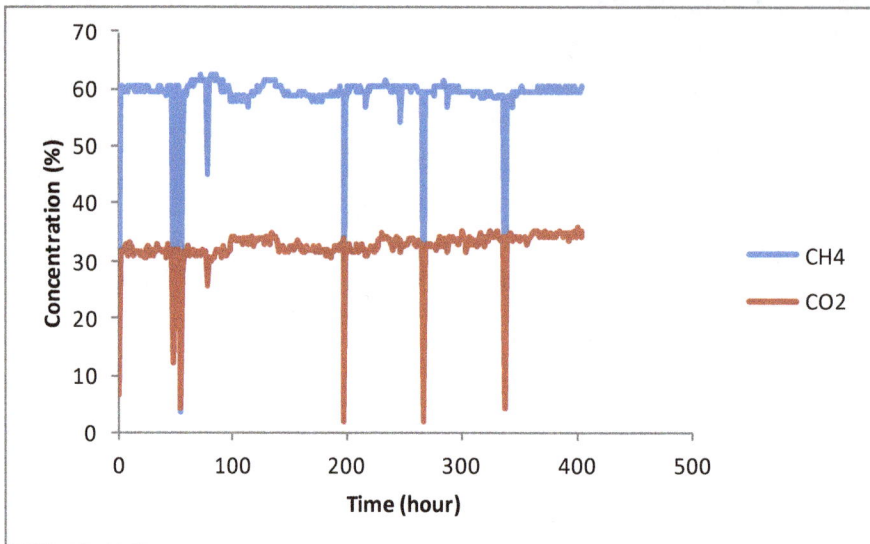

Figure 4. Hourly concentration of methane and carbon dioxide (Sept., 2011)

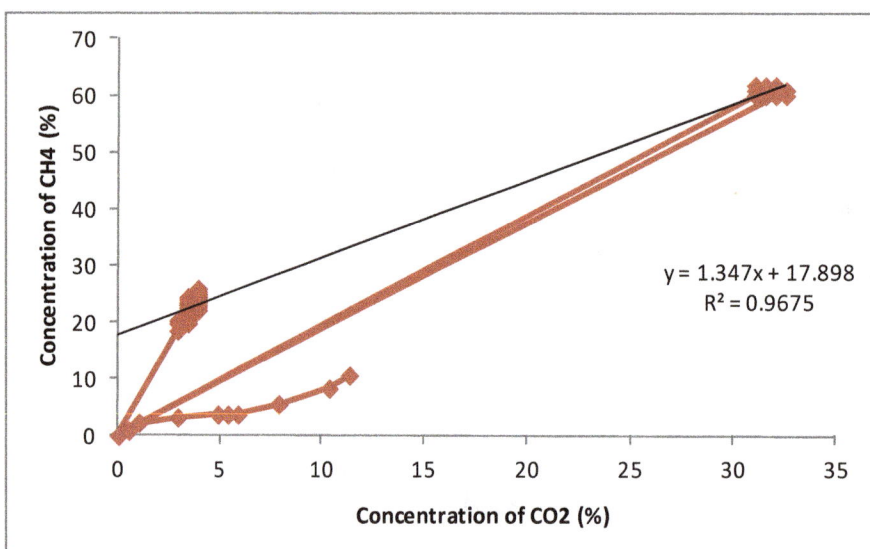

Figure 5. Graph of methane concentration against carbon dioxide concentration (June, 2011)

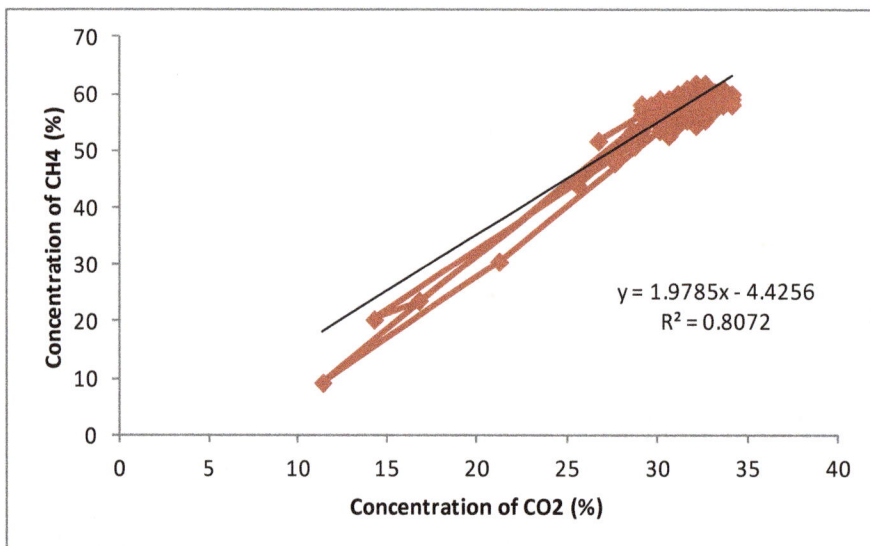

Figure 6. Graph of methane concentration against carbon dioxide concentration (July, 2011)

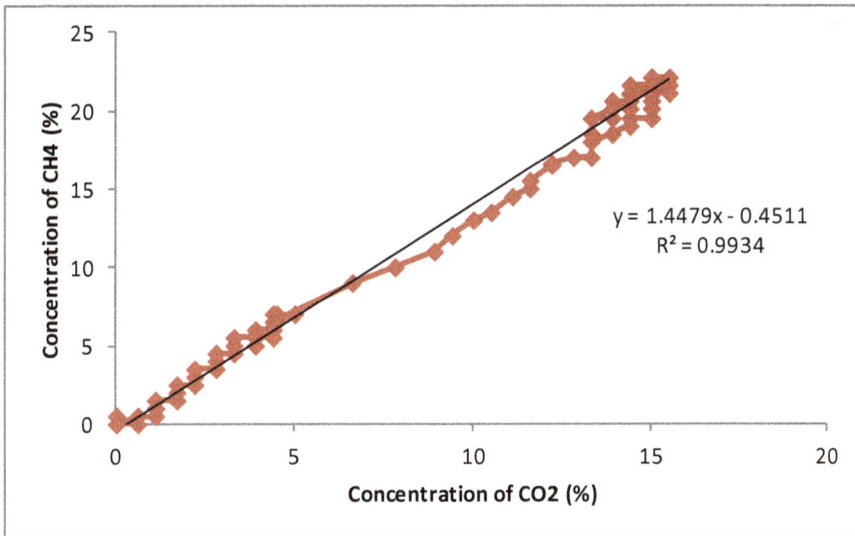

Figure 7. Graph of methane concentration against carbon dioxide concentration (August, 2011)

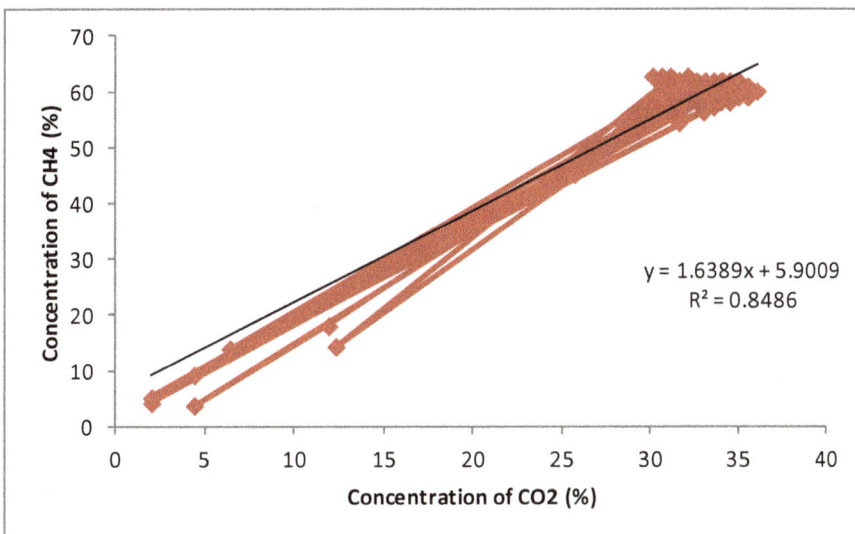

Figure 8. Graph of methane concentration against carbon dioxide concentration (Sept., 2011)

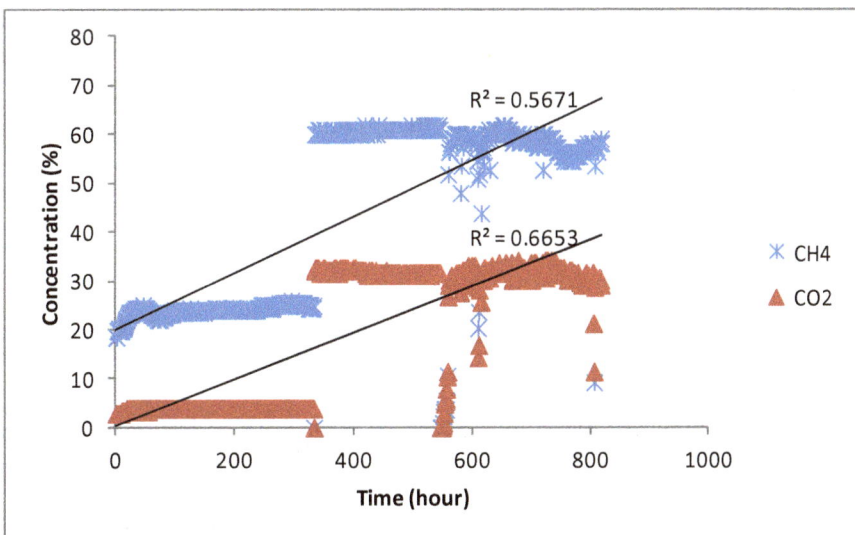

Figure 9. Graph of CH_4 and CO_2 concentrations against time. The correlation is for the entire monitoring period (June-September, 2011)

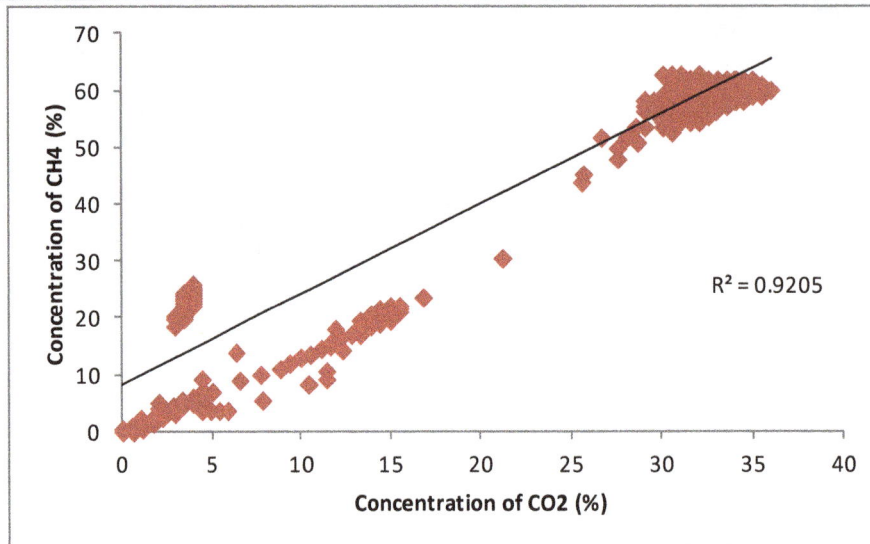

Figure 10. Graph of methane concentration against carbon dioxide concentration for the entire monitoring period (June-September, 2011)

Figure 3 shows the initial methane and carbon dioxide concentrations to be 0.5% and 0.6% respectively. They then remained constant for 6 hours and dropped to 0%. Methane concentration stayed at 0% for 70 hours and then went back to 0.5% and remained there for 41 hours before it gradually peaked at 21.6% and remained fairly there until the data was downloaded. Carbon dioxide on the other hand remained at 0% for 18 hours before going back to 0.6%. It then remained constantly there for 57 hours before gradually going up to 15.5% and fluctuated around this until the end of the monitoring period.

The trend of methane and carbon dioxide concentration in Figure 4 resemble those of Figure 2, however; the frequency of the rising and falling of their concentrations is higher than those of Figure 2. Methane had the lowest and highest concentrations of 13.9% and 62.7% respectively, while carbon dioxide had 6.4% and 35.5% as its lowest and highest concentrations respectively.

Figures 5-8 show graphs of methane against carbon dioxide. They display the correlations of methane with carbon dioxide. Methane had correlations of 0.9675, 0.8072, 0.9934, and 0.8486 with carbon dioxide in Figures 5-8 respectively. Methane and carbon dioxide had correlations of 0.5671 and 0.6653 (Figure 9) respectively with time over the entire monitoring period and also a correlation of 0.9205 (Figure 10) with each other for the same time scale.

5. Conclusions

- The very high concentrations of both the methane and carbon dioxide are significance of increased emission of these gases from landfills into both indoor and ambient atmospheres. This trend could be attributed to their environmental controls.
- The structure of time variation of the gases is a proof that higher concentrations should be expected in the nearest future.
- The high positive correlations between methane and carbon dioxide are indications that the emission of methane incites a concomitant emission of carbon dioxide and vice versa.

References

[1]   Mustafa MA (2010). Environmental Modelling and Health Risk Analysis (ACTS/RISK). Springer Dordrecht Heidelberg, London New York. ISBN 978-90-481-8607-5, DOI 10.1007/978-90-481-8608-2.

[2]   Scheutz C, Bogner J, Chanton JP, Blake D, Morcet M, Aran C, Kjeldsen P. (2008) Atmospheric emissions and attenuation of non-methane organic compounds in cover soils at a French landfill. Waste Management, vol. 28, pp1892-1908.

[3]   Scheutz C, Kjeldsen P, Bogner JE, Visscher AD, Gebert J, Hilger HA, Huber-Humer M, Spokas K. (2009) Microbial methane oxidation processes and technologies for mitigation of landfill gas emissions. Waste Management and Research, 27, pp 409-455.

[4]   Streese J, and Stegmann R. (2003) Microbial oxidation of methane from old landfill in biofilters. Waste Management, 23, pp573-580.

[5]   Eklund B, Anderson EP, Walker BL, and Burrows DB. (1998) Characterization of landfill gas composition at the Fresh Kills Municipal Solid-Waste Landfill. Environmental Science Technology, vol. 32, pp 2233-2237.

[6]   USEPA (1995). Air emissions from Municipal Solid Waste Landfills – Background Information for Proposed Standards and Guidelines. Emission Standards Division, U.S. Environmental Protection Agency, Office of Air and Radiation Office of Air Quality Planning and Standards Research Triangle Park, North Carolina 27711.

[7]   West OR, Siegrist RL, Mitchell TJ, and Jenkins RA (1995). Measurement error and spatial variability effects on the characterisation of volatile organic in the subsurface. Environmental Science and Technology, 1995, 29 (3), 647-656.

[8]   Katy B, Helen H, Lara P, Don B, and Cecilia M (2009).The VOCs Handbook: Investigation, assessing, and managing risks from inhalation of VOCs at land affected by contamination. CIRIA Report 766.

[9]   NHBC, (2007). Guidance on evaluation of development proposals on sites where methane and carbon dioxide are present.

[10] Thorneloe SA, Barlaz MA (1993). Global methane emissions from waste management. Atmospheric methane: sources, sinks, and role in global change, NATO ASI Series, vol. 13. Springer, New York.

[11] Thorneloe SA, Doorm M (1994). Methane emissions from landfills and open dumps. In: EPA report to congress on international anthropogenic methane emissions: estimates for 1990, USEPA, Office of Policy, Planning and Evaluation. EPA-230-R-93-010.

[12] Environmental Agency, (2008). Methane http://www.environment-agency.gov.uk/business/topics/pollution/185.aspx.

[13] Raich JW, Schlesinger WH, (1992). The global carbon flux in soil respiration and its relationship to vegetation and climate. Tellus, 44B, 81-91.

[14] Schlesinger WH, (1997). Carbon balance in terrestrial detritus. Annual Review of Ecology and Systematics, 8, 51-81.

[15] Epron D, Bosc A, Bonal D, Freycon V (2006). Spatial variation of soil respiration across a topographic gradient in tropical rainforest in French Guiana. Journal of Tropical Ecology, 22, 565-574.

[16] Wilson S, Card G, and Haines S. (2008). The local authority guide to ground-gas. The Chartered Institute of Environmental Health: London.

[17] Boucher O, Friedlingstein P, Collins B, Shine KP. (2009). "The indirect global warming potential and the global temperature change due to methane oxidation". Environmental research letters 4:044007.

[18] Agency for Toxic Substances Disease Registry (ATSDR), (2001). Landfill gas primer. An overview for environmental health professionals [available online] URL: http://www.atsdr.cdc.gov/HAC/landfill/html/intro.html. Accessed 7th May, 2009.

[19] International Panel on Climate Change (IPPC), (2001). Climate Change: The Scientific Basis. Cambridge University Press, Cambridge.

[20] Blasting, TJ (2009). Recent Greenhouse Gas Concentration. Carbon Dioxide Information Analysis Centre (CDIAC). URL: http://cdiac.ornl.gov

[21] International Panel on Climate Change (IPCC), (2007). The physical science basis, Fourth Assessment Report, Working Group I. Cambridge: Cambridge University; http://www.ipcc.ch/ipccreports/ar4-wg1.htm.

Permissions

All chapters in this book were first published in IJEE, by International Energy and Environment Foundation (IEEF); hereby published with permission under the Creative Commons Attribution License or equivalent. Every chapter published in this book has been scrutinized by our experts. Their significance has been extensively debated. The topics covered herein carry significant findings which will fuel the growth of the discipline. They may even be implemented as practical applications or may be referred to as a beginning point for another development.

The contributors of this book come from diverse backgrounds, making this book a truly international effort. This book will bring forth new frontiers with its revolutionizing research information and detailed analysis of the nascent developments around the world.

We would like to thank all the contributing authors for lending their expertise to make the book truly unique. They have played a crucial role in the development of this book. Without their invaluable contributions this book wouldn't have been possible. They have made vital efforts to compile up to date information on the varied aspects of this subject to make this book a valuable addition to the collection of many professionals and students.

This book was conceptualized with the vision of imparting up-to-date information and advanced data in this field. To ensure the same, a matchless editorial board was set up. Every individual on the board went through rigorous rounds of assessment to prove their worth. After which they invested a large part of their time researching and compiling the most relevant data for our readers.

The editorial board has been involved in producing this book since its inception. They have spent rigorous hours researching and exploring the diverse topics which have resulted in the successful publishing of this book. They have passed on their knowledge of decades through this book. To expedite this challenging task, the publisher supported the team at every step. A small team of assistant editors was also appointed to further simplify the editing procedure and attain best results for the readers.

Apart from the editorial board, the designing team has also invested a significant amount of their time in understanding the subject and creating the most relevant covers. They scrutinized every image to scout for the most suitable representation of the subject and create an appropriate cover for the book.

The publishing team has been an ardent support to the editorial, designing and production team. Their endless efforts to recruit the best for this project, has resulted in the accomplishment of this book. They are a veteran in the field of academics and their pool of knowledge is as vast as their experience in printing. Their expertise and guidance has proved useful at every step. Their uncompromising quality standards have made this book an exceptional effort. Their encouragement from time to time has been an inspiration for everyone.

The publisher and the editorial board hope that this book will prove to be a valuable piece of knowledge for researchers, students, practitioners and scholars across the globe.

List of Contributors

João P. Ribau
LAETA, IDMEC, Instituto Superior Técnico, Universidade de Lisboa, 6 Av. Rovisco Pais, 1, 71049-001 Lisboa, Portugal

Ana F. Ferreira
LAETA, IDMEC, Instituto Superior Técnico, Universidade de Lisboa, 6 Av. Rovisco Pais, 1, 71049-001 Lisboa, Portugal

Amrit Adhikari
Telemark University College, Porsgrunn, Norway

André V. Gaathaug
Telemark University College, Porsgrunn, Norway

Dag Bjerketvedt
Telemark University College, Porsgrunn, Norway

Knut Vaagsaether
Telemark University College, Porsgrunn, Norway

Almaz Akhmetov
ENCA Management Ltd., 7 Pobedy Str., Esik, 040400, Kazakhstan Orizon Consulting, 6481 Elm Str., Suite 161, McLean, VA, USA

Denise Alves Fungaro
Instituto de Pesquisas Energéticas e Nucleares, IPEN–CNEN/SP- Av. Prof. Lineu Prestes, 2242, Cidade Universitária, CEP 05508-000 São Paulo SP, Brasil

Thais Vitória da Silva Reis
Instituto de Pesquisas Energéticas e Nucleares, IPEN–CNEN/SP- Av. Prof. Lineu Prestes, 2242, Cidade Universitária, CEP 05508-000 São Paulo SP, Brasil

Xiong Liu
Institute of Thermal Science and Power Engineering, Naval University of Engineering, Wuhan 430033, P. R. China
Military Key Laboratory for Naval Ship Power Engineering, Naval University of Engineering, Wuhan 430033, P. R. China
College of Power Engineering, Naval University of Engineering, Wuhan 430033, P. R. China

Xiaoyong Qin
Institute of Thermal Science and Power Engineering, Naval University of Engineering, Wuhan 430033, P. R. China

Military Key Laboratory for Naval Ship Power Engineering, Naval University of Engineering, Wuhan 430033, P. R. China
College of Power Engineering, Naval University of Engineering, Wuhan 430033, P. R. China

Lingen Chen
Institute of Thermal Science and Power Engineering, Naval University of Engineering, Wuhan 430033, P. R. China
Military Key Laboratory for Naval Ship Power Engineering, Naval University of Engineering, Wuhan 430033, P. R. China
College of Power Engineering, Naval University of Engineering, Wuhan 430033, P. R. China

Fengrui Sun
Institute of Thermal Science and Power Engineering, Naval University of Engineering, Wuhan 430033, P. R. China
Military Key Laboratory for Naval Ship Power Engineering, Naval University of Engineering, Wuhan 430033, P. R. China
College of Power Engineering, Naval University of Engineering, Wuhan 430033, P. R. China

Nantaporn Noosai
Department of Civil and Environmental Engineering, Florida International University, Miami, FL 33174, USA

Vineeth Vijayan
Department of Civil and Environmental Engineering, Florida International University, Miami, FL 33174, USA

Khokiat Kengskool
Department of Civil and Environmental Engineering, Florida International University, Miami, FL 33174, USA

Hao Yu
Department of Industrial Engineering, Narvik University College, Postboks 385 Lodve gate 2, 8505 Narvik, Norway

Wei Deng Solvang
Department of Industrial Engineering, Narvik University College, Postboks 385 Lodve gate 2, 8505 Narvik, Norway

Shiyun Li
Department of Industrial Engineering, Narvik University College, Postboks 385 Lodve gate 2, 8505 Narvik, Norway
College of Mechanical Engineering, Zhejiang University of Technology, No. 18 Caowang Road, 310016 Hangzhou, P.R.China

H. Bounaouara
CORIA UMR 6614 CNRS, Université et INSA de ROUEN, Avenue de l'Université, BP 12, 76801 Saint Etienne du Rouvray, Cedex, France
LESTE, Ecole Nationale d'Ingénieurs de Monastir, 5019 Monastir, Tunisie

J. C. Sautet
CORIA UMR 6614 CNRS, Université et INSA de ROUEN, Avenue de l'Université, BP 12, 76801 Saint Etienne du Rouvray, Cedex, France

H. Ben Ticha
LESTE, Ecole Nationale d'Ingénieurs de Monastir, 5019 Monastir, Tunisie

A. Mhimid
CORIA UMR 6614 CNRS, Université et INSA de ROUEN, Avenue de l'Université, BP 12, 76801 Saint Etienne du LESTE, Ecole Nationale d'Ingénieurs de Monastir, 5019 Monastir, Tunisie

G. Julien

Adounkpe
Laboratory of Applied Ecology, Faculty of Agronomic Sciences, University of Abomey Calavi, 03 BP 3908 Cotonou Republique du Benin

Clement Ahouannou
Département de Génie Mécanique et Energétique Ecole Polytechnique d'Abomey Calavi, Université d'Abomey Calavi, 03 BP 1175 Cotonou Republique du Benin

O. Lie Rufin Akiyo
Department of Geography of the University of Parakou, Republic of Benin BP 123 Université de Parakou Republic of Benin

Augustin Brice Sinsin
Laboratory of Applied Ecology, Faculty of Agronomic Sciences, University of Abomey Calavi, 03 BP 3908 Cotonou Republique du Benin

A. Papadaki
School of Mechanical Engineering, Department of Thermal Engineering, National Technical University of Athens, 9 Iroon Polytechniou Str., Zografou 15780, Athens, Greece

A. Stegou-Sagia
School of Mechanical Engineering, Department of Thermal Engineering, National Technical University of Athens, 9 Iroon Polytechniou Str., Zografou 15780, Athens, Greece

Georgina Nagy
Institute of Environmental Engineering, Faculty of Engineering, University of Pannonia, 10 Egyetem St., Veszprém, Hungary H-8200

Anna Merényi
Institute of Environmental Engineering, Faculty of Engineering, University of Pannonia, 10 Egyetem St., Veszprém, Hungary H-8200

Endre Domokos
Institute of Environmental Engineering, Faculty of Engineering, University of Pannonia, 10 Egyetem St., Veszprém, Hungary H-8200

Ákos Rédey
Institute of Environmental Engineering, Faculty of Engineering, University of Pannonia, 10 Egyetem St., Veszprém, Hungary H-8200

Tatiana Yuzhakova
Institute of Environmental Engineering, Faculty of Engineering, University of Pannonia, 10 Egyetem St., Veszprém, Hungary H-8200

Hashim R. Abdol Hamid
Environment Dept., International Energy and Environment Foundation, Najaf, P.O.Box 39, Iraq

Sulala M. Z. F. Al-Hamadani
Three Gorges Reservoir Area's Ecology and Environment Key Laboratory of Ministry of Education, Chongqing University, Chongqing, 400045, China

ZENG Xiao-lan
Three Gorges Reservoir Area's Ecology and Environment Key Laboratory of Ministry of Education, Chongqing University, Chongqing, 400045, China

M.M.Mian
Three Gorges Reservoir Area's Ecology and Environment Key Laboratory of Ministry of Education, Chongqing University, Chongqing, 400045, China

Zhongchuang Liu
Three Gorges Reservoir Area's Ecology and Environment Key Laboratory of Ministry of Education, Chongqing University, Chongqing, 400045, China
National Centre for International Research of Low-carbon and Green Buildings, Chongqing University, Chongqing, 400045, China

Ahmed F. Hassoon
Department of Atmospheric Sciences, College of Sciences, Al-Mustansiriyah University, Baghdad, Iraq

Maher A. R. Sadiq Al-Baghdadi
CFD Center, International Energy and Environment Foundation, Najaf, P.O.Box 39, Iraq

Tijo Joseph
School of Engineering, University of Guelph, Guelph, Ontario, Canada

Animesh Dutta
School of Engineering, University of Guelph, Guelph, Ontario, Canada

Kamal H. Lateef
Ministry of Science and Technology- Renewable Energy Directorate, Iraq

Azhaar K. Mishaal
Ministry of Science and Technology- Renewable Energy Directorate, Iraq

Ahmed M. Abud
Ministry of Science and Technology- Environment and Water Directorate, Iraq

Zainab Ziad Ismail
Department of Environmental Engineering, Baghdad University, Baghdad, Iraq

Ali Raad Talib
Department of Environmental Engineering, Baghdad University, Baghdad, Iraq

Nwachukwu Arthur Nwachukwu
Williamson Research Centre for Molecular Environmental Sciences, School of Earth, Atmospheric and Environmental Science, University of Manchester, M13 9PL, UK

www.ingramcontent.com/pod-product-compliance
Lightning Source LLC
Chambersburg PA
CBHW050455200326
41458CB00014B/5190